RACE AND RACISM

Essays in social geography

Edited by **Peter Jackson**

Department of Geography, University College London

London
ALLEN & UNWIN
Boston Sydney Wellington

Allen & Unwin, the academic imprint of

Unwin Hyman Ltd

PO Box 18, Park Lane, Hemel Hempstead, Herts HP2 4TE, UK
40 Museum Street, London WC1A 1LU, UK
37/39 Queen Elizabeth Street, London SE1 2QB, UK

Allen & Unwin Inc.,
8 Winchester Place, Winchester, Mass. 01890, USA

Allen & Unwin (Australia) Ltd,
8 Napier Street, North Sydney, NSW 2060, Australia

Allen & Unwin (New Zealand) Ltd
in association with the Port Nicholson Press Ltd,
60 Cambridge Terrace, Wellington, New Zealand

First published in 1987

British Library Cataloguing in Publication Data

Race and racism: essays in social geography.
1. Racism 2. Anthropo-geography
I. Jackson, Peter, 1955–
305.8 HT1521
ISBN 0–04–305002–6

Library of Congress Cataloging in Publication Data

Applied for

Typeset in 10 on 11 point Bembo by Latimer Trend & Co. Ltd., Plymouth and printed in Great Britain by Billing & Sons Ltd, London and Worcester

Preface

In September 1985 the Social Geography Study Group of the Institute of British Geographers held a three-day conference at Coventry (Lanchester) Polytechnic on the subject of 'Race and Racism'. It was a lively and stimulating meeting with geographers joining a range of other social scientists in vigorous but unacrimonious debate. The conference itself has been reported elsewhere (*Immigrants and Minorities* **4** (3), November 1985, pp. 85–9).

The present volume is a selection of essays derived from some of the papers that were given at the conference, together with one newly commissioned paper (by Susan Smith) and an introductory essay (by myself). Each of the papers has been revised in the light of issues raised at the conference and in response to editorial comments on successive drafts. The introduction attempts to place the essays in the disciplinary context of social geography and in terms of the wider social-science literature on 'race' and racism. The essays are then divided into a number of parts, each of which begins with a short introductory section, highlighting salient points and drawing out general themes.

Editorial work on this volume has been made much less onerous by the generous co-operation of numerous people to whom I would like to extend my thanks. Hugh Matthews gave invaluable local support in organizing the original conference. All of the contributors kept remarkably well to schedule in preparing their manuscripts for publication, responding positively and with good humour to my various requests and suggestions. Most of the maps were redrawn for publication by Lauren McClue in the Drawing Office at UCL. And finally, the following people deserve my special thanks for taking the time to comment helpfully on an earlier draft of the introduction: Guido Ambroso, Marian Mair, Bob Miles and Susan Smith.

PETER JACKSON

Acknowledgements

We are grateful to the following individuals and organizations who have kindly given permission for the reproduction of copyright material (figure numbers in parentheses):

The Open University for an extract from *The school in the multicultural society*, A. James & R. Jeffcoate (eds), Harper & Row 1981; Commission for Racial Equality for an extract from *Five views of multiracial Britain* in *Racism and reaction* by Stuart Hall; H. Giles and Cambridge University Press (6.1); Caribbean Review Inc. and Linda Marsten (13.1).

Contents

List of tables

List of contributors

John Cater, Department of Geography, Edge Hill College of Higher Education, Ormskirk, Lancashire, England

Judy Chance, Nuffield College, Oxford, England

Anne Dunlop, Department of Sociology, University of Glasgow, Scotland

Peter Jackson, Department of Geography, University College London, England

Mark R. D. Johnson, Centre for Research in Ethnic Relations, University of Warwick, Coventry, England

Trevor Jones, Department of Geography, Liverpool Polytechnic, Liverpool, England

Michael Keith, Department of Geography, University of Liverpool, England

Barry Kosmin, Council of Jewish Federations, New York, USA

Robert Miles, Department of Sociology, University of Glasgow, Scotland

Deborah Phillips, School of Geography, University of Leeds, England

Paul Rich, Department of Politics, University of Bristol, England

Vaughan Robinson, Department of Geography, University College, Swansea, Wales

Alisdair Rogers, School of Geography, University of Oxford, England

David Sibley, Department of Geography, University of Hull, England

John Silk, Department of Geography, University of Reading, England

Susan J. Smith, Centre for Housing Research, University of Glasgow, Scotland

Rika Uto, Department of Geography, University of California, Los Angeles, California, USA

Stanley Waterman, Department of Geography, University of Haifa, Israel

Introduction

The idea of 'race' and the geography of racism

PETER JACKSON

Racism is not confined to the beliefs of a few bigoted individuals who simply do not know any better. It is a set of interrelated ideologies and practices that have grave material effects, severely affecting black people's life-chances and threatening their present and future wellbeing.[1] Racism is deeply rooted in British society's unequal power structure and is perpetuated from day to day by the intended and unintended consequences of institutional policies and practices. Institutional racism is in turn sustained by the false representations of 'common-sense' racism and media stereotypes. Challenging racism, as this book seeks to do, therefore involves a range of complex and interacting issues. We should begin, though, as Kevin Brown argued in an earlier collection of essays on the social geography of ethnic segregation (Jackson & Smith 1981), by recognizing that:

> White academics with an interest in race must relinquish their self-appointed role as the 'translators' of black cultures, in favour of analyses of white society, i.e. of racism. (Brown 1981, p. 198)

A positive response to Kevin Brown's challenge involves a reappraisal of the academic and political significance of the concept of 'race' and of the 'race relations' industry in general. It suggests that geographers have paid too little attention to work in other branches of the social sciences, particularly concerning the radical critique of 'race relations' research. But, conversely, it suggests also that there are important territorial dimensions to the study of 'race' that make the geography of racism an important and relatively neglected field. In keeping with debates in other areas of human geography and social theory (e.g. Gregory & Urry 1985), this suggests that we need both to broaden our intellectual horizons to encompass a wider range of social-science perspectives while at the same time injecting a more adequately theorized conception of space and place into the general social-science literature on 'race' and racism.

The structure of this introduction follows the pattern suggested by these initial remarks, including both a commentary on recent debates over the social construction of 'race' and a discussion of the territorial expression of various forms of racism. For spatial structures are implicated in the production and reproduction of social relations in the sense that particular territorial forms both produce and reflect particular social processes. In order to substantiate these theoretical assertions and to explain the choice of sub-title, 'Essays in social geography', it is appropriate to begin with a short review of the disciplinary context in which this book has arisen.

Social geography and spatial sociology

This volume is intended to mark a firm departure from the established tradition of studies into the geography of racial and ethnic minorities that dates back at least to Emrys Jones' pioneering work, *A social geography of Belfast* (1960). That seminal work, together with parallel research in urban sociology in the United States, gave rise to a series of empirical studies of residential segregation that has been celebrated as one of the most successful examples of cumulative social science (Peach 1983, p. 124). Characterized as 'spatial sociology' by both supporters and critics of the genre, this work raised few questions about the meaning or significance of segregation but concentrated instead on describing the spatial pattern of minority-group concentration, with gestures towards an explanation in terms of the opposing forces of 'choice' and 'constraint' (see, for example, the essays in Peach 1975, Peach *et al.* 1981, Jackson & Smith 1981). Not surprisingly, this tradition of research drew often angry criticism from more radical scholars and from those involved in black political struggles who found this work guilty of 'narrow empiricism' at best and 'socio-cultural apologism for racial segregation' at worst (Bridges 1982, pp. 83–4).

The strength of such exercises in 'spatial sociology' has always been their relative sophistication concerning the quantitative measurement of segregation and of other patterns of social interaction, such as ethnic inter-marriage. Their weakness is a relative lack of theoretical sophistication, despite recent attempts to develop a theory around the notion of ethnic pluralism (Clarke *et al.* 1984). Alternatively, some authors followed the Weberian lines of regarding access to scarce housing resources as the crucial structural underpinning of minority-group segregation and deprivation, as in Rex and Moore's influential study of Sparkbrook (Rex & Moore 1967). Rather fewer followed the marxian logic of David Harvey's explanation of the process of ghetto formation (Harvey 1973), now elaborated further as part of his broader argument about the urbanization of capital (Harvey 1985).

Despite the significance of this critique, many geographers have continued to adhere to an outdated and problematic concept of ethnic 'assimilation' (e.g. Robinson 1982, Walter 1984), despite fundamental criticisms of the concept on political and theoretical grounds (e.g. Blaut 1983, Yinger 1981). 'Assimilation' is simplistically defined as the socially desirable converse of 'segregation', an historically inevitable outcome of a unilinear process of ethnic competition and upward social mobility. The advocates of minority-group 'assimilation' rarely pause to consider precisely whose interests such a process would serve, casually assuming it to be a universally desirable goal of social policy. It is particularly ironic that geographers have continued to adhere to such a view, as one of the most significant findings of the spatial sociologists was their revelation of the lack of empirical support for the assimilation thesis in terms of a consistent decline in ethnic segregation over time, corresponding with the development of the 'new ethnicity' in the United States (cf. Glazer & Moynihan 1975, Steinberg 1981). Rather belatedly, geographers have begun to realize that 'ethnicity' is a much more slippery concept than they had earlier assumed, necessitating new approaches to the concept's emergent properties and symbolic features (see, for example, Yancey *et al.* 1976, Gans 1979, Smith 1984).

There have, however, been some genuine advances within social geography in the past couple of decades. It is no longer 'unrealistic', for example, to expect geographers to leave their universities and to live with the subjects of their research as Bridget Leach once complained (Leach 1973, p. 236). There is a growing ethnographic tradition in geography which specifically aims to convey the subjective experience of different social groups, drawing on the discipline's long tradition of field research (cf. Jackson 1985). There have also been constructive moves within geography away from a unique focus on 'immigrant' groups as a problem or research topic in themselves, responding to Stuart Hall's comment that:

> Instead of thinking that confronting the questions of race is some sort of moral intellectual academic duty which white people with good feelings do for blacks, one has to remember that the issue of race provides one of the most important ways of understanding how this society actually works and how it has arrived where it is. (Hall 1981, p. 69)

This volume seeks to mark a further step in the transition from a social geography that is exclusively concerned with patterns of immigration, segregation and assimilation towards a more conscious attempt to deal with the political dimension of ethnic and racial studies, challenging the racism that is endemic in British society and in British academia. To this extent, social geography is lagging behind other branches of the social sciences where the study of 'race relations' has undergone a more searching radical critique. It is to this critique that we now turn.

The social construction of 'race'

Recent criticisms of the 'race relations' industry in Britain include the charge that academics have been guilty of abstracting and distorting black people's experience, and of complicity with successive governments in perpetuating the view that black people themselves are the 'problem', rather than the racism of a dominant white society. The case against the conventional sociology of 'race relations' is most succinctly put by two authors from the radical Institute of Race Relations:

> It was not black people who should be examined, but white society; it was not a question of educating blacks and whites for integration, but of fighting institutional racism; it was not race relations that was the field for study, but racism. (Bourne & Sivanandan 1980, p. 339)

To pursue the implications of this assertion within the social sciences requires that we recognize the problematic status of many of the concepts that have become familiar within the tradition of 'race relations' research. This includes not only such commonplace terms as 'assimilation', 'succession', 'pluralism' and so forth but also, more fundamentally, the concept of 'race' itself. If we are to advance an analysis that goes beyond the categories of common-sense understanding (cf. Lawrence 1982a), we must recognize that 'race' has no explanatory value and serves little, if any, analytical purpose. While it might perhaps be argued that 'race' is a valid concept in contemporary political debate, where black people's struggles cannot be reduced to simple class terms, its analytical value in the discourse of academic social science is much less certain. Instead, following Miles (1984), we shall speak of the idea of 'race', distancing ourselves from those who accord the concept explanatory status and focusing instead on its ideological effects in various domains (scientific, political, common-sense, etc.).

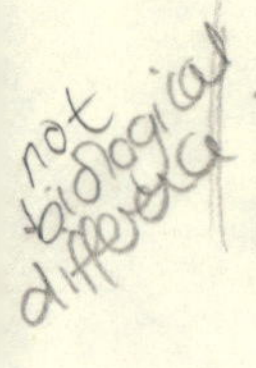

We begin by recognizing that 'race' is fundamentally a social construction rather than a natural division of humankind. With this in mind, Pierre van den Berghe defines 'race' as a group that is socially defined on the basis of physical criteria (van den Berghe 1967, p. 9), although as we shall argue below there are good grounds for recognizing some forms of racism that are couched in cultural rather than in purely physical terms. Van den Berghe defines racism similarly as any set of beliefs that organic, genetically transmitted differences (whether real or imagined) between human groups are intrinsically associated with the presence or absence of certain socially relevant characteristics (ibid., p. 11). He concludes that 'race' has no objective reality independent of its social definition (ibid., p. 148). Having quoted van den Berghe approvingly in this context it is disappointing to note his subsequent flirtation with sociobiology (e.g. van den Berghe 1978) which it is not possible to endorse here.

Although the urge to classify people into a finite number of 'races' has been widespread, it should not be understood as having its roots in an unalterable 'human nature'. The existence of so-called 'natural antipathies' between groups of people is a racist belief for which there is no secure scientific basis. The classification of people based on physical differences such as skin colour is even less 'natural', arising not from some innate human instinct but from specific historical circumstances. The process by which racist distinctions have been 'naturalized' is, in fact, one of the similarities between racism and sexism. For, as Stuart Hall has argued, both ideologies attempt to ground themselves in the evidence of nature:

> It is this transposition from historically and culturally created differences to fixed natural or biological or genetic differences which gives those two ideologies their deep-seated structure. (Hall 1981, p. 64)

Elsewhere, Hall has amplified this point, arguing that it is not helpful to define racism as a 'natural' and permanent feature of all societies, arising out of a universal 'human nature':

> It's not a permanent human or social deposit which is simply waiting there to be triggered off when the circumstances are right. It has no natural and universal law of development. It does not always assume the same shape. There have been many significantly different *racisms* – each historically specific and articulated in a different way with the societies in which they appear. Racism is always historically specific in this way, whatever common features it may appear to share with similar social phenomena. (Hall 1978, p. 26)

In order to counter arguments about the 'natural' origins of racist distinctions, the social construction of 'race' must therefore be understood historically. In Britain, this involves an examination of the colonial context, the legacy of slavery and the growth of the black population (cf. Fryer 1984). Yet, as recent research has shown, British attitudes to 'race' are more closely associated with domestic issues than has previously been assumed. Thus, Greta Jones has demonstrated how variations in British attitudes towards the Morant Bay rebellion in Jamaica reflected social tensions *within Britain itself* (Jones 1980, pp. 140–1). Similarly, Douglas Lorimer's work pays close attention to the overlap between 'race' and 'class' in the analysis of Victorian society (Lorimer 1979), an emphasis which Marian Mair has significantly extended in her recent dissertation on representations of black musicality in Britain since 1750 (Mair 1986).

The racism of previous generations may now appear self-evident. But one should beware of complacency in assuming that our own ideas are so much more enlightened. For, as several essays in the present volume argue (Chs 4, 12 & 13), contemporary social science can play a similar rôle to that of 19th-century natural science in providing academic legitimation for popular racist beliefs, to say nothing of the highly dubious rôle of sociobiology in this respect. More generally, though with specific reference to the evolution of American attitudes towards 'race', Jeffrey Prager has argued against our own smugness in accounting for the structural roots of contemporary inequality. He argues that current explanations that stress socio-economic or cultural reasons for the continued subordination of black people are no less ideological than previous biological or psychological explanations, as both prevent us from perceiving an appropriately individuated and differentiated black community (Prager 1982).

An historical perspective on 'race' also prevents the common misconception that Britain's 'race relations' problem began with the arrival of large numbers of black immigrants from the New Commonwealth in the post-war period. This attitude is frequently associated with the view that Britain has an enviable record of tolerance and fair play towards minority groups that faltered only with the 'flood' of immigrants who began to arrive after 1945. Colin Holmes demonstrates that this view is as historically unfounded as the idea that racism is a peculiarly working-class phenomenon: 'every major immigrant group since 1870 has been the target of some hostility and all sections of the receiving society have at times expressed opposition towards some immigrant group' (Holmes 1982, p. 13).

There are, though, two interrelated strands to the ideology of British tolerance and fair play. The first concerns the implication that there is some theoretical ceiling beyond which any further immigration will inevitably lead to conflict. Immigration can then be restricted on the allegedly humanitarian grounds of ensuring 'good race relations', a feature that was common to both Labour and Conservative party policies during the 1960s (cf. Ben-Tovim & Gabriel 1979, Reeves 1983). The argument is objectionable on the grounds that it ignores the human rights of prospective immigrants who may be perfectly entitled to live in Britain and on the grounds that it legitimizes racist attitudes. (The question of how to determine the magic number is also rarely confronted by those who advocate this 'theory' of immigration control.)

The second ideological strand concerns the use of sensationalist language about the 'flood' or 'tidal wave' of immigrants who are supposedly in danger of 'swamping' British culture. This view, which conveniently ignores the diversity that has always characterized 'British culture', is now so commonplace and apparently unexceptional that even the Prime Minister Margaret Thatcher can be quoted using its unmistakable vocabulary:

> The British character has done so much for democracy, for law, and done so much throughout the world, that if there is a fear that it might be swamped, people are going to react and be rather hostile to those coming in. (quoted in Miles & Phizacklea 1984, p. 5)

All these examples are guilty of the 'foreshortened historical vision' that Lawrence describes as fashionable in 'ethnicity studies' circles (Lawrence 1982b, p. 113). Lawrence goes on to provide a devastating critique of orthodox 'race relations' sociology, notwithstanding the impeccable liberalism of many of its central practitioners. The 'pathological' black family and the problems of 'black youth', allegedly 'torn between two cultures' or vulnerable to other forms of 'identity crisis' (cf. Cashmore & Troyna 1982) are all subjected to Lawrence's analytical probing, guided by the central idea that 'there are *power relations* in operation here which limit the range of choices black people can make' (Lawrence 1982b, p. 116).

Forms of contemporary racism

An historical approach is also capable of identifying the interplay between various forms of racism, contradicting the assumption that Victorian racism was exclusively 'scientific', for example, or that present-day racism is an entirely working-class phenomenon. One of the most important lessons of recent writings about 'race' is, in fact, the need to recognize the multiple forms that racism can take. For racism is not a unitary or static phenomenon:

> Racism does not stay still; it changes shape, size, contours, purpose, function – with changes in the economy, the social structure, the system and, above all, the challenges, the resistances to that system. (Sivanadan 1983, p. 2)

As Sivanandan goes on to argue, most forms of racism, unlike racialism or 'race prejudice', are *structured* (in the sense that they occur in the context of deeply entrenched, asymmetrical power relations) and *institutionalized* (in the sense that they are perpetuated, often unintentionally, through the policies and practices of public and private bodies). Thus, even so-called 'personal' racism is reinforced by institutionalized racism in housing, education and employment, and by the racist stereotypes that are regularly conveyed in the media. Measuring prejudice, as Vaughan Robinson sets out to do below (Ch. 7), therefore becomes a complex, multi-variate exercise that is not really amenable to empirical investigation.

The reproduction of racist ideologies similarly involves a range of social practices from overt aspects of public policy to more mundane features of everyday life. For even such an apparently inoffensive action as telling a racist or 'ethnic' joke serves to reinforce existing prejudices and actively reproduces the unequal social relations upon which more instrumental forms of racism are based. No better example could be found of the mutual interaction between agency and structure that characterizes contemporary structuration theory where 'the structural properties of social systems are both the medium and the outcome of the practices that constitute those systems' (Giddens 1979, p. 69). In this case, racism is a structural property of the British social system and racist jokes are one type of social practice that is both a medium for the reproduction of racist structures and an outcome of the structural racism that characterizes our society.

Institutional racism is perhaps the most invidious form of racism because it operates with the imprimatur of the state. It is, though, a form of racism that is often officially denied. Lord Scarman, for example, was willing to admit that 'police attitudes and methods have not yet sufficiently responded to the problem of policing our multi-racial society' (Scarman 1981, section 4.70), yet he was unwilling to see this as evidence of a deeper structural problem. Racial prejudice was interpreted as characteristic of a few (younger) police officers, while 'the direction and policies of the Metropolitan Police are not racist' (ibid., 8.20). More categorically still, Scarman insisted that 'institutional racism' does not exist in Britain (ibid., 9.1).

Not surprisingly, Scarman's remarks have become the focus of intense debate (Mason 1982, Williams 1985). The core of the debate concerns the issue of whether the charge of institutional racism can only be applied to policies and practices that are *intentionally* racist ('knowingly, as a matter of policy', in Scarman's words), or whether it can be extended also to those institutions that have racist *consequences* ('unwittingly discriminatory', in Scarman's terms), irrespective of the intentions of those who carry out the institution's policies. The debate has its roots in Weber's classic analysis of bureaucracy but the concept was forged in its contemporary sense in the course of the American civil rights movement. In this context:

> institutional racism can be defined as those established laws, customs and practices which systematically reflect and produce racial inequalities in American society. If racist consequences accrue to institutional laws, customs or practices the institution is racist whether or not the individuals maintaining those practices have racist intentions. (Jones 1972, p. 131)

A recent British example of the nature of institutional racism is provided by the Commission for Racial Equality's formal investigation of council-house provision in the London Borough of Hackney, which led to a non-discrimination notice being served under the Race Relations Act 1976. The CRE found that black people were being discriminated against by not being provided with housing of the same quality as that given to white people in similar circumstances (Commission for Racial Equality 1984). Hackney was not thought to be exceptional in this respect and a range of studies now exist that amply confirm the existence of racial discrimination in housing allocation policies in London and throughout Britain, as several papers in the present volume bear witness (Chs 3, 8 & 9). These studies show how discrimination has resulted from the normal policies and practices of local authorities which, irrespective of the intentions of particular officials, had given rise to systematic differences between similarly qualified black and white applicants.

Despite these comments on the irrelevance of official intentions concerning institutional racism, it is not difficult to provide other examples where the state and its agencies have been anything but innocent of the racist effects of their policies. Arnold Hirsch's research on Chicago housing, for example, shows the extent of the state's compliance in the making and remaking of racially segregated housing (Hirsch 1983). The construction of municipal housing ghettos is shown here to be an active process involving deliberate policy initiatives on the part of the Chicago Housing Authority, under the control of Mayor Richard J. Daley's Democratic 'machine'. Opposition by Chicago's white ethnic groups to what they politely term 'neighbourhood change' was given consistent government support and official sanction through the operation of the city's urban renewal and public housing agencies.

The effects of institutional racism are not, of course, limited to the housing market. Other aspects of social welfare provision show similar tendencies. Mark Johnson's essay (Ch. 10), for example, shows how public expenditure cutbacks have had particularly severe consequences for ethnic minorities, a factor that has increased the trend towards the 'privatization' of welfare services through the voluntary sector, as discussed below by Waterman and Kosmin for the case of London's Jewish community (Ch. 11).

While some forms of racism involve the assumption that social groups differ according to recognizable physical criteria, with at least an implication that these differences are innate (biologically inherited or otherwise passively received), there are also more 'modern' forms of racism that have an apparently cultural basis. These 'cultural' theories do not require traits to be inborn but do generally impute the existence of inescapable social differences which legitimize the assignment of inferiority and superiority. One such example is Oscar Lewis's 'culture of poverty' concept which ostensibly refers to the realm of 'culture' while actually involving the transmission of particular 'cultural' traits from generation to generation through its effects on children:

> By the time slum children are age six or seven they have usually absorbed the basic values and attitudes of their subculture and are not psychologically geared to take full advantage of changing conditions or increased opportunities which may occur in their lifetime. (Lewis 1965, p xlv)

Martin Barker suggests that the 'new racism' of the Tory right has a similar cultural basis, fusing the perspectives of ethology and sociobiology with a virulent strain of nationalism and xenophobia. According to Barker, the 'new racism' is being articulated through 'common-sense' political understanding in a particularly pernicious blend of 'pseudo-biological culturalism' (Barker 1981, p. 23). Studies that report the innate 'cultural disadvantage' of black children or the 'pathological' weaknesses of the black family are similarly guilty of cultural racism, ignoring the specific historical conditions that have given rise to such inequalities and the contemporary social forces that are responsible for perpetuating them.

Having considered the multiple forms that racism can assume and the need to develop a fully contextualized historical approach to the social construction of 'race' (cf. Banton, 1977, who treats 'race' more as an exercise in the history of ideas), it is now more feasible to attempt some definitions of this hydra-headed and intractable concept. Racism, as here conceived, involves the attempt by a dominant group to exclude a subordinate group from the material and symbolic rewards of status and power. It differs from other modes of exclusion in terms of the distinguishing features by which groups are identified for exclusion. However, racism need not have recourse to purely physical distinctions but can rest on the recognition of certain 'cultural' traits where these are thought to be an inherent and inviolable characteristic of particular social groups.

In this respect, racism shares many of the characteristics of nationalism. Both assert the existence of an essentially 'imagined community' (Anderson 1983). Indeed, in the course of exploring the particular history of different patterns of racialization in different parts of Britain, Miles has argued that 'English nationalism encapsulates racism': 'racism is the lining of the cloak of nationalism which surrounds and defines the boundaries of England as an imagined community' (Miles 1987, p. 38). Such distinctions are rarely clear-cut, however, and racism frequently parallels or intersects with other axes of discrimination such as those that operate along gender and class lines. Where several dimensions coincide, as with working-class Asians for example, it may be possible to talk about the double oppression of 'race' and class; Asian women may even be considered the subjects of a triple oppression (cf. Parmar 1982).

Rather than prioritizing either 'race' or 'class' in a reductive way, therefore, one must endeavour to find a more subtle means of exploring the complex relationship between the two dimensions. In his critique of *The Empire strikes back*, for example, Robert Miles argues that the 'black masses' are not a 'race' which has to be related to class, but, rather are 'persons whose forms of political struggle can be understood in terms of racialization within a particular set of production (class) relations' (Miles 1984, p. 230). Stuart Hall's position is not dissimilar. Arguing that black people's actions are *negotiated* under conditions of structural subordination, he claims that 'race' is the medium through which working-class blacks '... "live", experience, makes sense of and thus come to a consciousness of their structural subordination' (Hall *et al.* 1978, p. 347).

While some authors (such as Miles 1982) continue to restrict the term 'racism' to the ideological sphere, it is convenient here to extend its use to include a range of social practices that derive from racist beliefs. Some forms of racism involve direct *exclusionary practices*, either thinly disguised as in the case of residential zoning in the United States (Johnston 1984) or totally blatant as in the case of apartheid in South Africa (Smith 1982, 1983). The *ideological effects* of racism are, on the other hand, more insidious and even more pervasive. Racism is an ideology both in the general sense of a system of beliefs characteristic of a particular class or group (Williams 1977) and in the more restricted sense of a system of beliefs that serve to conceal the interests of a particular class or group (Urry 1981). Several of the essays in the present volume address this issue, revealing the mythical nature of Asian business success (Ch. 8), Irish stupidity (Ch. 6), Puerto Rican docility (Ch. 13) and the indolence of Southern blacks (Ch. 14).

As Gramsci's work demonstrates, however, dominant ideologies never achieve a position of unquestioned authority; they are always contested. Opposition to racism takes as many forms as racism itself, ranging from episodic violence and open confrontation to long-term symbolic forms of resistance such as those that have been documented by members of the Centre for Contemporary Cultural Studies at Birmingham (Hall & Jefferson 1976). The extent to which such 'rituals of resistance' remain confined to the symbolic level and the conditions under which they may achieve genuine political gains remains an open question (cf. Moore 1975; Sivanandan 1982, 1983).

The geography of racism

Many forms of racism have an explicitly territorial dimension that requires us to examine the complex interweaving of social relations and spatial structures. This is not to say that we should restrict our analyses of 'race' and racism to a single 'spatial dimension' (cf. Saunders 1985) but that we should be sensitive to the reciprocal relationship between society and space, recognizing both the spatial expression of social processes and the spatial constitution of society.

The geography of racism is discernible at a variety of scales. At the national scale, for example, it is clear that Britain is not a homogeneous political and ideological formation. Thus Susan Smith, below, writes of a specifically *English* racism in terms of the effects of successive government policies on particular geographical forms (Ch. 1), while Robert Miles and Anne Dunlop describe the way that Scottish politics has followed a very different pattern of racialization north of the border (Ch. 5). Similarly, in North America, as Alisdair Rogers and Rika Uto discuss from their work in southern California (Ch. 2), the changing map of residential segregation can be interpreted as a record of the changing relationship between home and work which in turn bears witness to changes in the spatial division of labour.

At the urban scale, Christopher Husbands' research in London's East End has revealed the consistent territorial basis of white working-class support for a succession of extreme right-wing political groups from Mosley's blackshirts to the contemporary National Front (Husbands 1982, 1983). This is not to say that racism is a peculiarly working-class phenomenon; evidence from the United States amply confirms the existence of similar exclusionary practices on the part of middle-class suburbanites who in this case have the additional power of the law and the courts on their side (Johnston 1984). Clearly, though, social relations are rather different in London's East End from the situation in other parts of the city where the local structure of housing and job markets, coupled with what Gerald Suttles has described as 'the cumulative texture of local urban culture' (Suttles 1984), have combined to produce quite different patterns of community life. Sandra Wallman's research in Battersea in South London would certainly support such an interpretation; here, 'the local area is the prime focus of identity and loyalty and there is a tolerance of political and social minorities such that racist or extreme political movements tend to be ignored or actively rejected' (Wallman 1982, p. 16). The implications of 'local style' and the interplay between different strands of personal identity clearly vary between neighbourhoods and have to be explored at the detailed level of ethnographic research, even down to individual households (Wallman 1984).

Other forms of racism have more obviously territorial foundations. The basis of present-day apartheid in the Group Areas Act of 1950 is a particularly extreme case and one that allowed a former South African prime minister to say that if he were to wake up one morning and find himself a black man 'the only major difference would be geographical' (B. J. Vorster, *Johannesburg Star*, 3 April 1973, quoted in Smith 1983). The geographical significance of other, less extreme forms of residential segregation has only recently begun to receive the attention it deserves. Richard Harris, for example, has posed some challenging questions about the sociological significance of residential segregation, particularly in terms of whether it serves to promote class consciousness within a particular residential community or whether it lowers the potential for conflict by reducing the possibility of interaction between different groups (Harris 1984).

Finally, research has only just begun to scratch the surface of the geography of racist violence and of organized resistance by black people. Despite the fact that racist attacks have a long history in Britain (Hiro 1971, Fryer 1984), public attention has been diverted by the media's preoccupation with the idea that black people are disproportionately involved as the perpetrators rather than the victims of violent crime. Recent examples of such 'moral panics' include the press's hysteria about mugging during the 1970s (Hall *et al.* 1978) and about the 'riots' during the early 1980s (Benyon 1984; Burgess 1985; Keith, Ch. 12 below).

Much less well documented are the incidents of racially motivated violence directed against black people that amount to an almost continuous and unrelenting battery of black (particularly Asian) people and their property (Doherty 1983). The police have been slow to respond to these attacks, often calling into doubt the evidence for a 'racial' motive, most notoriously in the case of the 'New Cross Massacre' (*Race Today* 1984). Even accepting official estimates, however, the level of racist violence is appalling. A Home Office survey in 1981, for example, found that Asians are 50 times and West Indians 36 times more likely to be the victims of racist attacks than whites. Only recently has the spatial incidence of such attacks been investigated and the evidence makes chilling reading. Although racial harassment is certainly a London-wide phenomenon, it has been most closely monitored in Tower Hamlets where particularly acute problems have been identified. Local MP Ian Mikardo complained to the Home Secretary in February 1983 of a 'catalogue of violence' by white people, mainly against Bangladeshis, that was met by the police with 'indifference, and sometimes even with hostility' (GLC 1984, p. 14). There is ample scope here for geographers to make a truly significant contribution towards the resolution of a pressing social problem.

The history of black people's resistance is as complex as racism itself. Nonetheless, there are good grounds to suspect that the movement has been geographically as well as historically varied. While there is little point in charting the geography of resistance for its own sake, there are good reasons for trying to relate the spatial incidence of resistance to the changing national and local circumstances in which black people's experience has been moulded. For example, Stuart Hall has hypothesized that the theory of 'assimilation' has to be tested against the real environment of the jobs and localities where black and white workers meet and live (Hall 1978, p. 27). He proceeds to relate the early phase of New Commonwealth immigration to the emergence of an open form of racism in the 'race riots' of Notting Hill and Nottingham in 1958. The Smethwick by-election of 1964 marks a second turning point in the history of British 'race relations', with Peter Griffiths' scurrilous use of the slogan 'If you want a nigger neighbour, vote Labour'. Powellism, too, had specific local roots albeit articulated within the context of a national debate on immigration.

Finally, it can be argued that the sites of black people's struggles have changed over time, focusing at some points on the workplace, at other times around the question of immigration, in the schools or, most recently, on the streets. None of these conflicts can be understood in isolation from the wider struggle of black people to gain a secure place within British society and within the historical context of their role as the suppliers of migrant labour (Miles 1982). It is vital that these conflicts be seen not as a response to a few people's irrational prejudices but as a reflection of a deeply entrenched pattern of racism that has its roots in real material conditions. Social geographers have their part to play in accounting for the present-day manifestations of racism and in understanding how the specific features of contemporary British racism have emerged.

Conclusion

Writing under the same title as the present volume in 1967, Pierre van den Berghe attempted to characterize what he then regarded as the dominant trends in the study of 'race relations' (van den Berghe 1967, pp. 2–8). First, he detected a 'new orthodoxy' concerning the influence of the social environment on perceived racial differences, as opposed to the influence of heredity and the physical environment. He noted the lack of a cross-cultural perspective and of an historical dimension. He criticized the isolation of 'race' from its social context, seeking a greater theoretical integration between 'race relations' and mainstream sociology. And finally, he pointed to the general complacency and optimism of liberal attitudes towards 'race', coupled with a tendency to see racism as a matter of individual prejudice that was amenable to ameliorative policies rather than as a matter of institutionalized practices that were deeply rooted in society's fundamentally unequal power structure.

While it is possible to record some progress during the past two decades, van den Berghe's critique remains apposite for the 1980s. To regard 'race' as a social construction rather than as a biological given may now be widely accepted in the social sciences but it has yet to  penetrate the public consciousness and to influence the realm of common-sense understanding. The need for cross-cultural research remains strong, yet most of the essays in the present volume comprise studies of a single society, leaving comparisons largely unstated. Similarly, the historical dimension is generally implicit rather than explicit, although in this case the present collection of essays fares rather better. The argument for integrating the study of 'race relations' within a more broadly conceived social science is also now quite generally accepted. The present volume attempts to extend this argument by injecting a more rigorously theorized conception of space and place into sociological theories about 'race' and racism. And finally, the complacency of liberal attitudes towards the issues of 'race' and racism is now definitely under attack. Several of the essays in this collection address this issue by providing a critique of the role of academic social science itself or by reviewing the problems that even well-intentioned anti-racialist policies face in counteracting the pervasive effects of institutional racism, whether in housing allocation or other aspects of welfare provision.

There is much that remains to be done, however, if future research is to combine the insights of a broadly structuralist critique of the roots of racism with a fully contextualized understanding of the experience of racism in particular localities at certain times (cf. Harvey 1984). The challenge then remains of carrying forward the critical sense of these essays into our own communities and professions, where racism remains pervasive and deeply entrenched.

Note

1 The term 'black' is used here to refer to all people of ethnic minority origin from Africa, Asia and the Caribbean. It refers as much to a state of consciousness as to physical appearance or skin colour.

References

Anderson, B. 1983. *Imagined communities: reflections on the origin and spread of nationalism*. London: Verso.

Banton, M. 1977. *The idea of race*. London: Tavistock.

Barker, M. 1981. *The new racism: conservatives and the ideology of the tribe*. London: Junction Books.

Ben-Tovim, G. & J. Gabriel 1979. The politics of race in Britain, 1962–79. *Sage Race Relations Abstracts* **4**(4), 1–56.

Benyon, M. (ed.) 1984. *Scarman and after: essays reflecting on Lord Scarman's report, the riots and their aftermath*. Oxford: Pergamon Press.

Blaut, J. M. 1983. Assimilation versus ghettoization. *Antipode* **15**, 35–41.

Bourne, J. & A. Sivanandan 1980. Cheerleaders and ombudsmen: the sociology of race relations in Britain. *Race and Class* **21**, 331–52.

Bridges, L. 1982. Review of C. Peach *et al.* (eds) *Ethnic segregation in cities* and P. Jackson & S. J. Smith (eds) *Social interaction and ethnic segregation*. *Race and Class* **24**, 83–6.

Brown, K. R. 1981. Race, class and culture: towards a theorization of the 'choice/constraint' concept. In *Social interaction and ethnic segregation*, P. Jackson & S. J. Smith (eds), 185–203. London: Academic Press.

Burgess, J. A. 1985. News from nowhere: the press, the riots and the myth of the inner city. In *Geography, the media and popular culture*, J. A. Burgess & J. R. Gold (eds), 192–228. London: Croom Helm.

Cashmore, E. & B. Troyna (eds) 1982. *Black youth in crisis*. London: Allen & Unwin.

Clarke, C. G., D. Ley & C. Peach (eds) 1984. *Geography and ethnic pluralism*. London: Allen & Unwin.

Commission for Racial Equality 1984. *Race and council housing in Hackney: report of a formal investigation*. London: Commission for Racial Equality.

Doherty, J. 1983. Racial conflict, industrial change and social control in post-war Britain. In *Redundant spaces in cities and regions? Studies in industrial decline and social change*, J. Anderson, S. Duncan & R. Hudson (eds), 201–39. London: Academic Press.

Fryer, P. 1984. *Staying power: the history of black people in Britain*. London: Pluto Press.

Gans, H. J. 1979. Symbolic ethnicity: the future of ethnic groups and cultures in America. *Ethnic and Racial Studies* **2**, 1–20.

Giddens, A. 1979. *Central problems in social theory*. London: Macmillan.

Glazer, N. & D. P. Moynihan (eds) 1975. *Ethnicity: theory and experience*. Cambridge, Mass.: Harvard University Press.

GLC 1984. *Racial harassment in London*. Report of a panel of inquiry set up by the GLC Police Committee.

Gregory, D. & J. Urry (eds) 1985. *Social relations and spatial structures*. London: Macmillan.

Hall, S. 1978. Racism and reaction. In *Five views of multi-racial Britain*, Commission for Racial Equality (eds), 23–35. London: Commission for Racial Equality.

Hall, S. 1981. Teaching race. In *The school in the multicultural society – a reader*, A. James & R. Jeffcoate (eds), 58–69. London: Harper & Row.

Hall, S., C. Critcher, T. Jefferson, J. Clarke, & B. Roberts 1978. *Policing the crisis: mugging, the state, and law and order*. London: Macmillan.

Hall, S. & T. Jefferson (eds) 1976. *Resistance through rituals: youth subcultures in post-war Britain*. London: Hutchinson.

Harris, R. 1984. Residential segregation and class formation in the capitalist city: a review and directions for research. *Progress in Human Geography* **8**, 26–49.

Harvey, D. 1973. *Social justice and the city*. London: Edward Arnold.

Harvey, D. 1984. On the history and present condition of geography: an historical materialist manifesto. *Professional Geographer* **46**, 1–11.

Harvey, D. 1985. *The urbanization of capital*. Oxford: Basil Blackwell.

Hiro, D. 1971. *Black British, white British*. London: Eyre & Spottiswoode.

Hirsch, A. R. 1983. *Making the second ghetto: race and housing in Chicago, 1940–1960*. Cambridge: Cambridge University Press.

Holmes, C. 1982. The Promised Land? Immigration into Britain 1870–1980. In *Demography of immigrants and minority groups in the United Kingdom*, D. A. Coleman (ed.), 1–21. London: Academic Press.

Husbands, C. T. 1982. East End racism, 1900–1980: geographical continuities in vigilantist and extreme right-wing political behaviour. *London Journal* **8**, 3–26.

Husbands, C. T. 1983. *Racial exclusionism and the city: the urban support of the National Front*. London: Allen & Unwin.

Jackson, P. 1985. Urban ethnography. *Progress in Human Geography* **9**, 157–76.

Jackson, P. & S. J. Smith (eds) 1981. *Social interaction and ethnic segregation*. London: Academic Press.

Johnston, R. J. 1984. *Residential segregation, the state, and constitutional conflict in American urban areas*. London: Academic Press.

Jones, E. 1960. *A social geography of Belfast*. Oxford: Oxford University Press.

Jones, G. 1980. *Social Darwinism and English thought: the interaction between biological and social theory*. Brighton: Harvester Press.

Jones, J. M. 1972. *Prejudice and racism*. Reading, Mass.: Addison-Wesley.

Lawrence, E. 1982a. Just plain common sense: the 'roots' of racism. In *The Empire strikes back: race and racism in 70s Britain*, Centre for Contemporary Cultural Studies (eds), 47–94. London: Hutchinson.

Lawrence, E. 1982b. In the abundance of water the fool is thirsty: sociology and black 'pathology'. In *The Empire strikes back: race and racism in 70s Britain*, Centre for Contemporary Cultural Studies (eds), 95–142. London: Hutchinson.

Leach, B. 1973. The social geographer and black people: can geography contribute to race relations? *Race* **15**, 230–41.

Lewis, O. 1965. *La vida: a Puerto Rican family in the culture of poverty – San Juan and New York*. New York: Random House.

Lorimer, D. A. 1979. *Colour, class and the Victorians: English attitudes to the Negro in the mid-nineteenth century*. Leicester: Leicester University Press.

Mair, M. 1986. *Black rhythm and British reserve: interpretations of black musicality in racist ideology since 1750*. Unpublished Ph.D. thesis, University of London.

Mason, D. 1982. After Scarman: a note on the concept of institutional racism. *New Community* **10**, 38–45.

Miles, R. 1982. *Racism and migrant labour*. London: Routledge & Kegan Paul.

Miles, R. 1984. Marxism versus the sociology of 'race relations'. *Ethnic and Racial Studies* **7**, 217–37.

Miles, R. 1987. Recent Marxist theories of nationalism and the issue of racism. *British Journal of Sociology*, **38**. 24–43.

Miles, R. & A. Phizacklea 1984. *White man's country: racism in British politics*. London: Pluto Press.

Moore, R. 1975. *Racism and black resistance in Britain*. London: Pluto Press.

Parmar, P. 1982. Gender, race and class: Asian women in resistance. In *The Empire strikes back: race and racism in 70s Britain*, Centre for Contemporary Cultural Studies (eds), 236–75. London: Hutchinson.

Peach, C. (ed.) 1975). *Urban social segregation*. London: Longman.

Peach, C., V. Robinson & S. Smith (eds) 1981. *Ethnic segregation in cities*. London: Croom Helm.

Peach, G. C. K. 1983. Ethnicity. In *Progress in urban geography*, M. Pacione (ed.), 103–27. London: Croom Helm.

Prager, J. 1982. American racial ideology as collective representation. *Ethnic and Racial Studies* **5**, 99–119.

Race Today 1984. *The New Cross massacre story: interviews with John La Rose*. London: Alliance of the Black Parents Movement, Black Youth Movement and the *Race Today* Collective.

Reeves, F. 1983. *British racial discourse*. Cambridge: Cambridge University Press.

Rex, J. & R. Moore 1967. *Race, community and conflict: a study of Sparkbrook*. London: Oxford University Press for the Institute of Race Relations.

Robinson, V. 1982. The assimilation of South and East African Asian immigrants in Britain. In *Demography of immigrants and minority groups in the United Kingdom*, D. A. Coleman (ed.), 143–68. London: Academic Press.

Saunders, P. 1985. Space, the city and urban sociology. In *Social relations and spatial structures*, D. Gregory & J. Urry (eds), 67–89. London: Macmillan.

Scarman, Lord 1981. *The Brixton disorders, 10–12 April 1981*. Cmnd 8427. London: HMSO.

Sivanandan, A. 1982. *A different hunger: writings on black resistance*. London: Pluto Press.

Sivanandan, A. 1983. Challenging racism: strategies for the '80s. *Race and Class* **25**, 1–11.

Smith, D. M. (ed.) 1982. *Living under apartheid: aspects of urbanization and social change in South Africa*. London: Allen & Unwin.

Smith, D. M. 1983. *Update: apartheid in South Africa*. Department of Geography & Earth Science, Queen Mary College, Special Publication No. 6.

Smith, S. J. 1984. Negotiating ethnicity in an uncertain environment. *Ethnic and Racial Studies* **7**, 360–73.

Steinberg, S. 1981. *The ethnic myth: race, ethnicity and class in America*. Boston: Mass.: Beacon Press.

Suttles, G. D. 1984. The cumulative texture of local urban culture. *American Journal of Sociology* **90**, 283–304.

Urry, J. 1981. *The anatomy of capitalist societies*. London: Macmillan.

van den Berghe, P. L. 1967. *Race and racism: a comparative perspective*. New York: Wiley.

van den Berghe, P. L. 1978. Race and ethnicity: a sociobiological perspective. *Ethnic and Racial Studies* **1**, 401–11.

Wallman, S. (ed.) 1982. *Living in South London*. London: Gower.

Wallman, S. 1984. *Eight London households*. London: Tavistock.

Walter, B. 1984. Tradition and ethnic interaction: second-wave Irish settlement in Luton and Bolton. In *Geography and ethnic pluralism*, C. G. Clarke, D. Ley & C. Peach (eds), 258–83. London: Allen & Unwin.

Williams, J. 1985. Redefining institutional racism. *Ethnic and Racial Studies* **8**, 323–48.

Williams, R. 1977. *Marxism and literature*. Oxford: Oxford University Press.

Yancey, W. L., E. P. Ericksen & R. N. Juliani 1976. Emergent ethnicity: a review and reformulation. *American Sociological Review* **41**, 391–403.

Yinger, J. M. 1981. Towards a theory of assimilation and dissimilation. *Ethnic and Racial Studies* **4**, 249–64.

PART I

Segregation reconsidered

The relationship between physical and social distance has exercised the imagination of geographers and sociologists for generations. While ample attention has been paid to the empirical question of mapping and measuring the residential patterns of different social groups, the theoretical significance of varying degrees of clustering and dispersal has received far less attention. The first three essays begin to address this issue, recognizing the *political* significance of segregation in both reflecting and reproducing the structure of social relations.

In the first essay, **Susan Smith** offers a political theory of residential segregation which she interprets as the geographical expression of a specifically English racism. She argues that, whatever their stated intention, housing and urban policies have had the practical effect of sustaining racial segregation. While issues of 'race' have rarely been the subject of direct government policy, she presents a persuasive argument to show how successive policies from slum clearance schemes to improvement grants and council house sales have contributed to the social reproduction of a spatially segregated black population.

Whereas Smith deals with the political context and social consequences of residential segregation, **Alisdair Rogers** and **Rika Uto** focus on the significance of the geographical separation of home and work in explaining the social geography of southern California. Emphasizing the centrality of production in the structuring of social relations, they employ post-Weberian location analysis and labour theory to explain the ethnic division of labour that has developed in association with the rise of high-tech industry in Los Angeles and Orange Counties. Drawing on the work of Michael Storper, Richard Walker and Allen Scott, in particular, Rogers and Uto seek to explain the persistence of the ghetto in social and spatial terms and the paradox of continued Latino in-migration at a time of high black unemployment. They explore the relationship between the spheres of production and reproduction, the 'racialization' of different kinds of labour, and the relatively neglected rôle of the state as an employer of minority labour.

David Sibley is also concerned with the rôle of the state, which he explores in relation to the development of settlement policies towards Gypsies. He shows how central and local government have opposed the Gypsy's characteristic mobility with specific residential policies involving sanctions on their location outside certain designated areas. Sibley further shows how the state legitimizes the use of such overt means of social control via the perpetuation of certain biological and cultural myths which serve to define the Gypsies as a 'deviant' group. Social marginality translates directly into spatial terms, as the ideological designation of the Gypsies as 'undesirable' permits the adoption of a crudely racist strategy of territorial containment.

1 *Residential segregation: a geography of English racism?*

SUSAN J. SMITH

Racial residential segregation is well documented empirically, but less often the focus of theoretical scrutiny. Arguing against such neglect, this chapter explores, from an historical perspective, some legislative and political aspects of the segregation of 'racial' minorities in Britain. Focusing primarily on those whose family histories originate in the New Commonwealth and Pakistan, I use the term 'black' to describe people of both South Asian and Afro-Caribbean appearance. Although this label masks important cultural differences – which Peach (1984) and Robinson (1981) show are important axes of residential differentiation *within* black communities – my starting point is in black people's common experience of segregation from the majority of white Britons.

It is impossible to specify accurately the size or characteristics of the black population (Peach 1982, Rees & Birkin 1984 itemize the difficulties), and the term 'race' is, of course, a social construction – real only in its effects and, except as a focus for combating racism, 'racist' in application. Nevertheless, Britain's economic and political history has defined dark-skinned immigrants and their descendants as a 'racial' category (see Miles 1982), whose incumbents experience more discrimination and disadvantage than do white immigrants (Brown 1984). From this starting point my aim is to assess how far racial residential segregation might be interpreted theoretically as an expression of the racism that permeates Britain's social and political life. To that end, I focus mainly on the effects of housing and urban policy in England (the different social and political context of the Scottish experience is outlined in Ch. 5). Obviously other political arenas, such as those encasing education and employment policies, also contribute to the social reproduction of a segregated black population: this chapter, however, is specifically concerned with the place of *residential* segregation in that wider process.

Race as a dimension of residential segregation

Ballard (1983) and Brown (1984) estimate the size of Britain's black population to be about 2.2 million (somewhat less than 4 per cent of the total population) in 1980–2. The OPCS labour-force survey (1983) estimates that of those whose 'ethnic origin' (by birth or descent) is in the West Indies/Guyana or South Asia, 46 per cent are Indian, 20 per cent Pakistani, 5 per cent Bangladeshi and 29 per cent West Indian/Guyanese. Although black people have lived in Britain since the 16th century – and before, according to Fryer's (1984) meticulous documentation – their present regional distribution was established during a brief period of sustained immigration (rather small-scale by international standards) during the late 1950s and the 1960s. The relationship between labour demand and the location of migrants is discussed by Peach (1966, 1968, 1978–9) and Robinson (1980b). This chapter, however, is less concerned with the economic influences that initiated segregation than with the political factors that help to sustain it.

The persisting spatial dissimilarity of black and white households on a regional scale is well documented elsewhere (Brown 1984, Jones 1978, Peach 1982). The majority of black people live in Greater London and the West Midlands (where 50 per cent of black households but only 20 per cent of white families live) and in the largest textile towns of Lancashire and Yorkshire; black households are less prominent in the coalfield regions and in the spatially peripheral heavy-industrial regions of Scotland, Wales, the North and the North West.

The black population is overwhelmingly urban – only about 3 per cent live in rural enumeration districts. As many as 43 per cent of West Indians and 23 per cent of Asians live in inner-city areas of London, Birmingham and Manchester, where only 6 per cent of the white population is to be found. Overall, three-quarters of the black population live in a set of urban enumeration districts which contain only one-tenth of the whites (Brown 1984). Black people constitute an average of 20 per cent of the population in these enumeration districts, though the figure can be much higher (Haynes 1983).

This broad locational dissimilarity between black and white Britain shows few signs of changing. The Policy Studies Institute survey found no significant dispersal of blacks into white-dominated wards between 1971 and 1982 (Brown 1984). Only in wards relatively densely populated by blacks has there been some residential dispersal from high- to low-concentration enumeration districts. Despite evidence of the limited dispersal of West Indians, associated with their movement into council housing (Lee 1977, Peach & Shah 1980), the PSI survey confirms the dominant theme of increased polarization of black and white residential space which runs through the work of Cater and Jones (1979) Peach (1966, 1982), and Jones (1978). Jones (1983), in fact, identifies a trend in some areas towards racially homogeneous neighbourhoods large enough to form the basis of a separate community life.

Although some of Britain's most deprived urban areas (including Glasgow, Belfast and Tyneside) have few black residents, overall in Britain black people are disproportionately likely to live in the most deprived enumeration districts (Brown 1984). Similarly, it is an enduring feature of successive surveys that, in each sector of the housing market, black people experience below-average housing conditions (see McKay 1977 for a summary; see also Clark 1977, Karn 1982). Although in absolute terms there has been some improvement to the quality of black people's housing in the past decade, black households are still under-represented as the occupants of detached and semi-detached homes, and they are more likely than whites to occupy pre-war properties, to share their home with another household, and to live at above-average densities. Given an association between residential differentiation and social disadvantage, there can be few grounds for complacency in the oft-quoted finding that intra-urban racial segregation in Britain is less marked than in the USA.

Obviously, segregation coupled with relative deprivation does not persist independently of economic inequality. The National Dwelling and Housing survey of 1977–8, for instance, revealed that black heads of households face disproportionate risks of unemployment. Black people born in the UK to West Indian parents are four to five times as likely as their white counterparts to be unemployed (Cross 1982). Moreover, white people, whether born in the UK or not, tend to be of higher status than any non-white immigrant group; and 'second generation' West Indians have, on average, an even lower status than their 'first generation' predecessors (Barber 1981). However, the locational dissimilarity and housing discrepancies between the majority of black and white households cannot be explained wholly in terms of income differentials (Clark 1977, Smith 1978). Indeed, McKay's study prompts him to conclude that in England, 'as far as racial minorities are concerned, residential location has been critical in determining their subordinate position in society' (McKay 1977, p. 16). It should not be assumed, therefore, that locational marginality simply reflects economic marginality. In part, this chapter is an attempt to elicit some of the political and social factors that mediate between the two.

What follows shows first how existing patterns of residential segregation reflecting racial inequality, which emerged from the economic exigencies of a short period of labour or refugee migration, have been sustained as a consequence – sometimes intentional, but more usually unanticipated – of a series of central and local government decisions implemented without specific reference to the distinctiveness of black people's disadvantage. Having considered the legislative basis of segregation, I then attempt to set these decisions in the wider context of the history of the idea of race and segregation in British politics. For however stimulating empirical debates about the form and intensity of segregation might be (see Jones & McEvoy 1978, 1979a,b; Lee 1978; Morgan 1980, 1983; Morgan & Norbury 1981; Peach 1979a,b, 1981; Robinson 1980a; Sims 1981; Woods 1976), a more fundamental, yet frequently neglected, question concerns its *meaning* within the British political economy. I suggest, therefore, that it is only within a theoretical framework directing attention to the politics of race – which of course reach beyond the governmental framework in which they are here located – that it is possible to explain how racial residential segregation is sustained and reproduced now that New Commonwealth immigration has all but ceased and as much as 40 per cent of the black population is British-born.

Public policies sustaining segregation

More than a decade ago, a report of the Select Committee on Race Relations and Immigration (1974–5) argued that few departments of the Home Office were equipped to deal effectively with the needs of black people. Concern was expressed regarding the rôles of the Departments of Employment, Education and Science, and Health and Social Security. The most serious reservations, however, related to the Department of the Environment, which is responsible for two areas of legislation carrying particularly important implications for race relations: housing policy and the Urban Programme. For the moment, I concentrate on the former. First, I suggest that although central government has always been reluctant to build the concept of 'race' into legislation combating disadvantage in housing and related fields, national housing policies have had a direct and unmistakable (if usually unintended) impact on the life-chances of the black population – most notably by working to sustain relatively high levels of intra-urban segregation. Local governments have had to work within the constraints imposed by the outcomes of centrally dispensed legislation. Nevertheless, the second part of this section provides an account of how local institutional practices have, usually inadvertently, encouraged racial segregation (in all sectors of the housing market) to be associated with the racial inequalities perpetuated in other spheres of social and economic life.

Housing policy: the rôle of central government

Black workers were originally forced to cluster into inner-city private rental accommodation as a consequence of the postwar housing shortage (Doherty 1983). Although Britain has never had a policy linking housing provision with labour migration, dispersal – initially conceived on a regional rather than intra-urban scale – was held to be the solution to immigrants' housing problems. The government's enthusiasm for dispersal came with a package of integrationist measures introduced, in the 1965 White Paper on Immigration from the Commonwealth, to offset public objections to tighter immigration controls (see Dummett & Dummett 1969). Subsequently, the dispersal ideal received further legitimation from the report of the Central Housing Advisory Committee (1969).

The government did not, however, commit itself to a comprehensive dispersal scheme. On the national scale, it was expected that dispersal, of both the black and the white populations, would be a natural outcome of successful regional economic development and planning. It was therefore the view of Maurice Foley, appointed in 1965 to co-ordinate the work of government departments in the integration of immigrants, that dispersal policies specifically directed towards the black population were unnecessary. At the regional level, it was argued in the same year by Robert Mellish that since immigrants would benefit with the rest of the population from overspill schemes, they required no special provision (Hansard 1964–5, v. 712, c. 35). On the intra-urban scale – the focus of this chapter – the responsibility for dispersal was assigned (by the White Paper *Race relations and housing*, Cmnd 6232, 1975) to the local authorities, which were expected to formulate a 'balanced view' on it. In so devolving responsibility for desegregation, the government effectively washed its hands of a contentious and divisive issue, avoiding confrontation with the white electorate (often perceived to be resistant to residential integration) and, by favouring the concept of dispersal, appearing to support black people's interests. Ironically, almost every major Housing Act both preceding and following this decision has had the consequence of sustaining racial segregation and reducing the residential options open to black households.

One of the earliest policy changes with such an effect was the shift from slum clearance and redevelopment to *in situ* improvement, introduced in the 1969 Housing Act. Between 1958 and 1968, slum clearance in England and Wales decanted some 160 000 to 180 000 people per year from inner-city slums into peripheral estates and high-rise flats. Most of those removed were white. The 'middle ring' Victorian and Edwardian apartment houses and terraces which then accommodated the majority of black households would have been the next to go had clearance policies remained in place. By the mid-1960s, however, there were many pressures working to change the emphasis of housing policy towards gradual *in situ* improvement. These pressures included local opposition to the scale and organization of redevelopment; a massive repairs problem in housing *not* scheduled for demolition; and government concern to control public spending in the face of economic decline (Bassett & Short 1980, p. 122). An additional pressure, though, was perceived social resistance to the rehousing of black people onto new council estates. Reginald Freeson raised this issue in Parliament in 1966, asking about the extent to which 'redevelopment plans are being held up by local authorities because they do not wish to accept responsibility for rehousing immigrants living in twilight zones of major city areas' (Hansard 1965–6, v. 725, c. 239). His question was dismissed as unfounded. Yet it was an allegation considered in a Political and Economic Planning survey in 1966 (Daniel 1968) and ratified in Rex and Moore's classic study of Sparkbrook (1967; see also Daniel 1968, McKay 1977, Smith & Whalley 1975). Smith and Whalley (1975, p. 82) further point out that even those relatively few black people who did live in clearance areas tended to be excluded from rehousing and were 'effectively corralled into the remaining area of suitable private housing'. The *timing* of the shift away from comprehensive redevelopment had the consequence, intended or otherwise, of retaining the black population in those relatively highly segregated areas of cities to which the migrant labour process had originally drawn them.

The General Improvement Areas (GIAs) provided for in the 1969 Housing Act, and the Housing Action Areas (HAAs) introduced from 1974, form the backbone of central government's area-based *in situ* improvement programme. It was always expected that such areas, especially HAAs, when designated in the major cities, would coincide with the main neighbourhoods of black residence (see the Department of the Environment's White Paper on *Race relations and housing*, Cmnd 6232, 1975). Certainly, both GIAs and HAAs tended, on declaration, to house more black families than had the slum-clearance districts. Additionally, as Rex (1981) points out, policies associated with the improvement programme actually enhanced those patterns of intra-urban segregation already preserved by a shift from redevelopment to improvement. The most potent of these policies are sketched below.

The first relates to housing association activity. In recent years, government withdrawal from public building, loss of council stock through sales, and rent increases in the public sector have all enhanced the rôle of housing associations in providing for the relatively large proportion of the black (particularly Afro-Caribbean) community who find that they must rent (see Niner 1984). Grant-assisted GIAs and HAAs offered an important incentive for housing association conversions. Because such tenancies were often more open to black households than were council lets, the process helped to intensify racial segregation in some areas.

Spatial variability in the distribution of local-authority mortgage finance had a similar effect. Council loans were available for the older, cheaper properties of the inner cities, in which building societies were loath to invest (see Williams 1977). Mortgages were allocated 'down market' (see Merrett 1982), often attracting black buyers to areas where black households were already statistically over-represented. As many as one-third of black owner-occupiers in Smith's (1976) survey relied on such finance. By 1982, the proportion had shrunk to 28 per cent for West Indians and 17 per cent for Asians; but by this time, since the granting of such loans peaked in 1973–5, the geography of council mortgages had already had its effect on segregation.

The late 1970s, and, especially, the years since the 1980 Housing Act, mark the culmination of a shift away from direct state provision of homes (through public-sector renting) towards policies stimulating high rates of owner-occupation together with private investment in neighbourhood improvement. Between 1978-9 and 1983-4, public expenditure on housing decreased by 55 per cent in real terms. Council-house building has been curtailed, existing stock is being sold, and tax incentives for mortgage borrowing remain high. Thus, while only 28 per cent of the country's housing stock was owner-occupied in 1953, the figure had risen to 54 per cent in 1976 and stands at over 60 per cent today.

Although Smith (1976) has shown that, for the black population, home-ownership does not mean better housing conditions, and Karn (1977) has established the extent to which the structure of housing finance discriminates against low-income owner-occupiers, a category into which the bulk of black homeowners fall, it is Robin Ward (1981, 1982) who best documents the effects on racial residential segregation of the so-called 'commodification' of housing encouraged by government policy (more accurately, the process has re-orientated rather than removed state intervention, e.g. by replacing housing subsidy in kind with tax relief on mortgage interest). Ward's argument is that as housing is increasingly traded according to market principles and interests, the social polarization between owners and renters, and between those owners whose homes provide a source of capital accumulation and those whose homes cannot store wealth, is intensified. As a consequence of the processes by which dwellings are now coming into owner-occupation, Ward shows further that much of this social polarization is aligned with racial differences and is expressed in the form of residential segregation.

On a national scale, Ward (1981) shows that the ability of black people to take advantage of the 'commodification' of housing varies regionally, reflecting the uneven pattern of economic change and urban development in postwar Britain. He goes on to show that, within cities, the precise pattern of council-house sales has helped to sustain high levels of racial segregation, restricting black households to inner-urban rings (Ward 1982). For the council homes most suitable for purchase and capital accumulation are located on suburban estates, and are almost exclusively occupied by white households. The more centrally located deck-access flats and maisonettes in which West Indians concentrate are much less appealing and are least able to store wealth. In fact, Maclennan and Ermisch (1986) suggest that as many as 85 per cent of council-home sales have been of houses rather than flats, and that, judging by their characteristics, only a small proportion could have been located in the inner cities.

As the council sector dwindles, therefore, and the best properties are sold to white tenants, the prospects for mobility amongst black council tenants (already concentrated in the inner city) are restricted to a limited range of inner-area destinations that can only reinforce present patterns of racial segregation. Those who do buy, or who are already owner-occupiers, face the locational inertia associated with a downward spiral of house conditions and relative house prices which, though also affecting poor whites, is most marked in Asian-dominated areas (Karn *et al.* 1985). Ward (1981, p. 15) concludes, therefore, that recent trends in the 'commodification' of housing are likely to 'increase *racial* polarization between white households in areas of desirable property and both Asians in inner areas of cheap owner-occupation who can expect a much lower return on their investment in housing and West Indians in the public sector'. For the future, patterns of property inheritance can only increase the economic differentials (and spatial dissimilarity) between black and white owner-occupiers; and black tenants moving into or within the public sector will have a decreasing range of locations from which to choose.

This brief survey suggests that it is central rather than local government whose decisions (both directly, and through their effects on the market) most firmly establish the limits of black residential space. Local institutions, have, nevertheless, played an important rôle with respect to the finer details of segregation, and they have helped to seal its relationship with racial disadvantage.

Policy in practice: the rôle of local institutions

Although central government retains control over housing standards and overall costs, local authorities have traditionally exercised considerable discretion in catering to housing need, in housing management and in improvement grant policies. In the past decade the scope of this discretion has steadily narrowed as local authorities become increasingly dependent on central government finance (see Kirby 1982). Ever more stringent cash limits have had the effect of penalizing the most expensive local governments – those in urban areas where the majority of the black population is concentrated (see Short 1984). Acknowledging such constraints, and those set by the policy outcomes sketched above, this section considers the achievements of local governments and local institutions (and the implications of these for segregation) in meeting the housing needs of black people.

During the 1960s, local authorities had little deliberate impact on the development or dissolution of racial segregation. As McKay (1977, p. 96) observes, 'many local governments (in areas of high immigration, many of them Labour-controlled), behaved as though blacks did not exist'. In these early post-immigration years, local authorities sustained the tradition of central government, failing to provide black renters with public housing and doing little to intervene in the processes sustaining segregation within the private sector. Below, the effects on racial segregation of the changing response of local institutions to the circumstances and needs of the black population are summarized for each major sector of the housing market in turn.

Private renting Both West Indian and Asian immigrants moved initially into privately rented accommodation. Although today they are under-represented in this sector (which houses only 6 per cent each of West Indian and Asian households but 9 per cent of white families), New Commonwealth and Pakistani tenants generally occupy below-average accommodation and live at above-average densities (Bovaird *et al.* 1985). Significantly, in 1982 as many as half the West Indian renters and one-third of Asians surveyed by PSI claimed to have experienced discrimination in obtaining their homes (Brown 1984).

One consequence has been the emergence of a dual rental market in which, by 1975 hardly any whites were renting from black landlords and less than 15 per cent of black tenants rented from white landlords (Smith 1976). The persistence of this trend is confirmed by Doling and Davies (1983) who find that black private tenants in Birmingham occupy poorer property than whites but pay higher rents. Amongst other things, this reflects black tenants' lesser use of the Rent Acts to obtain fair renting agreements. Black renters appear to face both direct and indirect discrimination sufficient to sustain segregation and confine them to homes of inferior quality.

The traditional rôle of the privately rented sector (based on individual or commercial landlords) is increasingly assumed by housing associations, which are publicly funded but privately administered. Some sections of the black population cluster disproportionately into these properties, which accommodate only 2 per cent of all householders but 8 per cent of West Indian families, 4 per cent of Bangladeshis and 3 per cent of Hindu families. Thus, despite the relatively scanty documentation of housing association policies and practices towards black people, Niner (1984) argues that their significance for racial minorities is considerably greater than their small contribution to the property market suggests. Although she is able to provide information on the treatment of black applicants by three large urban housing associations – one apparently not discriminating at all; a second discriminating against blacks in terms of allocation criteria, waiting time, choice and type of dwelling allocated; and the third locating blacks in racially segregated areas of poor environmental quality – it is too soon to make definitive statements about the impact of this sector on the overall pattern of racial segregation and housing disadvantage.

Owner-occupation – a low-income solution? Between 1966 and 1971, black households moved rapidly out of private renting into home-ownership (which, according to PSI, now accounts for 72 per cent of Asian families but, because of their more rapid entry into the council sector, only 41 per cent of West Indian households). As private renting contracted, building societies' responses to the nation's changing economic fortunes were important in shaping patterns of racial segregation. Moreover, the discriminatory practices which contribute to this are shown by Stevens *et al.* (1982) to reflect local branch discretion rather than overall company policy. At a local level, therefore, building-society activity has had important consequences for the differentiation of residential space.

On the one hand, the property boom of the early 1970s allowed building societies to attract massive investment (as their interest rates became more competitive), increasing the availability of mortgage finance and encouraging more flexible lending criteria which opened up the inner-ring housing market to marginal buyers. The consequences included increased racial segregation (as a corollary of accelerated white outmigration), an increase in black owner-occupation, and the birth of a generation of Asian estate agents whose rôle will be considered shortly (see Cater 1981).

Karn (1982), on the other hand, looks at the more obvious effects of fiscal restraint and housing shortage on building-society activity, establishing that it is more difficult under these circumstances for black people than for whites to obtain funding for house purchase (see also Brown 1984). She shows that mortgage refusals discriminate against black buyers (Haynes, 1983, too notes that only 50 per cent black buyers in contrast to 75 per cent whites have a building-society mortgage); that the policy of not lending to non-savers is more often waived for whites than for blacks; and that when loans are granted, stereotyped beliefs about area preferences 'steer' buyers so as to sustain existing patterns of segregation.

The inadequacy of conventional mortgage provision encouraged not only local authorities but also banks and finance houses to intervene. The propensity of the latter to lend on the cheap pre-1919 properties of the inner city (i.e. those within the financial grasp of the black population) was qualified by high interest rates, short loan periods, and punitive clauses to safeguard investment. This all worked to the disadvantage of a broad cross-section of the black population as compared with only a minority of white buyers. According to Karn (1982, p. 46) the dependence of black, especially Asian, owners on such alternatives to conventional lending has contributed to their very marked concentration in the inner cities. Ironically, this means that many black borrowers pay more for their credit and – given the structure of tax relief on mortgage interest and restrictions on the option mortgage scheme – receive less subsidy than higher-income owners with more expensive homes (Karn 1977).

The activity – or non-activity – of building societies has perhaps been most significant for the segregation of black owner-occupiers, but a second source of local influence can be traced to the practices of estate agents. Discrimination here has been apparent since the mid-1960s (Daniel 1968), and Cater (1981) shows that vendor conventions can help to sustain segregation. His research in Bradford indicates that most sales *to* Asians are made through Asian estate agents operating in a residual market and selling dwellings *for* Asians. The process is augmented by white agents who 'steer' their few Asian clients towards cheap older properties, again in Asian-dominated areas. Additionally, as many as 50 per cent of Asian-to-Asian transfers may take place informally, involving simply an exchange of dwellings in the least desirable areas. All three practices help perpetuate an already marked level of Asian segregation within some of the city's worst properties.

Having bought into relatively segregated neighbourhoods, usually in the least attractive, ageing urban areas, black owners appear to have less opportunity than whites to improve their properties (Brown 1984). Although a higher proportion of black than white owners apply for local authority improvement grants, of those surveyed by PSI, only 45 per cent of West Indian applicants (in contrast with 65 per cent of Asians and Whites) were successful.

As in the privately rented sector there is evidence amongst owner-occupiers not only that social and economic policies can have the unanticipated outcome of reinforcing segregation, but also that local institutional practices tend to increase the probability of this being coupled with relative disadvantage.

Public-sector housing The opening of the public sector to black tenants in the past decade has undoubtedly contributed to an overall improvement in black people's housing conditions. This is also responsible for the small amount of desegregation indicated by some measures. From housing just 1 per cent of black households in the mid-1960s, local authorities accommodated 4 per cent of Asian and 26 per cent of West Indian households in 1974 (Smith 1976), 10 per cent and 45 per cent in 1977 (according to the National Dwelling and Housing Survey of that year), and 19 per cent and 46 per cent in 1982 (Brown 1984).

Surveys of racial discrimination published in the mid-1960s found that black people often failed to register for council housing because they believed they would be unlikely to qualify. Evidence of the discriminatory effects of residence requirements, owner-occupier disqualification, and bias against unmarried couples and extended families, suggests the belief was justified. Henderson and Karn (1984), however, show that it is not merely difficulty of access that gives cause for concern, but also the tendency for blacks to receive the worst-quality properties and to be concentrated into certain estates – usually those of the poorest quality in the inner areas of cities. Black tenants (especially Asians) tend to be allocated older housing than their white counterparts; and Brown (1984) indicates that even controlling for number of council homes lived in, time of allocation, and characteristics of household, black families live in smaller, more crowded homes than white tenants, they tend more often to be allocated flats, and their flats tend to be located on the higher floors of multistorey blocks. Local reports in London (Phillips 1986), Hackney (CRE 1984a), Liverpool (CRE 1984b) and Walsall (CRE 1985), as well as the survey of the Association of Metropolitan Authorities (1985), find similar evidence of disadvantage amongst black tenants. Not surprisingly, in the recent British social attitudes survey, only 21 per cent of black respondents as compared with 36 per cent of the population as a whole find council estates pleasant to live in; and this, as Jowell and Airey (1984, p. 91) point out, must contribute to the 'growing evidence that the council estates on which ethnic minorities are concentrated *are* the most unpleasant of council estates'.

Karn (1981, 1982) offers some penetrating accounts of how black people's disadvantage in council housing arises partly out of the routine expediency of the home allocation bureaucracy. Although there exist finely tuned procedures which aim to grade applicants according to housing need, Karn argues that it is only through an extra informal grading of tenants' 'respectability' that an effective system for rationing homes – a scarce resource of variable quality – can be sustained. Using a variety of examples, she demonstrates how racial stereotypes, as well as images of class and gender, become associated with scales of disrespectability. A 'common-sense' association of blacks (as well as women and the lower classes) with low respectability provides some working criteria by which to meet management imperatives to let homes quickly while avoiding trouble with the (white) majority of tenants and applicants (see also Phillips, Ch. 9 below). As a consequence of giving the worst homes in the worst areas to households who fare badly according to a number of qualitative criteria such as housekeeping standards, public-sector housing is allocated according to hierarchical rather than egalitarian principles and 'geographical segregation of West Indians and Asians and the poorest white families in the private sector is being repeated and reinforced in the public sector' (Karn 1981, p. 21). This pattern of reinforcement is neatly summarized by Flett (1979) in a brief review of council housing in Greater London, Manchester and Birmingham (where the majority of black tenants are housed). The allocation procedures in all three places resulted in the net suburbanization of white households and the increased concentration of black people in the inner city.

Early-established racial disadvantage in council housing appears to be exacerbated by the operation of transfer systems (Brown 1984, Karn 1981). McKay's (1977) analysis suggests that this is inevitable. The coupling of segregation with disadvantage is a contingency of racially discriminatory housing allocation procedures which are likely to continue since 'Britain's public housing has been and continues to be geared to the "respectable" working class rather than to the economically marginal or to a black population largely unloved and possibly feared by many white public-housing tenants' (McKay 1977, p. 182).

Although the allocation of public housing has usually mirrored patterns of racial segregation in the private sector, it is through their allocation procedures that local authorities have theoretically had scope to shape the relationship between black and white residential space. Dispersal policies, however, which became popular following the Cullingworth report (Central Housing Advisory Committee 1969), lacked central guidelines and often became 'little more than a rather vague and general attitude' amongst housing managers (Smith & Whalley 1975). Where dispersal has been practised the results were often unexpected. Flett (1979), for instance, shows how, despite attempted dispersal, by 1976, 70 per cent of black families housed by the GLC lived in the four boroughs of Tower Hamlets, Southwark, Lambeth and

Hackney (a paradoxical trend towards greater borough-specific concentration compared with previous years). Flett provides evidence to show that this does not reflect tenants' choices. Re-sorting occurred without desegregation and, as a consequence of local authority practice, 'whilst blacks were staying in inner London, and in that sense not dispersing, many had lost one of the great advantages of concentration: the support of their own network of friends and acquaintances' (Flett 1979, p. 187).

Perhaps the best documented dispersal policy – that implemented in Birmingham – was initiated partly, at least, in response to white tenants' complaints about the too-rapid entry of black families. The quota policy – introduced to reduce the degree of segregation between West Indians in postwar inner-ring properties, Asians in pre-1919 middle-ring terraces and white families dominating the postwar outer estates – not only proved unlawful under the Race Relations Act 1976 (Section 1), as it denied people homes on the grounds of colour, but failed to achieve dispersal and made little progress in redressing the inequalities experienced by black people in the public housing system. Flett *et al.* (1979) show that despite the stated intention of dispersal, blacks who said that they would prefer to live on a white estate were less likely to have their preferences met than were white applicants requesting a white estate. Overall, the preferences of all applicants were most likely to be met when they specified an area in which the applicant's own racial group was already dominant.

Birmingham's experience questioned the concept of dispersal, and its fate was sealed by the CRC's report on *Housing choice and ethnic concentration* (Community Relations Commission 1977) which shows that black people often prefer to live in relatively segregated areas. Certainly, dispersal ideology diverts attention away from the need to eliminate the inequalities of wealth and status that deny black people a strong position within the housing market (see Jones 1980, Lee 1977, Rex 1981). Today, therefore, the emphasis is shifting away from ideas of dispersal quotas and towards the notion of 'equality targets' as described by Seager and Singh (1984). However, for any interpretation of the development of segregation in England over the past 40 years, the fact is that irrespective of intentionality, the thrust of national housing policies has been towards racial segregation, the effects of most local institutions have been to protect the housing environment of privileged whites from the entry of blacks, and the outcome is that racial segregation is associated with black people's disadvantage.

The highest stage of white supremacy?

The housing policies of national and local governments in Britain have succeeded neither in dispersing the black population nor in reducing their relative deprivation. To understand this, it is necessary to move beyond the concept of residential 'choice' and beyond the purview of government intent. Accordingly, the remainder of the chapter briefly assesses how interpretations of racial segregation might be informed by analysing the ambivalence of public policy towards racial minorities, by exposing the history of ideas about race and segregation in British politics, and by explaining the impact of these ideas in terms of the position of black people in a democratic system that is inherently unresponsive to their needs.

My argument has been that the major decisions sustaining racial segregation in England have been taken at the level of national government, notwithstanding the important local institutional practices through which segregation is reproduced. Quite why such decisions have encouraged segregation to persist – in association with racial disadvantage and despite expressed intentions to the contrary – might be appreciated on two interrelated levels. The first concerns the ambivalence of urban policy towards the specific needs of black people.

Given the assumption of successive governments that the black population would disperse and integrate once its members had lived in Britain long enough to overcome the disadvantages associated with immigrant status, it is hardly surprising that there has always been reluctance to build a concept of race into legislation combating discrimination and disadvantage. The same liberal indignation that kept Fenner Brockway's Racial Discrimination Bill at bay for a decade continues to sustain resistance to the introduction of measures to alleviate racial disadvantage in housing and related fields. It is assumed that black people's problems differ in degree rather than in kind from white people's and that their interests will best be served by policies addressed to general, rather than specifically racial, disadvantage.

The dilemma of striving for racial justice without institutionalizing the concept of racial difference across the full range of economic and social policy is therefore resolved (in theory) by assigning responsibility for the eradication of racial discrimination to the executors of race relations legislation – principally, the relatively toothless Commission for Racial Equality (CRE) and, on occasion, the police – and assuming that this takes care of the 'racial' dimension of other policies. Racial *disadvantage* is not recognized in law, because (tacitly) its components are thought of as divisible into racial discrimination on the one hand, and the general hardship of all deprived groups on the other. Thus, the housing policies discussed above could afford to be 'colour blind' because racial discrimination in housing is unlawful and can be tackled through the courts. Other factors which place racial minorities at a disadvantage in the housing market are expected to have been dealt with by more general policies attacking urban deprivation.

This reasoning fails, however, because anti-discrimination legislation cannot deal with the disadvantages which are a legacy of past racist practices. The consequences of ignoring this are well-illustrated in Cross's (1982) analysis of the targeting of the Urban Programme. This venture was launched amid a wave of concern for race relations, and it spearheaded the attack on urban deprivation in the years during which racial segregation became entrenched. Cross (1982) shows that just 3 per cent of Programme funds in 1980–1 (rising to 12 per cent in 1985–6, according to the Department of the Environment 1985) were allocated to 'ethnic projects'. Despite the imagery associated with the Programme, few parts of it address race-related issues directly (Edwards & Batley 1978, Haynes 1983, Higgins *et al.* 1983, Department of the Environment 1985). Cross (1982) estimates, in fact, that the majority of the black population (60 per cent) live outside the Partnership and Programme authorities, and he shows that even within them, black organizations fare worse than their white equivalents in attracting funds.

Here, as in so many areas of urban policy, efforts remain limited to tackling general disadvantage, ignoring specifically racial disadvantage, and justifying this by appeal to the safeguards built into anti-discrimination legislation. But the utilitarian underpinnings of such policies are directed towards majority (white people's) disadvantage, whereas black people are a disadvantaged minority with a weak political voice which is too easily ignored in the distribution of scarce resources. Legislation has failed to ensure that, in catering to the demands of the disadvantaged amongst a (white) majority, initiatives such as the Urban Programme would adequately serve the interests of the black minority. Certainly, policies directed towards general disadvantage have not afforded the black population, on average, a stronger position in the housing market; nor have they significantly alleviated the relative deprivation associated with racial segregation. I suggest, therefore, that, within the present form of British democracy, as long as specifically racial disadvantage (i.e. that conferred by a history of directly and indirectly racist practices) remains unacknowledged in legislation combating urban deprivation, racial segregation will continue to be associated with black people's exclusion from a fair share of social and economic resources.

In so far as key policies shaping racial segregation have done so largely as the unanticipated outcome of political decisions made in response to a range of social and economic pressures, racial segregation (and even its association with black people's disadvantage) might, at this first level, be interpreted as the 'innocent' by-product of pragmatic and utilitarian politics. However, the consistency over a quarter of a century of the outcomes of an apparent jumble of urban-economic policies and executive practices suggests an alternative explanation which provides the beginnings of a political theory of racial segregation in England. My argument in introducing a second level at which the link between residential segregation and racial inequality might be interpreted is that the 'unexpected' negative outcomes of 'colour-blind' housing policy seems less unpredictable in the light of the status of 'race' and the concept of segregation in British politics more generally.

To grasp fully the meaning and significance of segregation, it is necessary to move beyond the expressed aims and intentions of legislators to capture the political inspiration for the policies and practices that have allowed black people to remain confined to those same 'specific locales, parts of the housing stocks and corners of the economy' which were always marginal and which currently show few signs of economic regeneration (Brown 1984, p. 323). To this end a touchstone for the interpretation of racial segregation is to be found in the tenor of political discourse which, by virtue of the legislation into which it feeds and because of the mass-media attention it attracts, has some bearing on most levels of public life and institutional practice.

An evaluation of segregationist ideology in British political thought over the past 40 years, based on a reading of parliamentary debates and papers, is the subject of another essay (Smith, forthcoming). Here, one or two themes are picked out to illustrate the relevance of a political perspective to an explanation of racial segregation. Crudely, postwar parliamentary debate about 'race' and segregation can be divided into three overlapping periods, distinguished according to politicians' definitions of and responses to the presence of black people in urban Britain.

In the immediate postwar period, the segregationist ideologies which Rich (1985) shows were imported to Victorian England from South Africa and the American South lay dormant in the political process. A concept of racial difference obtained, but it was based on the ideas of cultural difference (backwardness), and on the assumption (or hope) that black immigration would be short-lived. In time, therefore, 'colonial immigrants' might be expected to disperse from the inner areas of the major cities and to become assimilated into a British way of life. In short, black people in Britain were thought of as culturally backward, 'childlike' and morally inferior (see Ben-Tovim 1978); the 'problem' was seen in terms of an immigrant history; and the solution was presumed to lie in immigrants' spatial integration and 'natural' absorption into English culture. The real problems of 'race' and racism were thought to lie overseas.

Later, particularly in the mid- to late 1960s, racism became more explicit in British politics. Much of the rhetoric was segregationist (Peter Griffiths, for instance, won his Smethwick seat in the 1964 general election with the slogan 'If you want a nigger neighbour, vote Labour'), but this was incidental, not fundamental, to the fairly free expression of racist sentiment that prevailed for a short time. Black people were redefined as illegitimate competitors for scarce resources and as a threat to the material wellbeing of whites (see Joshi & Carter, 1984, for an account of this shift from 'colonial' to 'indigenous' racism). Enoch Powell successfully reconceptualized the 'problem' of 'immigrants' in Britain in segregationist terms by conjuring up the image of 'alien territories' occupied by black outsiders – a definition more or less accepted by moderates and extremists alike during the steady retreat of politicians in all parties towards the views of a racist right wing (Foot 1965, p. 234). The key differences of opinion, therefore, lay in the 'problem's' solution, which was seen in terms of dispersal, repatriation or enforced removal. Although most politicians favoured the first alternative, Powell's influential rhetoric had raised, for the first time, the possibility that dispersal might be associated not with 'assimilation' but with a modification (i.e. swamping) of English culture (see Phillips 1977). The panacea of integration was therefore reformulated as a threat, and the seeds of a new ambivalence towards black people in Britain were sown.

Currently, political ambiguity towards 'race' and racism remains, but British political discourse is cloaked increasingly in a new subtlety and apparent respectability in its treatment of 'race' and racial difference. Whereas segregationist ideas were present in the relatively overt racist politics of the late 1960s, it is only now, in the wake of what Barker (1981) terms the 'New Racism', developed in the political context of the New Right, that racial segregation begins to prove acceptable as a cultural norm at the same time as specifically racial inequalities remain unrecognized by the economic individualism that underlies monetarist public policy. This new racism sustains essentially racist social boundaries by presenting them as a 'value-free' acknowledgement of cultural difference rather than as an assertion of white superiority. Extending this argument, fears of 'swamping' soon seem reasonable, and a notion of separate but (eventually) equal has worked its way into common-sense vocabularies. The 'problem' of racial segregation is now no longer linked, as it was in the 1960s, with notions of dispersal and integration. Rather, it is linked with an assumption about the immutable differences of a population whose presence is often discussed as if it violates the 'reasonable' rights – both material (concerning access to resources) and symbolic (concerning citizenship) – of white Britons. Cushioned by the languages of Reeves' (1983) 'sanitary coding', racism in British politics is increasingly, if euphemistically, expressed in segregationist terms, and echoes of Powell's 'alien territories' loom again in the debates of the 1980s.

This brief outline of changing political sentiment serves to show that the policies most obviously working to sustain segregation were largely initiated in a period – the second described above – when segregationism was incidental but not fundamental to the reproduction of racism in British politics. At a time when a new form of *de facto* segregationism seems on the ascendency, therefore, and might be fundamental to the development of a new, more subtle form of racism in parliamentary politics, the beneficial effects of even the most altruistic of policies might be hard to secure; and, given their limited penetration of national politics, the prospects for black Britons of breaking the association between segregation and disadvantage seem relatively slim.

The most obvious feature of political discourse and public policy as discussed above is the virtual exclusion of black representatives from decision-making power. Notwithstanding the importance of extra-parliamentary black 'resistance', in terms of public policy, Britain's black population has always been legislated *for* by white politicians responding to the demands and perceived needs of a predominantly white electorate. While this has taken its toll at local government level, it is in central government, where the policies inadvertently sustaining segregation were formed, that this absence has been most telling. In short, the effects of legislation on racial segregation were unanticipated precisely (but not only) because black people's interests remained peripheral to the decision-making procedures. In this sense, the association of residential segregation with racial inequality is explicable when viewed as a consequence of the limited voice of the black population in British politics.

This weak political voice is inevitable, given the small proportion of the electorate that is black and in view of their consequent vulnerability within parties serving the interests of a white majority within the inertia of consensus politics. Thus, successive governments have co-operated to depoliticize race-related issues, which, because they divide politicians along non-partisan lines, are a potential threat to party unity (Ashford 1981). Similarly, the general agreement over the 'immigrant problem' expressed in the racist politics of the late 1960s 'involved not only a convergence of policy between the two parties but also a willingness to forgo the opportunity to appeal directly to the immigrants themselves for political support.' (Freeman 1979, p. 102.)

I suggest, therefore, that in so far as segregation reflects and reproduces racial inequalities in the distribution of opportunities, rewards and life-chances, its persistence, its relationship with black disadvantage, and its rôle in the reproduction of social (racial) relations, must be explained within a theoretical framework (a) explicitly linking the general tenor of political discourse with the intended and unintended outcomes of *specific* legislative and executive decisions, and (b) locating the black minority within a democratic system inherently unresponsive to its needs.

The formal presentation of such a theory is beyond the scope of this chapter, but it is significant that this political perspective is favoured by the analysts who have perhaps done most to further the development of theories of racial segregation outside Europe. Segregation is, for both John Cell (1982) and George Fredrickson (1981), a political phenomenon, created and enforced by political power. Apartheid in South Africa is practised to sustain white political dominance, and the cost is not always offset by the availability of cheap black labour (Fredrickson 1981). The Jim Crow laws of the American South are remembered for the United States' brief flirtation with *de jure* segregation, but the key to the system was disfranchisement. Segregation, then, is a form of exclusion, but not in Godard and Pendaries' (1978) straightforward sense of restricted access to housing due to the weak market position of the poor and the manipulative techniques of managers and investors. Rather, it is a form of political exclusion which can isolate and marginalize the demands of a minority too small to make any serious political challenge (and, in the case of blacks in Britain, internally fragmented as to class, culture and housing tenure). From this point of view, whatever *else* segregation represents – social support for a cultural minority or an economic niche for black entrepreneurs, for instance – at a political level there are grounds for interpreting racial segregation in Britain as an expression of 'white supremacy' – a geography of English racism.

References

Ashford, D. E. 1981. *Policy and politics in Britain*. Oxford: Basil Blackwell.

Association of Metropolitan Authorities 1985. *Housing and race. Policy and practice in local authorities*. London: AMA.

Ballard, R. 1983. Race and the census: what an ethnic question would show. *New Society*, 12 May, 212–14.

Barber, A. 1981. Labour force information from the National Dwelling and Housing Survey. *Research Paper* 17. London: Department of Employment.

Barker, M. 1981. *The new racism. Conservatives and the ideology of the tribe*. London: Junction Books.

Bassett, K. & J. R. Short 1980. *Housing and residential structure*. London: Routledge & Kegan Paul.

Ben-Tovim, G. 1978. The struggle against racism: theoretical and strategic perspectives. *Marxism Today* **22**, 203–13.

Bovaird, A., M. Harloe & C. M. E. Whitehead 1985. Private rented housing: its current role. *Journal of Social Policy* **14**, 1–23.

Brown, C. 1984. *Black and white Britain*. London: Heinemann.

Cater, J. 1981. The impact of Asian estate agents on patterns of ethnic residence: a case study in Bradford. In *Social interaction and ethnic segregation*, P. Jackson & S. J. Smith (eds), 163–83. London: Academic Press.

Cater, J. C. & T. P. Jones 1979. Ethnic residential space: the case of Asians in Bradford. *Tijdschrift voor Economische en Sociale Geografie* **70**, 86–97.

Cell, J. W. 1982. *The highest stage of white supremacy*. Cambridge: Cambridge University Press.

Central Housing Advisory Committee 1969. *Report on council housing: purposes, procedures and priorities*. London: HMSO.

Clark, D. 1977. *Immigrant responses to the British housing market: a case study in the West Midlands conurbation*. Research Unit on Ethnic Relations Working Paper 7.

Commission for Racial Equality 1984a. *Race and council housing in Hackney*. London: CRE.

Commission for Racial Equality 1984b. *Race and housing in Liverpool: a research report*. London: CRE.

Commission for Racial Equality 1985. *Walsall Metropolitan Borough Council: policies of housing allocation*. London: CRE.

Community Relations Commission 1977. *Housing choice and ethnic concentration*. London: CRC.

Cross, M. 1982. The manufacture of marginality. In *Black youth in crisis*, E. Cashmore & B. Troyna (eds), 35–52. London: Allen & Unwin.

Daniel, W. W. 1968. *Racial discrimination in England*. Harmondsworth: Penguin.

Department of the Environment 1975. *Race relations and housing*. Cmnd 6232. London: HMSO.

Department of the Environment 1985. *Urban Programme: minutes of evidence* (Monday, 25 November 1985). HoC Committee of Public Accounts. London: HMSO.

Doherty, J. 1983. Racial conflict, industrial change and social control in post-war Britain. In *Redundant spaces in cities and regions?* J. Anderson, S. Duncan & R. Hudson (eds), 201–39. London: Academic Press.

Doling. J. & M. Davies 1983. Ethnic minorities and the protection of the rent acts. *New Community* **3**, 487–95.

Dummett, M. & A. Dummett 1969. The role of government in Britain's racial crisis. In *Justice First*, L. Donnelly (ed.), 25–78. London: Sheed and Ward.

Edwards, J. & R. Batley 1978. *The politics of positive discrimination: an evaluation of the Urban Programme 1967–77*. London: Tavistock.

Foot, P. 1965. *Immigration and race in British politics*. Harmondsworth: Penguin.

Flett, H. 1979. Dispersal policies in council housing: arguments and evidence. *New Community* **7**, 184–94.

Flett, H., J. Henderson & B. Brown 1979. The practice of racial dispersal in Birmingham, 1969–1975. *Journal of Social Policy* **8**, 289–309.

Fredrickson, G. M. 1981. *White supremacy*. New York and Oxford: Oxford University Press.

Freeman, G. P. 1979. *Immigrant labor and racial conflict in industrial societies*. New Jersey: Princeton University Press.

Fryer, P. 1984. *Staying power. The history of black people in Britain*. London and Sydney: Pluto Press.

Godard, F. & J. Pendariés 1978. Rapports de propriété, ségrégation et pratique de l'éspace résidential. *International Journal of Urban and Regional Research* **2**, 78–100.

Haynes, A. 1983. *The state of black Britain*. London: Root Books.

Henderson, J. & V. Karn 1984. Race, class and the allocation of public housing in Britain. *Urban Studies* **21**, 115–28.

Higgins, J., N. Deakin, J. Edwards & M. Wicks 1983. *Government and urban poverty*. Oxford: Basil Blackwell.

Hiro, D. 1971. *Black British, white British*. London: Eyre & Spottiswoode.

Jones, P. N. 1978. The distribution and diffusion of the coloured population of England and Wales. *Transactions of the Institute of British Geographers* New Series **3**, 515–32.

Jones, P. N. 1980. *Ethnic segregation, urban planning and the question of choice: the Birmingham case*. Paper presented to the Symposium on Ethnic Segregation in Cities, St. Antony's College, Oxford.

Jones, T. P. 1983. Residential segregation and ethnic autonomy. *New Community* **11**, 10–22.

Jones, T. P. & D. McEvoy 1978. Race and space in cloud cuckoo land. *Area* **10**, 162–6.

Jones, T. P. & D. McEvoy 1979a. Reply to Peach. *Area* **11**, 84–5.

Jones, T. P. & D. McEvoy 1979b. Reply to Peach. *Area* **11**, 222–3.

Joshi, S. & B. Carter 1984. The role of Labour in the creation of a racist Britain. *Race and Class* **23**, 53–70.

Jowell, R. & C. Airey 1984. *British social attitudes: the 1984 report*. Aldershot: Gower.

Karn, V. 1977. *The impact of housing finance on low-income owner-occupiers*. Working Paper 55, Centre for Urban and Regional Studies, University of Birmingham.

Karn, V. 1981. *Race and council housing allocations*. Paper presented to the Policy Seminar on Race Relations, Nuffield College, Oxford.

Karn, V. 1982. *Race and housing in Britain: the role of the major institutions*. Paper presented to the Anglo-American seminar on Ethnic Minorities and Public Policy, Middle Aston House, Oxfordshire.

Karn, V., J. Kemeny & P. Williams 1985. *Home-ownership in the inner city. Salvation or despair.* Aldershot: Gower.

Kirby, A. 1982. The external relations of the local state in Britain: some examples. In *Conflict, politics and the urban scene*, K. R. Cox & R. J. Johnston (eds), 88–104. London: Longman.

Lee, T. R. 1977. *Race and residence. The concentration and dispersal of immigrants in London.* Oxford: Clarendon Press.

Lee, T. R. 1978. Race, space and scale. *Area* **10**, 365–7.

Maclennan, D. & J. Ermisch 1986. Housing policy and inner city change. In *Social and economic change in inner cities*, V. Hausner (ed.). Oxford: Oxford University Press.

McKay, D. H. 1977. *Housing and race in industrial society. Civil rights and urban policy in Britain and the United States.* London: Croom Helm.

Merrett, S. 1982. *Owner occupation in Britain.* London: Routledge & Kegan Paul.

Miles, R. 1982. *Racism and migrant labour.* London: Routledge & Kegan Paul.

Morgan, B. S. 1980. *The measurement of residential segregation.* Geography Department, King's College London, Occasional Paper No. 11.

Morgan, B. S. 1983. A distance-based interaction index to measure residential segregation. *Area* **15**, 211–17.

Morgan, B. S. & J. Norbury 1981. Some further observations on the index of residential differentiation. *Demography* **18**, 251–6.

Niner, P. 1984. Housing associations and ethnic minorities. *New Community* **11**, 238–48.

OPCS 1983. Labour force survey 1981: country of birth and ethnic origin. *OPCS Monitor* LFS 83/1 and PP 83/1. London: OPCS.

Peach, C. 1966. Factors affecting the distribution of West Indians in Great Britain. *Transactions of the Institute of British Geographers* **38**, 151–63.

Peach, C. 1968. *West Indian migration to Britain: a social geography.* London: Oxford University Press for Institute of Race Relations.

Peach, C. 1978–9. British unemployment cycles and West Indian immigration. *New Community* **7**, 40–3.

Peach, C. 1979a. Race and space: a comment. *Area* **11**, 82–4.

Peach, C. 1979b. More on race and space. *Area* **11**, 221–2.

Peach, C. 1981. Conflicting interpretations of segregation. In *Social interaction and ethnic segregation*, P. Jackson & S. J. Smith (eds), 19–33. London: Academic Press.

Peach, C. 1982. The growth and distribution of the black population in Britain, 1945–1980. In *Demography of immigrants and minority groups in the United Kingdom*, D. A. Coleman (ed.), 23–42. London: Academic Press.

Peach, C. 1984. The force of West Indian island identity in Britain. In *Geography and ethnic pluralism*, C. Clarke, D. Ley & C. Peach (eds), 23–42. London: Allen & Unwin.

Peach, C. & S. Shah 1980. The contribution of council house allocation to West Indian desegregation in London, 1961–71. *Urban Studies* **17**, 333–42.

Phillips, D. 1986. *What price equality?* GLC Housing Research and Policy Report No. 9. London: Greater London Council.

Phillips, K. 1977. The nature of Powellism. In *The British right*, R. King & N. Nugent (eds), 99–129. Westmead: Saxon House.

Rees, P. H. & M. Birkin 1984. Census-based information systems for ethnic groups: a study of Leeds and Bradford. *Environment and Planning A* **16**, 1551–71.

Reeves, F. 1983. *British racial discourse*. Cambridge: Cambridge University Press.

Rex, J. 1981. Urban segregation and inner city policy in Great Britain. In *Ethnic segregation in cities*, C. Peach, V. Robinson & S. J. Smith (eds), 25–42. London: Croom Helm.

Rex, J. & R. Moore 1967. *Race, community and conflict*. London: Oxford University Press for Institute of Race Relations.

Rich, P. 1985. Doctrines of racial segregation in Britain: 1800–1914. *New Community* **12**, 75–88.

Robinson, V. 1980a. Lieberson's P* index: a case study evaluation. *Area* **12**, 307–12.

Robinson, V. 1980b. Correlates of Asian immigration to Britain, 1959–74. *New Community* **8**, 115–23.

Robinson, V. 1981. *The dynamics of ethnic succession: A British case study*. Working Paper No. 2, University of Oxford School of Geography.

Seager, R. & G. Singh 1984. Race, housing and equality targets. *Housing* **20**, 25–6.

Select Committee on Race Relations and Immigration 1974–5. *The organisation of race relations administration* HC 448–I. London: HMSO.

Short, J. R. 1984. *The urban arena*. London: Macmillan.

Sims, R. 1981. Spatial separation between Asian religious minorities: an aid to explanation or obfuscation? In *Social interaction and ethnic segregation*, P. Jackson & S. J. Smith (eds), 123–35. London: Academic Press.

Smith, D. 1976. *The facts of racial disadvantage*. London: PEP.

Smith, D. J. 1978. The housing of racial minorities: its unusual nature. *New Community* **6**, 18–26.

Smith, D. & A. Whalley 1975. *Racial minorities and public housing*. London: PEP.

Smith, S. J. forthcoming. *White supremacy in Britain? Critical interpretations of racial segregation*. Cambridge: Cambridge University Press.

Stevens, L. *et al.* 1982. *Race and building society lending in Leeds*. Leeds Community Relations Council.

Ward, R. 1981. *Race, housing and wealth*. Paper presented to the policy seminar on Race Relations, Nuffield College, Oxford, 27 November 1981.

Ward, R. 1982. Race, housing and wealth. *New Community* **10**, 3–15.
Williams, P. 1977. *Building societies and the inner city*. Working Paper 54, Centre for Urban and Regional Studies, University of Birmingham.
Woods, R. I. 1976. Aspects of the scale problem in the calculation of segregation indices: London and Birmingham, 1961 and 1971. *Tijdschrift voor Economische en Sociale Geografie* **67**, 169–74.

2 *Residential segregation retheorized: a view from southern California*

ALISDAIR ROGERS AND RIKA UTO

How is ethnic residential segregation to be constituted as a valid problem within geography? Existing interpretations are so greatly influenced by the Chicago School, which was successful at defining the objects of study and analysis, that it seems difficult to engage in a genuinely different form of inquiry. This essay is an attempt to sever links with that tradition, at least with regard to the metropolitan scale. A realignment of empirical work, a well-defined method of abstraction and a theoretically sound objective are required for any workable research programme. We suggest a recovery of 'work' and 'labour' as central themes in social geography, and a parallel re-emphasis on the production, circulation and consumption of commodities in urban and economic geography. As a result, there must be a shift away from the rigorous plotting of night-time location towards an understanding of the geographical separation of workplace and residence as a key feature in the structuring of ethnicity and the formation of class.

Our account begins with a brief résumé of some of the limitations of past approaches, which indicates the direction of our realignment. The theoretical basis of such a formulation is outlined, and is then partly substantiated by case studies from Los Angeles and Orange Counties, California.

A re-examination of residential segregation

The study of residential segregation is concerned with the uneven distribution of social groups across urban space, and the concomitant formation of socially homogeneous residential areas. While there is no single method or approach, most studies use concepts of space, 'race' or ethnicity or both, and class. We contend that the first step of realignment is a critical re-examination of these concepts in terms of commodity production, the labour process and territory. Rational abstraction and the development of meaningful concepts are essential to any programme (Sayer 1984).

In studies of residential segregation 'race' and ethnicity have frequently been taken as pre-given and/or the analytical and explanatory equivalents of class.[1] Based on such supposed equivalence there have been attempts to prove that 'race' is more important than, or contains, class. This can only be so if 'race' and class are both treated as descriptive generalizations drawn from a single-levelled reality. 'Race' is indeed a generalization of empirically given phenotypical characteristics. However, class may be a descriptive concept (i.e. stratification); but it may also be an explanatory concept, based on abstract and theorization (Miles 1982, 1984; Walker 1985).[2] In this sense 'race' cannot contain class. Furthermore, one may wish to examine racism rather than 'race' to shift focus away from a particular social category in itself, towards historical and ideological formulations. Our problem is to specify class relations and the real material effects of racism as an ideology. In social geography this may begin with indicating the differentiation of labour by racism (among other factors) and its differential incorporation and reproduction. In this respect, it is not sufficient to treat 'racial' differences at the level of distribution alone (Rex & Tomlinson 1979, Parkin 1979).[3] The social relations of production give the terrain upon which (re)distributive strategies act, and if class is seen as central to the transformation of society in a way which 'race' or religion say, are not, then one must begin with those relations of production.[4]

Spatial sociology has separated space and society in a theoretical sense. This is exemplified by the calculation of indices of segregation which represent in abstract form specific spatial configurations. Such indices may be correlated with or regressed against socio-economic variables to decompose 'racial' segregation into x per cent 'race' and y per cent class effects. Once again, 'race' and class are put into a false zero-sum equation which assumes their equivalence. Secondly, such exercises treat as empirical 'facts' aspects of social formations which are constituted spatially and temporally, not directly reducible to aspatial and atemporal snapshots of social stratification. The spatial configuration does not spring fully formed from the social formation and so does not merely record or reflect it as a 'fact' to be decomposed. Thirdly, by treating the distribution of all-of-group-*A* across a set of spatial units as a 'fact', indices direct attention away from a possible plurality of causes and consequences. Spatial patterns, in the form of a ghetto, natural area or all-of-group-*A* are not satisfactory criteria for rational abstraction (Sayer

1985a). To regard a black ghetto as a single 'fact' to be explained one assumes some internal and necessary relation between all persons classified as 'black', regardless of their other characteristics. 'Race' is not a strong enough concept to bear this load. Spatial contiguity is not sufficient to assume analytical continuity.

Neither 'race' nor space can satisfactorily define discrete objects of inquiry *a priori*. In urban areas the primary forces responsible for the organization of objects in space are those of commodity production, circulation and consumption, which include the connections between workplace and residence (Scott 1980). The city is not a *tabula rasa* upon which the social distance preferences of individuals or predefined social groups are enacted in a magical process of sifting and sorting. What the Chicago School 'tradition' failed to explore is that the built environment and social space are constantly defined by systems of workplace–residence embedded in the larger association between production and reproduction. Production was 'lost' in Chicago and thereafter. For this reason, we address not the distribution of residential space among social groups, but the social production of urban space and the simultaneous reproduction of social groups implicated in it.

A reformulation of residential segregation

Since neither space nor 'race' has independent analytical validity, one must search for new objects of analysis. In post-Weberian location theory these objects are the organization of production processes in predefined sectors, represented by firms (Scott 1982, 1983a; Storper & Walker 1983; Walker & Storper 1981). Although the locational dynamics of certain sectors may be examined, the intent is rarely to account for the entirety of any metropolitan or regional assemblage of industry. Space does not define the inquiry. This analysis could and should be extended to the question of labour, its supply and reproduction. Since almost everyone works, one is again left with the problem of identifying discrete objects of inquiry.[5] 'Racial' and ethnic definitions have been excluded, so production sectors must provide the objects, particularly if they can be assumed to have identifiable labour markets associated with them. Therefore, the objective is not to account for the entirety of the location of any one ethnic or 'racial' group. Instead, ethnic and 'racial' categories are the results in part of ideological processes which are primarily (but not exclusively) significant in relation to production, and to its particular sectors.

The scale of the analysis is therefore shifted to one defined by the sets of practices associated with production and the reproduction of labour, chiefly journey-to-work regions or local labour markets. In the case of services or retailing this need not be confined to any one part of the metropolitan area.

Of equal significance is the centrality of work and labour in everyday life, which has been systematically downplayed by social geography. As a subdiscipline it has tended to focus on non-work, e.g. residence, public behaviour and grafitti. The separation of living and working is reflected in the divisions between social and economic geography, or urban and industrial sociology. Under the impetus of marxist discourse production has moved to a more central position in urban theory. The strength of the concepts of work and labour is that they bridge the gap between workplace and residence that is ingrained in capitalist societies, and is all too often regarded as non-problematic (Harvey 1977, 1985; Katznelson 1981). Social geography may examine this separation both in terms of its importance in shaping the physical distinction and interrelation between production and reproduction spaces, and in terms of its significance for class consciousness. Katznelson has argued that this division partly accounts for some of the peculiarities of US class history. The American working class acted like a class at work and like an ethnic group at home. Lastly, the labour process itself describes a junction between the contextual and the compositional, or the general and the specific (Thrift 1983).

Using insights from post-Weberian location theory we intend to examine both the causes and significance of social residential segregation. If it is not found to be significant then there is no reason to search for its causes. Rather than review the extensive list of apposite texts, we shall identify certain common features.

The first area of agreement is that residential differentiation is implicated in the process of class formation, and that it has a significance in the anatomy of capitalist societies. From this follow a number of points. The connections between living and working, residence and workplace, reproduction and production are historically and geographically varied, and have significance in the political activity of classes and everyday life (Dear 1981; Harris 1984; Harvey 1975, 1977, 1985; Katznelson 1981; Scott 1980; Walker 1981). As a result, the immediate experience or practical understanding of class may become compartmentalized (Walker 1985). In addition, the non-economic forces of class formation find space for expression. Social categories such as class, 'race' and gender are reproduced at work, at home and in the community (Castells & Murphy 1982, Dear 1981, McDowell 1983).

The second convergence is the recognition that urbanization takes on forms outlined by the capitalist mode of production in general and the dynamics of commodity production, circulation and consumption in particular. The historical periodization of capital accumulation results in the spatial imbrication of the urban form, and as a consequence capital creates barriers to itself in the form of the built environment and the spatial division of labour (Harvey 1982). From these two common observations it follows that residential differentiation incorporates the contradictions of capitalism, and that these resolve chiefly around labour, and the use and exchange values of fixed capital. Lastly, because of periodic restructuring, there will be local variations of social class formations (Massey 1984, Urry 1981).

There are also important differences, two of which need recalling. Some authors study the general motion of capital through its accumulation crises and translate these events into urban space either directly or through financial and related institutions (Harvey 1978, Walker 1981). Others begin with a dissection of the diagnostic features of capitalism such as commodity production or the labour process, then establish contingent spatial properties (Scott 1983a, Storper & Walker 1983). Within this approach there is an important divergence, regarding the status of labour as either secondary to the location of production or central to the explanation. This would seem to be a fruitful area for research. It may be that the reproduction of labour becomes a periodic obstacle to production.

There are several possible syntheses of these texts around the topic of residential segregation. Most authors have only dealt with economic classes, the simplest case. However, a combination of insights into production, location and class structuration does not exclude non-economic class formation. In fact, the contrary may be true. We may be able to understand the reproduction of categories such as 'race' and ethnicity only in the context outlined below. This is because they are above all ideological expressions implicated in the fragmentation, subordination, supply and reproduction of labour.

To illustrate the general type of analysis or procedure implied by these observations, we have prepared a diagram of how ethnic residential segregation may be constituted as a problem (Fig. 2.1). The linkages should be read as procedural and not causal, at least not in any simple one-directional sense. Each component in reality consists of institutions and human agents, such that there is always the possibility of change. In this scheme, residential segregation appears as a 'hinge' between the theory of production location and the concept of class structuration. We begin with the former to address questions of cause, and the latter for questions of significance.

In brief, residential segregation both expresses and impacts upon social and spatial divisions of labour, which in turn have their origins in the labour processes associated with commodity production, circulation and

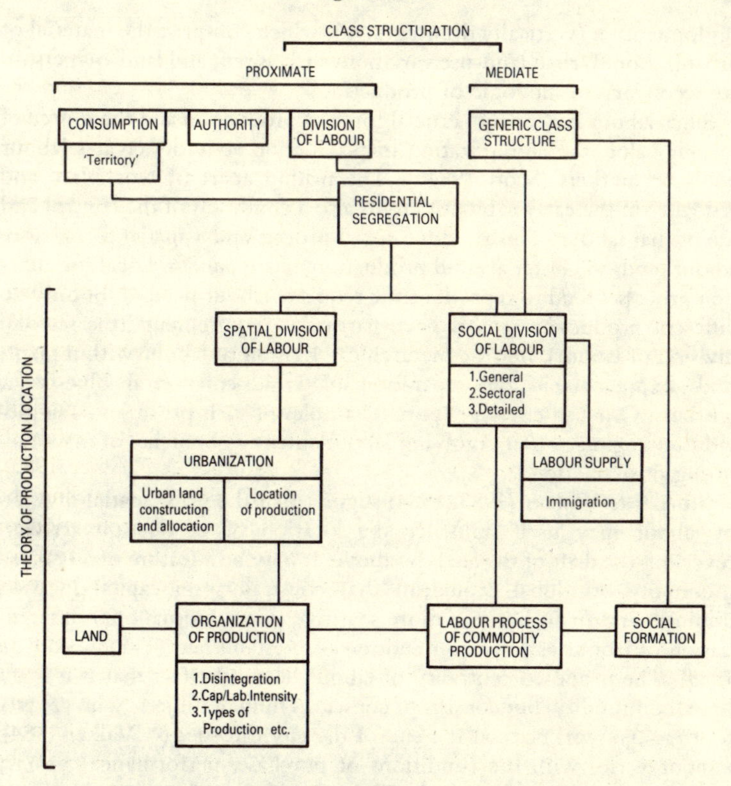

Figure 2.1 Theorizing ethnic residential segregation.

consumption. The co-ordinates of this process are provided by the social formation, which also defines the terrain of class relations. Commodity production, as a diagnostic of capitalism, is the first stage of active research.

One may identify various types of production activity defined by their dominant technology and its means of conversion and transfer (Storper & Walker (1983) identify six types, for example). Production is organized into vertically disintegrated or integrated forms, depending upon capital or labour intensity, the routinization of work, cost of inputs and so on (Scott 1983a). The crucial point is that each type of production has its own intra-metropolitan locational tendencies, stemming from organizational characteristics such as inter-plant linkages, subcontracting arrangements, and uncertainty (Scott 1983b, 1984a). These can be resolved into tendencies towards dispersal (vertically integrated) and agglomeration (vertically disintegrated), which comprise the material of urbanization. Urban land-use variations such as rent and land-ownership are secondary to the logic of production.

Since labour is a (or the) crucial force of production and the source of surplus value, the configuration of production activities creates labour pools or markets (Scott 1984b). The prising apart of workplace and residence in the early capitalist city is also a condition of the control and use of that labour. This is both a social process and a spatial form, since labour tends to cluster around production, *ceteris paribus*. Local socialization processes tend to reproduce the requisite labour needs. Not only do different production sectors have particular requirements (the sectoral division of labour), but the hierarchical division of labour within plants finds expression in the separation of white-collar and blue-collar workers. One can envisage 'pure' examples of such processes in single-industry regions, often involving labour-intensive branches of manufacturing or extraction.

Storper & Walker (1983, 1984) suggest that the prior spatial division of labour may itself influence the location of production activity, reversing the drift of the analysis above. It may be a feature of advanced monopoly or global capitalism that constraints on capital location diminish and firms become more sensitive to an international differentiation of labour as price competition is re-established (Trachte & Ross 1985). The proposed centrality of labour lies in the fact that it is not a 'true' commodity, but consists of conscious human subjects who are free to leave the workplace at the end of the day (Storper & Walker 1984). Labour varies with the conditions of purchase, performance capacity and circumstances of control. The creation of labour pools sets in motion ties of local affinity and organization which extend a web of linkages between a community and a workplace. These may influence the behaviour of labour at work and its suitability for employment. The strike record, wage rates, employee reliability, the nature of recruiting networks and other factors enter into the firm's calculations.

The availability of labour itself is a recurrent problem for capital owners, who must induce other individuals to sell their labour power either by coercion or by persuasion. This necessity is periodically resolved by relocation, by the employment of new labour sources (e.g. immigrants and women), or by conversion to machinery. Conversely, labour has the options of seeking work elsewhere as individuals, or of organizing collectively to improve conditions *in situ*. The history of American capitalism is a history of immigration. In turn, immigration and migration are defining characteristics of ethnicity and 'race' (Hershberg 1981, Katznelson 1981, Ward 1982). The processes of the communal categorization of individuals and the supply of labour are intertwined. Miles states that:

> The process of racial categorisation can . . . be viewed as affecting the allocation of persons to different positions in the production process and the allocation of material and other rewards and disadvantages to groups so categorised within the class boundaries established by the dominant mode of production. (Miles 1982, p. 159.)

The relevance of 'race' and ethnicity is twofold. In the first place migrant or immigrant labour is always of a different national origin to domestic labour. Since capital can defray the costs of the reproduction of labour to the country of origin, such workers are usually cheaper. They may also be in a weaker position in the workplace. Secondly, racism is an ideology that both justifies and reproduces this initially subordinant position, and extends control of labour outside the workplace itself. Just as production draws on variegated labour, so its continuation may etch this differentiation into society itself.

The discussion of labour opens up the wider ramifications of residential segregation. An explanation defined solely in terms of the spatial implications of production and the centrality of labour can only reveal the crudest armature. It is clear that urban areas are not commonly demarcated into discrete labour pools grouped around types of production activity. Neither are cities completely the expression of series of imbricated production–reproduction cells. To extend the analysis one must consider the significance of residential segregation for the structuration of class.

Several authors have discussed the connections between segregation and class formation, mostly drawing on the theoretical expositions of Giddens (1973).[6] He reasons that the mediate or general forces of class structuration given by the social formation are conditioned by proximate and local factors. These include the division of labour, the nature of authority relations and distributive groupings involving the consumption of economic goods. These groupings may reinforce divisions in 'market capacity' – that is, status with regard to property ownership, credentials and possession of manual ability. The most significant of these is the tendency towards community or neighbourhood segregation. Therefore, in a general sense, residential segregation is to be understood in terms of the smooth (or otherwise) reproduction of market capacity, hence labour. The community is the major site of child-raising and socialization practices, social networks and affinitive links such as marriage, and image and status. As such, Scott (1985) has hypothesized that social segregation may be part of the agglomeration economies of urbanization.

Not only is the community the site of the reproduction of labour, it is also a locus of labour's struggle against capital (Harvey 1977). Local cultures and social institutions may become the bases of local political radicalism (Cooke 1985). The resilience of labour in communities is added to the calculations of the spatial division of labour.

If the above holds true, then it is reasonable to assume that residentially segmented neighbourhoods aid in the reproduction of 'race' and ethnicity as meaningful categories with real effects at all levels of the social formation. Far from predicting the demise of 'race' and ethnicity therefore, marxist analysis may appreciate their maintenance in terms of the repeated fragmentations of the experience of class. Classes in the generic sense are modified by the proximate forces of the workplace–residence division and spatial segmentation into residential areas.

The outcome of this discussion is that the valid object of inquiry is no longer the location of every member of a group *A*, since 'race' and space have no independent analytical validity. We attempt to explain residential segregation of non-occupational groups by combining a theory of production location with a theory of territorial development, stressing the reproduction of labour as the key interpretative factor. In other words there would be 'racial' and ethnic segregation stemming from the locational dynamics of production and the various labour relations of its sectors, independently of any discrimination based in the distribution or exchange sector of housing. There has not been space to expand the discussion into relevant political and ideological factors. These include the rôle of the state in controlling and differentiating labour, and the ideological representation of workers, e.g. dextrous Chicanas and Filipinas, green-fingered Japanese and tough Chicanos. To demonstrate the real difference our theorization might make to research we present some preliminary work conducted in Southern California. One of the most promising modes of inquiry is comparative analysis, since it may uncover what is general and what is contingent or local.

The ghetto and the *barrio* in Los Angeles

Los Angeles County was the place of residence for 7 477 503 people in 1980, of whom 2 066 013 were recorded by the Census Bureau as of Spanish origin and 943 968 categorized as 'black'. In US urban politics these two groups have moved to centre stage in the 1980s. However, any research which tried to discover and describe the social and spatial boundaries between them would encounter problems. Both groups are significantly differentiated by socio-economic status, and the term 'Latino' embraces people as diverse as Kanjobal Indians from the hills of Guatemala who speak no Spanish to the descendants of the original Californios.[7] To begin an analysis with the raw and uninterpreted data of spatial distributions would be to translate a nebulous spatial phenomenon into a real analytical one. Bearing this in mind, we lay aside the long-established Mexican *barrio* of East Los Angeles and focus on one illustration of the social and spatial relation between blacks and Latinos to examine the specificity of racial categorization, the so-called Latinization of the ghetto (Garcia 1985, Oliver & Johnson 1983; Fig. 2.2). This process is apparently accompanied by ethnic conflict and competition over jobs, housing and welfare. In this example two sets of questions may occur: why has the ghetto persisted for so long, and why is there sizable Latino labour migration coincident with high rates of black unemployment?

The context of these issues is provided by a series of changes which have been described as the transition from a state-monopoly capitalist city to a global capitalist city (Soja *et al.* 1983, Trachte & Ross 1985). This restructuring is occurring during the pathological recovery of US

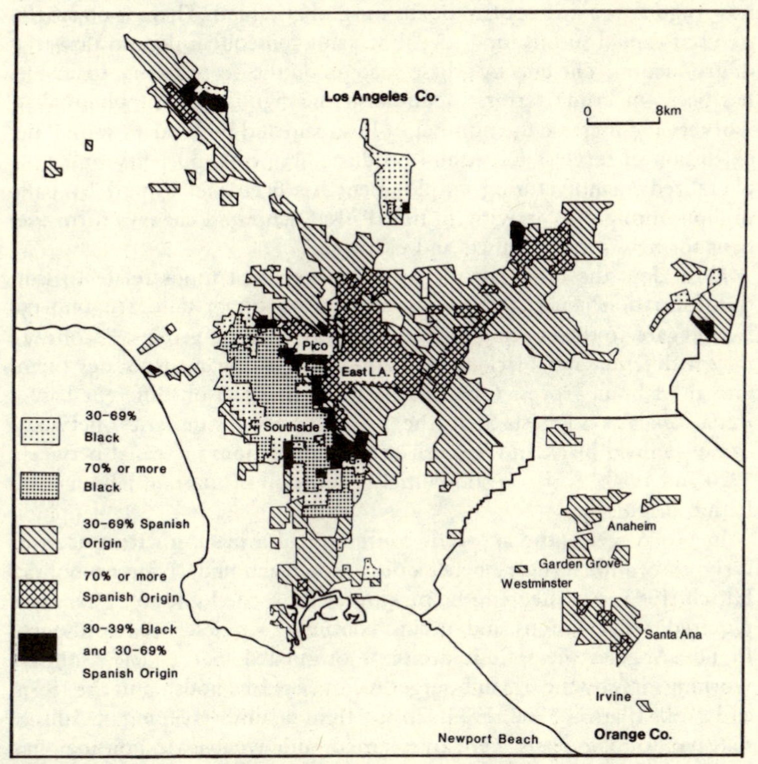

Figure 2.2 Blacks and Latinos in Los Angeles and Orange County, California (from US Census, 1980).

capitalism in a process of deepening and widening capital accumulation (Davis 1984). There is a resultant industrial, occupational and therefore spatial differentiation which cannot easily be described as either a recovery or a recession (Soja 1984). As a result, the proximate forces or local factors of class structuration are particularly intense and vividly geographical. The established social order of the city is changing with the creation of a 'New Ellis Island' (Anderson 1983).

The industrial structure has become more diverse and decentralized since the 1960s, but general growth has concealed sectoral divergences (Soja *et al.* 1983). Within manufacturing there has been a transition from automobiles, rubber and aircraft to durable goods and non-durable goods such as apparel, but also electronics and high-tech (Report to the Coalition to Stop Plant Closings 1981). There is a growth of service employment, clerical and managerial, emphasizing the expansion of the financial, insurance and real-estate sector and the recent emergence of Los Angeles as a key control nexus for global capital. There is no single trend of capital substituting for labour and consequent decentralization of production. The effect of these changes on the occupational structure has been, in crude terms, to diminish the significance of blue-collar workers and increase the number of low-paid and high-salary jobs. The expansion of services (especially producer services) and petty or 'peripheralized' manufacturing employment has been met in part by two million immigrants arriving in the 1970s. Such rapid changes form the basis for new social conflicts and cleavages.

How does the boundary between blacks and Latinos relate to this transformation? First, the re-ordering of job opportunities is uneven with regard to the existing spatial distribution of the groups. Secondly, racism differentiates black from Latino labour and incorporates them into the labour process (or excludes them from it) on different bases. Initial analysis suggests that the occupational dissimilarity between census-defined black and Spanish-origin populations increased between 1970 and 1980.[8] Lastly, a substantial proportion of migrant labour is of Latino origin.

In Los Angeles, the apparent continuity of a black ghetto since the early years of the century belies a discontinuity in underlying causes. As Hirsch (1983) commented, the making of the second ghetto in Chicago required fresh decisions, and not just continued acquiescence to old ones. In Los Angeles these fresh decisions originated in the defeat of the working class by the urban bourgeoisie over public housing in the 1930s and 1950s (Parson 1982, 1984). In the fight against 'creeping socialism' massive housing plans were overturned, and what little housing was built reinforced 'racial' boundaries. The black population expanded through migration from the South when the war industries demanded labour (Collins 1980, De Graaf 1970). Postwar lay-offs destroyed the nascent connections between steady work and secure residence which had created stable white ethnic communities in Long Beach and elsewhere. Mass suburbanization in these years split the unionized, mainly white working class from minority workers both spatially and politically. Residences freed by the departure of the whites in South-Central Los Angeles relieved a chronic housing shortage, and facilitated the coalescence of a single black ghetto. It was not a community bound together by stable employment. Later, union and employer resistance in the factories of the immense urban manufacturing belt began to relent, and public-sector employment was opened up. The thaw was insufficient to prevent the Watts commodity riots in 1965, and subsequent private and public investment failed to reverse socio-economic disadvantage. By the 1970s Southside had a dire investment vacuum, and exported workers to other parts of the metropolis.

Manufacturing employment in Los Angeles reached a peak in 1969, and from 1979 de-industrialization became severe. This affected minority workers who had finally succeeded in entering unions and industrial plants. Between 1978 and 1982 at least 75 000 jobs were lost to plant closures, notably in the auto-related industries clustered in South Los Angeles (Soja *et al.* 1983). Southside lost 40 000 jobs and 20 000 people, with the result that median family income more than halved (Soja 1984). The stable core of black occupational structure was devastated.

Nonetheless, the ghetto did form the basis of cultural and political power. Tom Bradley, now in a fourth term, was elected mayor in 1973 with a solid black vote. By the 1980s the status divisions contingent upon local economic changes became more pronounced. Southside is polarized between a high-income 'buppie' (black urban professional) west and a low-income 'underclass' south-east. One can no longer presume a single 'black' experience. Whereas 32 per cent of all black persons are below the projected poverty level, Rose (1981) indicates that Los Angeles is exceptional in national terms in that two-thirds of all black professionals participated in non-black housing markets. More research needs to be done on the emergence of a local black bourgeoisie.

This brief review suggests that in apparent contrast to our theoretical formulations, the residential segregation of blacks has not been consistently inscribed in clear-cut workplace–residential interaction. It is evident that much of what constitutes black racialization has resulted from more direct contact with the institutions of the state (through public employment and welfare provision, for example). Moreover, it results from considerable exclusion from waged labour rather than inclusion. In this context it is important to note that Marx's concept of labour may include the condition of non-waged employment, often conceptualized as a reserve army or 'underclass'. Both terms are ambiguous and unsatisfactory. Black racialization is typified by a long history of the failure or inability to create model workplace–residence relations established by the white working class, which were then transformed by the decentralization of production with increasing capital intensity (Yancey *et al.* 1985). Black unemployment in Los Angeles has remained consistently high, but this does not imply that the ghetto is simply an internal colony devised to reproduce a reserve army. The relationship between spatial form and social process is far too complex to be reduced to functional relations between 'the ghetto as colony' and US capitalism alone. The task is to unravel the layers of work–residence and non-work–residence which have both overlapped and become separated to form a contiguous spatial form.

The Latinization of the ghetto is occurring in a context of the differentiation of the black population in social and spatial terms. Between 1970 and 1980 the enumerated Spanish-origin population of the South-Central planning region doubled from 50 000 to 100 000, accounting for 21 per cent of the residents in 1980 (US Bureau of the Census). This is almost certainly an underestimate, given the failure of the census to count all undocumented workers. During the same period the black population fell by 8 per cent. Compared with the metropolitan area as a whole, a disproportionate number of these Latinos were foreign-born, perhaps 60 per cent, or of other than Mexican origin. In addition, they are poorer than local blacks and far less likely to own houses.[9] Taken together with fieldwork data this preliminary analysis indicates that the Latinization is led by a newly arrived immigrant workforce, including large numbers of Central Americans. There must be a significant proportion of the County's undocumented population in these neighbourhoods. Bordering Southside, close to downtown Los Angeles, the Pico area is reckoned to contain 300–400 000 Central Americans, many of them refugees without asylum. The vast majority have arrived within the past ten years. What explains the presence of this community, and what significance does it have?

Similar events have been observed in New York City by Sassen-Koob, which she has interpreted as a 'peripheralization at the core' (1981, 1982, 1985). As part of the internationalization of capital, areas of US cities which were in decline are being re-employed, with the infusion of largely foreign capital and Third World labour. In New York City such recombination was made possible by the disciplining of domestic labour during the fiscal crisis of the 1970s. While Los Angeles experienced nothing so drastic, this restructuring is an attractive hypothesis to be examined (although the explanation may be different). There are at least three diagnostic trends.

The first is a polarization of the inner-area occupation structure, involving a growth in the high-status professional, managerial and technical employees of financial and commercial firms, and a corresponding and complementary increase in low-wage service workers, especially in producer services. These individuals service the needs and life-styles of the professional, managerial and technical class in restaurants, hotels, laundries and so forth. A second trend is employment in labour-intensive, low-wage, vertically disintegrated and highly clustered industries, such as apparel. Scott (1984a) has identified a convergence of firms in this sector on downtown. The $3.5 billion garment industry in Los Angeles employs 125 000 workers, mainly in women's dress production (Report to the Coalition to Stop Plant Closings 1981, Wolin 1981). In contrast to New York and most of the rest of the country, it is a growth sector.[10] Ninety per cent of the workers are estimated to be undocumented, 80 per cent female and only 20 000 unionized (Wolin 1981). The vast majority are Latino and Asian; there is an unknown quantity of home-work and health-and-safety code violations are endemic. The processes of state-defined illegality, racism, sexism and exploitation combine to create a flexible, low-wage labour force. Together with service jobs they form a centralized complex of employment opportunities, essentially a peripheral and horizontal (i.e. of limited upward mobility) labour market. Given the intermittent, low-wage and unstable employment offered, and the importance of informal recruiting, both labour and employers tend to maximize access to each other by spatial convergence.

The third feature of this peripheralization is therefore the creation of an immigrant community and the recovery of part of the built environment, in a partial and transitional revalorization of inner-area substandard dwellings. Sassen-Koob (1981) describes New York's immigrant community as a holding operation, a low-cost complement to upper-class gentrification and the rapid spread of condominiums. It includes, in both cities, many small businesses catering to ethnic clientele, generating a local petty bourgeoisie.

We hypothesize that the economic 'glue' that holds together such inner-city ethnic communities is given by the spatial tendencies of local labour demands, rather than by the characteristics of housing markets. The dissolution of such communities in the past has been associated with the emergence of more stable workplace–residence relations, which allow a greater distance between the two components. There is, however, no inevitability about such shifts. They are contingent upon changes in the local organization of production. At present the inner-area community and the local labour demand are integral to the reproduction of a low-wage, non-unionized and non-Anglophone workforce. Yet the space it occupies is also demanded by the expansion of corporate capital, resulting in a series of community struggles since the 1970s (Haas & Heskin 1981). The intensification of these conflicts may arise as the community begins to support militant action in the workplace, as it has begun to do. Or it may revolve around the second generation, neither immigrants nor *indocumentados*, who will neither be slotted into sweatshop labour nor be educated enough to feed the demands of skilled office work. This is a process that social geography should interpret and inform.

The Latinization of the ghetto is therefore an apparently spatial process which is rooted in the reordering of production, which reaches out to draw labour from beyond domestic borders and which disproportionately links one group to an expanding low-wage sector and the other to increasingly insecure public and industrial sectors. The results of this occupational, social and spatial restructuring are as yet unpredictable. The effect of residential transition has been to challenge the political, economic and territorial power of blacks in the city. Residential mixing has not drawn the two groups together to any significant social degree, despite the objective grounds of common subordination. The political and cultural evidence points to the obvious empirical salience of residential segregation. Yet this account has tried to show how such spatial forms can be interpreted, not as the simple fragmentation (and consequent irrelevance) of class, but as the segmentation of the experience of class. Class is never experienced in a 'pure' form; it is always mediated by the communal and territorial realities of everyday life (Walker 1985). In Los Angeles, this has implied an erosion of the ghetto, not at the margins but from inside, and the emergence of a *barrio* counterposing a peripheralized labour force to an affluent élite and a racialized minority.

High-technology production in Orange County

Orange County is located immediately to the south of Los Angeles County, and its significance lies in the clear dominance of economic forces in its urbanization. In contrast to Los Angeles, where there is a multiplicity of production–reproduction relations expressed largely in terms of the inter-sectoral division of labour, Orange County has a sharp intra-sectoral division of labour based primarily on 'high-tech' industry. This is the basis of a distinct class polarization which is reflected relatively unambiguously in reproduction space in the form of occupational residential segregation. In this respect it resembles Silicon Valley (Santa Clara Co., California) as the 'post-industrial' equivalent of coal-mining or steel-mill regions. Furthermore, residential segregation adopts a 'racial' significance because of the particular requirements of the dominant economic sector.

Industrialization began relatively late in Orange County, as branch plants in the aircraft industry began to decentralize from Los Angeles in the 1950s (Scott 1986). The region was (and still is) primarily agricultural, divided into a few large ranches of which one, the present Irvine Company, still owns one-sixth of the county's area. There were a few scattered enclaves of Mexican agricultural workers but little urbanization. From 1960 to the present the history of the region can be described in terms of the formation and consolidation of a high-tech complex, which exemplifies the image of the Sunbelt. There was plentiful and cheap land and labour, as well as the environmental amenities of beaches and a good climate. The population of the county grew from 704 000 in 1960 to 1 900 000 in 1980, while high-tech employment rose from 25 000 in 1959 to 167 000 in 1981 (Scott 1986, US Bureau of the Census).[11]

The form of urbanization in Orange County is a direct result of the locational tendencies and labour requirements of the high-tech sector. The labour profile demonstrates considerable polarization. On the one hand there are highly-skilled technicians and engineers involved in research and development, and a skilled management staff responsible for control functions. On the other, there is a low-skilled blue-collar workforce concerned with producing assembled goods. The relationship is top-heavy, with the ratio of non-production workers to production workers in manufacturing being 6.1 : 10.0. The US average is 3.5 : 10.0 (Scott 1986).

The blue-collar labour force contains a substantial number of Latino and Asian immigrant workers, providing cheap labour in competition with overseas employers. Between 1970 and 1980 the Spanish-origin population of the county rose from 160 000 to 285 000, an increase of 78 per cent (US Bureau of the Census). Estimates of the number of *indocumentados* are as high as 50 000.[12] Many Mexicans arrived as migrant labourers working in agriculture and in the rapid construction of the county, typically as clean-up crews. One-sixth of the whole country's Vietnamese population is in Orange County, centred on Westminster and Garden Grove. It is estimated that one-fifth of all benchwork is done by Vietnamese. Notable by its almost complete absence is the black population. This poses an intriguing question. Could it be that the peculiar labour requirements of the high-tech sector have squeezed out the need for higher-waged working-class employees?

The locational dynamics of the high-tech sector, with its vertical disintegration and declining plant sizes, have been towards polarization and an intensification of the original complex (Scott 1986). Two main clusters have formed, one around Anaheim and Fullerton and another further south in Irvine, a city incorporated in 1971. This location continuity has reinforced the residential patterns which began to emerge in the 1960s. Residential communities to the immediate south and east of the Irvine complex as well as the coastal communities of southern Orange County, such as Newport Beach and Laguna Beach, have attracted the wealthier managers and engineers. The poorer workers are found in scattered inland locations to the north, most notably Santa Ana and Garden Grove, in close proximity to the high-tech clusters. Given the location of high-tech firms in virtually undeveloped areas and the 'clean' characteristics of the industry, we would expect the distance relationship between work and home to be relatively short for most workers. A preliminary analysis of journey-to-work times for Orange County workers in 1980 has shown that more than 60 per cent of all workers travel less than 20 minutes to work.[13] There is no significant difference in travel time between Anglos[14] and Mexicans, although slightly shorter time is travelled by Anglos, accounted for mostly by the short times travelled by women engaged in clerical work. Thirty-five per cent of Anglo women workers live within 10 minutes of their workplace compared to 25.2 per cent for men. If we consider only those industries previously defined as high-tech core industries, however, results show that slightly shorter times are travelled by Mexicans. Sixty-seven per cent of Mexican workers travelled 20 minutes or less, compared with 62 per cent of white workers. A further study of journey-to-work patterns would provide important insights into the workplace–residence relationships of a newly urbanized growth centre. We would expect such relationships to differ from those found in urban areas with other forms of industrialization.

The proximity of residence to major centres of employment provides an important amenity to the recruitment of employees. The competitive bidding for highly-skilled labour by the high-tech sector results in serious efforts to produce both attractive working and living environments. Tennis courts and exercise rooms inside the plant may be complemented by Mediterranean-style condominium complexes along the coast.

In contrast, some *barrios* developed from old agricultural settlements. These are now becoming seriously over-crowded and under-serviced, and suffer from problems similar to those found in the old Eastside neighbourhoods in Los Angeles. The marginal position of such *barrios* was vividly illustrated by Garden Grove's attempt to 'dump' Buena Clinton, a small district of 4000 Latinos, on Santa Ana in return for offering them rights to Garden Grove's golf course (*Los Angeles Times*, 28 December 1983).

In recent years there have been new trends towards vertical disintegration resulting in a transfer of labour from unionized to non-unionized plants. There is also a concomitant growth in the number of blue-collar workers, which is against the national trend, and a decline in the number of bureaucrats and managers (Scott 1986). Furthermore, land-price inflation has limited the willingness of plants and technicians to move to the county, so that the central business district has switched to finance and control functions. A CBD is appearing in the otherwise dispersed and low-density regions. Traffic congestion and the disruption caused by continuous construction have extended journey-to-work times for commuters. Perhaps because of such diseconomies some firms are seeking locations elsewhere in the county or outside.[15]

As the contradictions of rapid and uncontrolled urbanization deepen, the prospects for the poor minorities of Orange County are not promising. If plants leave, will the prosperity of the region continue for long enough to allow some benefits to trickle down to the semi-skilled workers? Will service employees replace manufacturing? In a single-industry region such endogenous factors as technological change and such exogenous factors as international competition and Federal defence and procurement contracts could undermine the emergent minority communities. The relations between Asians and Latinos and the high-tech sector require further elucidation rather than more theoretical inference. Is there a place for them in the county which has most consistently supported Ronald Reagan? It appears that in the restructuring of 'old' Los Angeles and 'new' Orange County the future of racialized communities is increasingly circumscribed by events of international dimensions.

Conclusion

We began by indicating that the present constitution of residential segregation as a problem in social geography was founded upon inadequate conceptions of 'race' and space. A failure to treat the forces of urbanization and incorporation fully resulted in a misidentification of 'race' and ethnicity as given social forms. This led to some chaotic conceptualization by indices and spatial forms of what are varied and contingent phenomena, which have a possible multiplicity of real causes. 'Race' is not an explanation; racism is a real ideological construct. Both 'race' and space could usefully be revised in terms of the connections between spheres of production and reproduction.

We proposed that any reformulation could begin on the higher ground of post-Weberian location analysis and labour theory. Above all we stressed the centrality of work in the everyday life of human beings, a fact strangely forgotten in social geography. By deducing that the social and spatial divisions of labour originate in the labour process of commodity production, circulation and consumption, we sought to explain the existence of residential differentiation. To account for the obvious differences between a pure production-centred city and actual urban places, or between the model of a labour pool and the multi-textured reality of a community we invoked notions of class structuration. The existence of residentially homogeneous and segmented areas was significant in terms of the reproduction of labour, which could also take non-economic forms. Our analysis was therefore directed towards local class forces and the structuration of generic class structures incorporating 'racial' and ethnic categorization.

From our account it should be clear that there is no one theory of segregation that encompasses such diverse social groups as the aged, the mentally ill, or gays. Our focus has been quite narrowly fixed on labour as a social relation, and on the metropolitan scale rather than a more localized one. On such a canvas the capital–labour relations seems pre-eminent over affective ties. We have tried to indicate nonetheless that such local forces may have an effective impact upon metropolitan structure in its details, though evidence for this is somewhat thin.

Using two case studies we attempted to develop this retheorization, placing great importance on the quantitative examination of social systems rather than the inference of theorization. Los Angeles and Orange Counties provided cases of the production-centred structuring of residential space. We have made a series of simplifying assumptions in order to tease out the specific relations between production and reproduction. These simplifications are less realistic in Los Angeles County than in Orange County, which is relatively recently urbanized. Nonetheless, it is these relations which we believe must form the basis for any discussion of residential segregation.

Future research lies not only in detailed explication of the relations we have adduced, but also of the assumptions we have made. The peculiarity of black racialization, as constituted through both work and 'non-work', suggests an analysis of the specific points of intervention of the state, and coercion outside the sphere of work itself. In Orange County, the extraordinary dominance and success of economic forces, and the relatively smooth reproduction of a labour force, has forestalled the emergence of ethnic conflict in the local state. In both cases, there is scope for the detailed study of work as a routinized daily activity, the everyday meaning of ethnicity and 'race'. Our retheorization draws upon the strong tradition of geography for describing and explaining the regional constitution of society. Finally, we agree with Sayer (1985b) in recognizing the political and social imperative to understand local complexities and specificities rather than imposing the preformed imagos of a marxist discourse upon them.

Notes

1 For the purposes of this chapter we make no distinction between 'race' and ethnicity. In the American context at least the difference is difficult to maintain and has been pre-empted by an *a priori* empirical differentiation of black 'race' and European ethnicity. See Miles (1982, Ch. 3).

2 The difference between an explanatory concept obtained by abstraction and a descriptive concept obtained by generalization is at the centre of marxist (and realist) analysis. There is no real problem with presenting a table of income categories against racial categories, if it reveals the required information. This is descriptive usage. However, a regression equation that represents 'race' as the residual term after 'class' has been accounted for, thus proving that 'race' 'explains' x per cent of the variation, is an inappropriate usage. It accords 'race' causal efficacy.

3 These comments apply to all notions of closure, strategy, voluntary/involuntary segregation, choice and constraint, positive/negative forces of segregation and so on that draw on Weberian themes, explicitly or not. See Brown (1981) and Miles (1982).

4 Parkin's bourgeois-class theory does not hold class to be central to the transformation of the social formation. He states, '. . . class conflict may be without cease, but it is not inevitably fought to a conclusion' (1979, p. 112). There are many sources of potential conflict which is inevitable and irresolvable, and power may be funnelled through class, status and party. See also Giddens (1973).

5 Pahl (1984) has shown that 'work' is an activity which can only be defined properly by context, with regard to source of labour, for whom the task is done, where and so on. We take 'work' to include waged and unwaged activity, formal and informal activities or both production and reproduction spheres. However, in the case studies we employ a more restrictive usage of 'work', as waged employment.

6 See Walker (1985) for an extension and critique of Giddens' position which attempts to make clearer the distinction between class and division of labour.

7 The US Bureau of the Census is not a completely accurate guide to the Spanish-origin population of Los Angeles. It gives 2 066 013 in 1980, of whom 1 650 934 are Mexican, 36 662 Puerto Rican, 44 289 Cuban and 334 218 other. Other estimates vary from 2.1 million to 3.5 million, including 200–250 000 Salvadorans, 50–75 000 Guatemalans, 25–50 000 Nicaraguans, 50–100 000 others. Estimates of undocumented workers, over half of whom are likely to be of Spanish origin, vary from 200 000 to 1.1 million depending on interest or method of calculation (Anderson 1983; Cornelius, Chavez & Castro 1982; *Los Angeles Herald Examiner* 24 July 1983, 8 May 1984; *Los Angeles Times* 20 February 1984).

8 Calculations are based on the 15-class industrial classification from the US Census, which is not strictly comparable between the two years. In 1970 the Index of Dissimilarity was 21.9, in 1980 19.5. In manufacturing the Spanish-origin population was well over-represented in the non-durable sectors of food, apparel, furniture, fabricated metals and machinery. Blacks were greatly over-represented in the public sector, professional services and transportation.

9 Black home-ownership is 30 per cent and Spanish-origin 7 per cent. Black median family income is $11 744, Spanish-origin is $11 014 (US Bureau of the Census).

10 Between 1977 and 1982 the number of production workers employed in women's dress manufacture rose 70 per cent in Los Angeles–Long Beach, and fell 54 per cent in New York City. Wages in the former were two-thirds of those in the latter, making apparel the lowest-waged sector of production (Annual Wage Survey 1982).

11 Scott (1986) defines 'high-tech' as a 'core' of four SICs (machinery except electronics, electric and electrical equipment, transportation equipment and instruments and related products) and a 'penumbra' of three other SICs. The core provides 74 per cent of all manufacturing employment in the county.

12 The census records 286 339 Spanish-origin people in the County, including 232 472 Mexicans, 5534 Puerto Ricans, 4820 Cubans and 43 313 others. They comprise 14.8 per cent of the county's population. Asians and blacks make up 4.5 per cent and 1.2 per cent respectively (*Los Angeles Times*, Orange County edn. 24 July 1983).

13 The data for this study are obtained from the Public Use Microdata File of the US Census.

14 'Anglos' can be defined as white people of non-Spanish origin.

15 Recent estimates suggest that Orange County's demographic growth has been slower in the 1980s than the 1970s, in contrast to Los Angeles County (*Los Angeles Times*, 18 February 1986).

References

Anderson, K. 1983. The New Ellis Island. In *Time*, 13 June 1983.

Brown, K. R. 1981. Race, class and culture: towards a retheorization of the choice/constraint concept. In *Social interaction and ethnic segregation*, P. Jackson & S. J. Smith (eds). London: Academic Press.

Castells, M. & K. Murphy 1982. Cultural identity and urban structure: the spatial organisation of San Francisco's Gay community. In *Urban policy under capitalism*, N. Fainstein & S. Fainstein (eds). Beverly Hills: Sage.

Collins, K. E. 1980. *Black Los Angeles: the maturing of the ghetto 1940–50*. Saratoga: Century 21.

Cooke, P. 1985. Class practices as regional markers: a contribution to labour geography. In *Social relations and spatial structures*, D. Gregory & J. Urry (eds). London: Macmillan.

Cornelius, W. A., L. R. Chavez & J. G. Castro 1982. *Mexican immigrants and southern California: a summary of current knowledge*. Research Report Series 36, Center for US–Mexican Studies, University of California at San Diego.

Davis, M. 1984. The political economy of Late-Imperial America. *New Left Review* **143**, 6–38.

Dear, M. 1981. Social and spatial reproduction of the mentally ill. In *Urbanization and urban planning in capitalist society*, M. Dear & A. J. Scott (eds). London: Methuen.

De Graaf, L. B. 1970. The city of Black Angels: emergence of the Los Angeles ghetto 1890–1970. *Pacific Historical Review* **39**(3), 323–52.

Garcia, P. 1985. Immigration issues in urban ecology: the case of Los Angeles. In *Urban ethnicity in the United States: new immigrants and old minorities*, L. Maldonado & J. Moore (eds). Beverly Hills: Sage.

Giddens, A. 1973. *The class structure of advanced societies*. London: Hutchinson.

Haas, G. & A. D. Heskin 1981. Community struggles in Los Angeles. *International Journal of Urban and Regional Research* **5**(4), 546–64.

Harris, R. 1984. Residential segregation and class formation in the capitalist city: a review and directions for research. *Progress in Human Geography* **7**(1), 26–49.

Harvey, D. 1975. Class structure in a capitalist society and the theory of residential differentiation. In *Processes in physical and human geography*, R. Peel, M. Chisholm & P. Haggett (eds). London: Heinemann.

Harvey, D. 1977. Labour, capital and class struggle around the built environment in advanced capitalist societies. *Politics and Society* **5**(3), 265–79, 288–93.

Harvey, D. 1978. The urban process under capitalism: a framework for analysis. *International Journal of Urban and Regional Research* **2**(1), 101–31.

Harvey, D. 1982. *The limits to capital*. Oxford: Blackwell.

Harvey, D. 1985. *Consciousness and the urban experience*. Oxford: Blackwell.

Hershberg, T. (ed.) 1981. *Philadelphia: workplace, family and the group experience in the 19th century*. New York: Oxford University Press.

Hirsch, A. R. 1983. *Making the second ghetto*. Cambridge: Cambridge University Press.

Katznelson, I. 1981. *City trenches*. Chicago: Chicago University Press.

Massey, D. 1984. *Spatial divisions of labour*. London: Macmillan.

McDowell, L. 1983. Towards an understanding of the gender division of urban space. *Environment and Planning D: Society and Space* **1**(1), 59–72.

Miles, R. 1982. *Racism and migrant labour*. London: Routledge & Kegan Paul.

Miles, R. 1984. Marxism versus the sociology of 'race relations'? *Ethnic and Racial Studies* **7**(2), 217–37.

Oliver, M. & J. Johnson, Jr. 1983. Interethnic conflict in an urban ghetto: the case of Blacks and Latinos in Los Angeles. In *Research in social movements, conflict and change VI*, R. Ratcliff & L. Kriesberg (eds). Greenwich: Jai Press.

Pahl, R. E. 1984. *Division of labour*. Oxford: Blackwell.

Parkin, F. 1979. *Marxism and class theory: a bourgeois critique*. London: Tavistock.

Parson, D. 1982. The development of redevelopment: public housing and urban renewal in Los Angeles. *International Journal of Urban and Regional Research* **6**(3), 393–413.

Parson, D. 1984. Organized labour and the housing question: public housing, suburbanization and the urban renewal. *Environment and Planning D: Society and Space* **2**(1), 75–86.

Rex, J. & S. Tomlinson, 1979. *Colonial immigrants in a British city*. London: Routledge & Kegan Paul.

Report to the Coalition to Stop Plant Closings 1981. Graduate School of Architecture and Urban Planning, Report 56, University of California at Los Angeles.

Rose, H. M. 1981. The black professional and residential segregation in the American city. In *Ethnic segregation in cities*, C. Peach, V. Robinson & S. Smith (eds). London: Croom Helm.

Sassen-Koob, S. 1981. *Exporting capital and importing labour: the role of Caribbean migration to New York City*, Occasional Paper 28. New York Research Program in Inter-American Affairs, New York University.

Sassen-Koob, S. 1982. Recomposition and peripheralization at the core. *Contemporary Marxism* **5**, 88–100.

Sassen-Koob, S. 1985. The new labour demand in global cities. In *Cities in transformation*, M. P. Smith (ed.). Beverly Hill: Sage.

Sayer, A. 1984. *Method in social science*. London: Hutchinson.

Sayer, A. 1985a. The difference that space makes. In *Social relations and spatial structures*, D. Gregory & J. Urry (eds). London: Macmillan.

Sayer, A. 1985b. Industry and space: a sympathetic critique. *Environment and Planning D: Society and Space* **3**(1), 3–29.

Scott, A. J. 1980. *The urban land nexus and the state*. London: Pion.

Scott, A. J. 1982. Production system dynamics and metropolitan development. *Annals of the American Association of Geographers* **72**(2), 185–200.

Scott, A. J. 1983a. Industrial organisation and the logic of intra-metropolitan location I: theoretical considerations. *Economic Geography* **59**, 233–50.

Scott, A. J. 1983b. Industrial organisation and the logic of intra-metropolitan location II: a case study of the printed-circuits industry in the Greater Los Angeles region. *Economic Geography* **59**(4), 343–67.

Scott, A. J. 1984a. Industrial organisation and the logic of intra-metropolitan location III: a case study of the women's dress industry in the Greater Los Angeles region. *Economic Geography* **60**(1), 3–27.

Scott, A. J. 1984b. Territorial reproduction and transformation in a local labour market: the animated film workers of Los Angeles. *Environment and Planning D: Society and Space* **2**, 277–307.

Scott, A. J. 1985. Location processes, urbanization and territorial development: an exploratory essay. *Environment and Planning A* **17**, 479–501.

Scott, A. J. 1986. High technology industry and territorial development: the rise of the Orange County complex, 1955–1984. *Urban Geography* **7**(1), 3–45.

Soja, E. W. 1984. *LA's the place: economic restructuring and the internationalization of the Los Angeles region.* Paper presented at the Annual Meeting of the American Sociological Association, 27–31 August 1984, San Antonio.

Soja, E. W., R. Morales & R. Wolff 1983. Urban restructuring: an analysis of social and spatial change in Los Angeles. *Economic Geography* **59**, 195–230.

Storper, M. & R. Walker 1983. The theory of labour and the theory of location. *International Journal of Urban and Regional Research* **7**(1), 1–43.

Storper, M. & R. Walker 1984. The spatial division of labour: labour and the location of industry. In *Sunbelt/frostbelt: urban development and regional restructuring*, L. Sawyers & W. K. Tabb (eds). New York: Oxford University Press.

Thrift, N. 1983. On the determination of social action in space and time. *Environment and Planning D: Society and Space* **1**(1), 23–57.

Trachte, K. & R. Ross 1985. The crisis of Detroit and the emergence of global capitalism. *International Journal of Urban and Regional Research* **9**, 186–217.

Urry, J. 1981. Localities, regions and social class. *International Journal of Urban and Regional Research* **5**(4), 455–74.

Walker, R. 1981. A theory of suburbanization: capitalism and the construction of urban space in the United States. In *Urbanization and urban planning in capitalist societies*, M. Dear & A. J. Scott (eds). London: Methuen.

Walker, R. 1985. Class, division of labour and employment in space. In *Social relations and spatial structures*, D. Gregory & J. Urry (eds). London: Macmillan.

Walker, R. & M. Storper 1981. Capital and industrial location. *Progress in Human Geography* **5**(4), 473–509.

Ward, D. 1982. The ethnic ghetto in the United States: past and present. *Transactions of the Institute of British Geographers* **NS7**(3), 257–75.

Wolin, M. L. 1981. Sweatshop: undercover in the garment industry. *Los Angeles Herald-Examiner*, 14 January–1 February.

Yancey, W. L., E. P. Ericksen & G. H. Leon 1985. The structure of pluralism: 'We're all Italian around here, aren't we Mrs. O'Brien?' *Ethnic and Racial Studies* **8**(1), 94–116.

3 *Racism and settlement policy; the state's response to a semi-nomadic minority*

DAVID SIBLEY

The principal concern of this essay is racism as it is manifested in the response of central and local government to British Gypsies. In order to put this question into perspective, however, the discussion is prefaced with some comments on the state's perception of minority groups as 'problems', particularly in regard to the spatial concentration of minorities in cities. The contrast between the responses to black minorities, for example, which have an integral rôle in capitalist economies, and to a semi-nomadic, peripheral minority may be illuminating.

In ethnically-mixed societies that pursue ostensibly pluralist policies, direct control of the residential locations of a particular group is unusual. It is evident that the residential choices of the black British population are highly constrained as a result of institutional racism, which compounds the effects of class to create an unequal distribution of rewards in the job market, education and the housing market, but the control of residential location is not an explicit feature of policy. Here, racism is structurally evident in practice but is generally unstated and masked (Mason 1982, cited by Williams 1985, p. 329).

Possibly because of the deeply imbedded nature of racism in British society, there is a tendency to reserve the label 'racist' for regimes where coercion is part of the daily routine of control, including control of settlement and migration, notably in South Africa where, as Western (1982, p. 218) has observed, 'The social geography . . . of cities is a product of witting intent; the hand is not hidden here.' While there is clearly a difference between the practice of racism in South Africa and in Western liberal democracies, it is the case that the residential distribution of minority populations has been seen as a problem in societies where people are supposed to be able to exercise choice in both the public and private sectors of the housing market. A number of dubious proposals for housing ethnic minority groups appeared in Britain, for example, during the 1970s. While these ideas were not taken up by policy makers at the national level, they probably reflect beliefs that were widely held. Perry (1973) put it like this: 'Dispersal may be concerned primarily with getting people out of decaying city centres, or with breaking up tight-knit racial groups, with getting coloured people accepted in suburban neighbourhoods, or with a combination of these three.' She went on to suggest that, for practical reasons, many immigrants (sic) will continue to live in inner cities for some time. The claim that dispersal was necessary – that the spatial concentration of the black population was undesirable – appeared in an influential report by Cullingworth in 1969 (Central Housing Advisory Committee 1969) and he reaffirmed this view in the foreword to a Community Relations Commission report in 1977. Cullingworth qualified his position in this and later writing but in 1979 he was still able to suggest that 'there is a presumption – to put it no more strongly – that there is a scale of homogeneity which is problem-creating' (Cullingworth 1979, p. 163).

Two instances of Dutch policy may be cited which support the view that white societies can see little positive about black communities in urban areas. First, the housing allocation policy of Rotterdam City Council includes the 'clustered deconcentration' of the black population, in order to provide access both to better housing and public services (Mik 1983), but this policy is opposed by some organizations in the black community (Roseval 1981, personal communication). The second was the practice (until the mid-1970s) of directing Surinamese immigrants to small towns in order to ease assimilation and avoid ghetto-formation in large cities. This kind of response is part of the process of what Solomos (1984, p. 9, cited by Williams 1985) has termed 'the construction of black communities as problems' in which a concentration of black people becomes synonymous with deprivation and deviant behaviour. The tendency of some academics to use the presence of a black community in an area as a social indicator contributes unwittingly to the same negative image. As Kantrowitz (1981, p. 54) has argued, in relation to white ethnic rather than black minorities but the point seems equally apposite, the academic establishment has contributed to a negative view of ethnic segregation – segregation is bad. Dispersal is the obvious liberal solution, neglecting all the benefits of concentration, notably, the existence of thresholds for the provision of community services, the possibility of acquiring a constituency and gaining political power and the positive reinforcement of identity within a racist society. These positive features of spatial concentration have been recognized by a few writers, for example, Deakin and Cohen (1970) and Boal (1976), the last-named emphasizing the advantages of residential concentration for protection, the avoidance of white racism and the preservation of minority culture. There is rather more evidence in the social-policy literature, however, of negative, racist attitudes.

In Britain, at least, this kind of thinking has not been translated into practices affecting the larger black minorities except at the local level by some housing authorities. This is not the case, however, for British and other European Gypsies and for peripheral groups like the indigenous populations of North America and Australia, for whom settlement control has been a central feature of policy, part of a strategy of social control in the sense that Cohen (1985, p. 2) uses the term: that is, 'planned and programmed responses to expected and realized deviance rather than ... the general institutions of society which produce conformity.' In the location, design and management of settlements for the Inuit and Dene in Canada, the Australian Aborigine population and Gypsies and Travelling people in Britain, Ireland and Holland, the desire to disperse, contain and transform is evident, either because these minorities are seen as a threat to the capitalist economy or because their labour power is required, or both. The difference in the dominant society's response to these groups and to the larger, primarily urban, black minorities is probably related to the economic distinctiveness of peripheral minorities, which compounds the issue of racism. In their use of land in particular these groups provoke a reaction which is similar to that experienced by the plot-land residents in Britain in the 1930s (Hardy & Ward 1984) or the residents of squatter settlements in metropolitan areas in the Third World (see, for example, Collier 1976, p. 47). In this essay, however, I want to examine the ways in which the racism of the larger society has shaped the response to peripheral minorities, recognizing that racism is one dimension of the larger problem of exploitation within a capitalist society, albeit a very important one. I will concentrate on the case of British Gypsies but extend the argument to Gypsies elsewhere and to other peripheral minorities.

Gypsies and racism

The Swann Report, published in 1985, refers to the 'universal hostility and hatred' experienced by the Travelling community and, with particular reference to Gypsy children, the 'racism and discrimination, myths, stereotyping and misinformation, the inappropriateness and inflexibility of the education system' (Commission for Racial Equality 1985, p. 18). While the terminology is rather confusing, there is, at least, an admission that racism is a problem for the Gypsy community where, before, discussions of racism in Britain in the context of policies for minority groups have neglected racist attitudes to Gypsies. As with other minorities, we have to distinguish between popular expressions of hostility which can be identified unequivocally as racist and the not so obvious racism which is manifested in government responses and academic analyses. The latter is more important because it has direct and immediate consequences for the economy and social wellbeing of Travelling people.

There are, however, some difficulties involved in establishing the racist nature of the relationship between the dominant society and Gypsies. Wallerstein (1983, p. 78) has argued that 'the beliefs that certain groups were "superior" to others in certain characteristics relevant to performance in the economic arena always came into being after, rather than before, the location of these groups in the work force. Racism has always been "post-hoc".' This economic rationalization of racism does not have an obvious application to Gypsies because of their marginal relationship to the dominant economy. In particular, their avoidance of wage labour, except where family labour is used in seasonal agricultural work, has insulated Travellers from direct economic exploitation. In fact, the usual relationship of dominance or subordination is reversed in Gypsy/*gauje* (non-Gypsy) encounters, with exploitation of the *gauje* being a source of ethnic pride. It must be admitted, however, that the scope for exploitation is very restricted given the limited control over resources which Gypsies can exercise in the marginal and interstitial areas of the space-economy. The issue of racism is further confused by the nomadism of the minority group, whether this nomadism is real or wrongly attributed to a particular group of Travellers. There is an enduring conflict between settled peoples and nomads which is usually resolved to the disadvantage of the latter (Rapoport 1978). Inevitably, Travelling people violate concepts of property and pose a threat to the smooth operation of the dominant economy. Movement in itself is viewed negatively in a society that is attached to property. The threat posed by a migratory group is often expressed as a fear of 'a Gypsy invasion', regardless of the actual mobility of Travellers – the stereotype is of a minority group of prodigious mobility.

There is a further conflict between the order of the larger society, for example in regard to accepted relationships between land uses, and the disorder of the Gypsies' way of life – although the latter is only a different kind of order, reflecting the integrated nature of Gypsy culture. The idea of a spatial separation of work, residence and recreational activities is alien to Gypsies but the integration of these activities in space is a form of deviance according to a dominant world-view. As Parekh (1974) has suggested, this concern with order may be a particularly

British or European problem for the same obsession is lacking in Indian culture, for example.

Although hostility to Gypsies may be bound up with their real or mythical nomadism, there is a clear racist dimension to the problem in that the assumed inferiority of the group is supported by reference to spurious biological and cultural arguments. If a nomadic people are also of a different race, they become that much more of a pariah group. It may be useful, therefore, to trace the origins of racist myths applied to Gypsies and to see how they are manifested in current attitudes.

Biological and cultural myths

The main value of an historical analysis of the problem is that it may help us to make connections between past and present attitudes and practices. As Williams (1985, p. 337) has suggested, 'it is important to trace varieties of ideological discourses, from early emphasis upon [blacks] as in need of civilizing, from scientific racism and biological determinism to contemporary liberal discourses of deprivation.' Although there may be a legitimate historical interest in racism as an ideological issue (Lorimer 1978, Livingstone 1984), this focus could prove diversionary if the relevance to current attitudes, including those embodied in social theory, were not established. Thus, the following examples are included to demonstrate continuity in racial stereotyping. Although not all are directed specifically to Gypsies, they have contributed to the formation of a racial stereotype of Gypsies (and indigenous minorities).

Eighteenth- and nineteenth-century attitudes are particularly interesting. Although these are often labelled 'scientific' or 'theological' it is probably more useful to think of both as expressions of the Victorian imperialist culture of the upper classes. The flavour of the debate in clerical circles is suggested by the views of the progressionists and degenerationists, which were in fact equally racist. John Wesley, as an example of the latter, maintained that some societies existed in a state of nature, a degenerate state from which they would only be saved by civilization and religion (Douglas 1978, p. 12). This sentiment is conveyed nicely in Emma Marshall's Victorian novel, *Houses on wheels*, in which a Gypsy boy, Lennie, dies from consumption. His *gauje* friend Bernard reflects on his death:

> 'Thank God', Bernard said. And as he turned sadly away, he prayed that the Good Shepherd might seek out and save many more of these poor lost sheep by our highways and hedges – those dwellers in houses

> on wheels – and put it into the hearts of many who have only thought of them . . . as hopelessly sunk in sin and ignorance to stretch out a hand of help to promote any wise scheme for their rescue from the deeps of wickedness, and misery, and sin. (p. 385.)

While the more liberal progressionists such as Henry Burnett Tylor and Adam Smith saw evidence of change from a primitive to a civilized state without religious intervention, they were still attached to notional scales that sought to distinguish between the lowest (savages) and the highest (white Europeans).

The views of the scientific establishment, notably Galton, were essentially the same. Galton, for example, referred to 'a wild, untameable restlessness . . . innate with savages', and to a veneer of civilization, including 'numerous instances in England where the restless nature of a gypsy half-blood asserts itself with irresistible force' (Biddiss 1979, pp. 68–9). Like the theologians, he devised a ranking which put 'the average standard of the Negro race . . . two grades below our own; that of the Australian native . . . at least one grade below the African' (Blacker 1952, p. 325). More sinister were the arguments of those who professed a materialist biology, notably Haekel who argued that 'the careful rooting out of weeds among good and useful plants would make easier the struggle for life among the better portions of mankind' (Billig 1982, pp. 70–1), a sentiment which received approval from Darwin, Engels and Lenin, and which paved the way for genocide. If there had been a subsequent rejection of these views and a reconstruction of social theory in an explicitly anti-racist form, we could dismiss the 19th-century legacy. However, similar rationalizations of prejudice appear in a diluted and mystified form in social theories which still have some currency, notably modernization theory. Problems like 'the Gypsy problem' are couched in terms that fit such theory and, in this sense, racism permeates theory and practice.

Racism and the modernization of peripheral minorities

A premise of many ethnographic studies of indigenous minorities and Gypsies is that change in the minority culture is inevitable, either because increasing contact with the dominant society is interpreted as an attempt to enter the modern world, with the promise of greater material wellbeing, or because the expansionary and disruptive capitalist economy appears to leave no room for small-scale, semi-autonomous economies. Mooney (1976, p. 391) has summarized this perspective as follows: 'native Indians or other contacted peoples are supposed to be in the process of an inevitable one-way change from a "traditional", "primitive", or "native-oriented" state. Underdevelopment ... is assumed to be the original condition of the acculturating society; full development will come with complete acculturation, that is, with integration into white society'. For indigenous populations in industrialized societies, modernization involves urbanization, either through urban development in the periphery or as a result of population movements from the periphery to the urban core. A different gloss might be put on the process by substituting 'internal colonialism' for 'modernization', but from either point of view, in accounts of the acculturation or incorporation of indigenous minorities or nomadic peoples we are presented with two contrasting images. The first is of a rural people, *at one with nature*, subsisting with a traditional economy, involving hunting, gathering and some trading in the market economy, or – in the case of Gypsies – crafts and the provision of services to an agricultural population; the second is of a dislocated group, attempting to cope with an alien urban environment, exhibiting what is often interpreted as deviant behaviour, such as alcoholism and minor criminality. The rôle of the researcher in many instances has been to map the transition from the first state to the second state.

The opposition of rural and urban in studies of peripheral minorities corresponds to what Douglas (1975) has termed a contrast between 'man' and 'not man' (sic). By this, she means that some (racial minority) groups appear from a mainstream perspective to be outside civil society, belonging instead to 'The Wild', in the sense that they are a part of nature. Thus, in popular mythology we have 'real Indians', 'real Eskimos', 'real Gypsies', 'true Romanies' – either inhabiting areas that are distant and inhospitable or remembered as inhabitants of a rural past. The importance of an imputed racial purity is that the people actually encountered by members of the larger society, often in conflict situations and particularly in cities, can be dismissed because they do not conform to the romantic racial stereotype. In the case of British Gypsies the use of terms like 'tinker', 'itinerant' and 'diddikai' all suggest a failure to meet the standards implied in the stereotyped view – they effectively dehumanize and legitimate oppressive policies. Similarly, Brody (1971) noted that native North Americans on skid row were popularly considered to be of mixed race, unlike 'real Indians' who would not have problems of alcoholism. The use of the stereotype in responses to Gypsies at several levels – in popular protests, in local and central government policies and in some academic writing – suggests that this prejudice is deeply rooted in the British social consciousness. The following two examples of informed comment are characteristic:

> ... it soon turned out that the so-called 'Gypsies' did not have painted wooden caravans or boast dark lustrous eyes; nor did they make a living by selling heather and stealing the odd chicken. They were Irish scrap dealers.
>
> It was these travellers apparently who later formed part of the invasion of Hampstead Heath, drawing attention to a bizarre and intractable problem: the emergence in the middle of a modern city of a shifting and growing group of social outcasts. (*The Observer*, 3 February 1985.)

> In relation to a complaint against Ladbroke Racing Ltd, brought by the Commission for Racial Equality on behalf of two Gypsies, the regional complaints officer said that the CRE would strongly assert that true Romanies were an ethnic group within the meaning of the Act; travelling people were 'a grey area'; if an Irish tinker was being discriminated against because he was Irish, he would have grounds for complaint. (*The Guardian*, 20 April 1985.)

These kinds of categorizations have serious consequences. Because the stereotype of the 'real Gypsy' is by definition mythical, all Gypsies are likely to be considered deviant and in need of corrective treatment. Whatever the reality, it will never correspond to the romantic portrayal of Gypsy life because the existence of a nomadic community in a highly structured industrial society is simply unacceptable. Thus, it could be argued that the response to Gypsies in a capitalist society demonstrates Wallerstein's assertion that racism provides a post-hoc justification for economic oppression, where oppression takes the form of exclusion from the main centres of production and reproduction and relegation to marginal space.

The containment of Gypsies

In Europe there is a long history of expulsions and, at the same time, attempts to control through assimilation. In France, for example, between 1912 and 1969 all nomadic Gypsies over 14 years of age had to carry a '*carnet anthropometrique*' detailing a number of physical characteristics. This had to be presented to the police on arrival and departure from a town to facilitate surveillance and eviction (Williams 1982). In 1928–9, Holland, Belgium and Germany proposed that 'the gypsy plague', that is the movement of Gypsies across national frontiers, should be tackled by the League of Nations. Bavaria had compiled a register of male Gypsies in 1925, and in 1930 the German state established a central agency for controlling Gypsies (*Zentralstelle für die Zigeunerbekampfung in Deutschland*). These controls, inspired by Nazi racist ideology, culminated in mass extermination of European Gypsies – in Poland and the Soviet Union in 1941, and in Auschwitz and other concentration camps in 1943–4. Reflecting on these events, Sijes (1979, p. 173) detects an echo of the Nazi occupation of the Netherlands, when many Gypsies were confined to camps prior to deportation to Germany, and in the Dutch Caravan Sites Act of 1968 which identified native Dutch Travellers (*Woonwagenbevoners*) as a special category who could only live in prescribed locations.

In England and Wales, the historical experience of Gypsies has been similar to that of communities in continental Europe. There have been penal sanctions against Gypsies since the Tudor period, but the first systematic attempt to restrict settlement was the Caravan Sites Act of 1968 (Eliz. 2. 1968, ch. 52). At the time, this was widely interpreted as liberal legislation which would enable the Traveller community to continue a nomadic way of life without harassment but I have argued elsewhere (Sibley 1981, pp. 93–100) that it would be more appropriate to see the Act as 'a programmed response to deviance'. Although one objective of the legislation was to establish a national network of sites, it was not intended that large cities should be included in the network, ostensibly because there was no space or there were no Gypsies. Thus, a number of urban authorities were exempted from the requirement to provide sites or were allowed (in the case of the inner London boroughs) to make minimal provision. Since many Gypsies were living in cities in the early 1970s, it might be assumed that the racist stereotype of the real, rural Gypsy was affecting the vision of administrators, who either could not see urban Gypsies or dismissed them as 'itinerants' who were somehow beyond the scope of the Act. There was certainly a strong disposition not to accept Gypsies as part of urban society, and exclusion is still an objective, as a recent case in Bradford suggests:

> A solicitor claimed that Bradford was trying to run 23 Gypsies out of town by means of a wide-ranging court injunction. The injunction sought would stop them entering or remaining on Council land or premises, or any other land or premises, without written permission.
>
> 'The Metropolitan district council has asked for an order that would amount to a pass law . . . I have never come across a case that has asked for such sweeping powers. It is the equivalent of building a legal fence around Bradford. Where do these Gypsies go? Many have been living in Bradford for 15 to 20 years.' (*The Guardian*, 11 May 1985.)

Settlement control

Apart from local measures, the main instrument of control is designation, which was included in the 1968 legislation in order to encourage local authorities to provide sites for those families 'residing in or resorting to' their area. Under Section 10(3) of the Act, any family stopping in a designated area (county or county borough) but not on an official site would be subject to a fine 'increasing for every day on which the offence is so continued'. The phrase 'resorting to' is not defined in terms of frequency of visit or length of stay and is thus open to different interpretations. Because local authorities see Gypsies as a problem they will generally try to make minimal provision. If the minister then decides that provision is adequate, designation can be granted, giving the authority power to prosecute Travellers not accommodated on sites. In Peterborough, for example, the minister agreed that the upgrading of an existing site to provide 72 pitches would qualify the authority for designation even though local Gypsy support groups estimated that the number of families was consistently above 100. Arguments about numbers thus became crucial to decisions on whether or not to apply a policy which will have fundamental consequences for the movement and residence of Travellers. Peterborough City Council went to the trouble to put a full-page advertisement in a local newspaper explaining the benefits of designation. Beneath a cartoon showing an illegal encampment with scrap cars, domestic refuse and a 'no litter' sign, giving a clear negative message, a description of the proposed site was coupled with the following enthusiastic comment on designation: 'When this is granted, camping on sites other than official ones will become a criminal act. The City Council, or indeed others, will be able to invoke the Criminal Law to ensure speedy eviction of anyone illegally camping' (*Peterborough Standard*, 26 April 1984).

The effect of designation on the settlement of Travellers is unclear, however, first because designated authorities may respond to illegal encampments with differing degrees of vigour and secondly because the extent of designated territory will vary – where contiguous districts are designated the problem is potentially more serious than where the district is isolated and families are able to stop in adjacent areas.

The case of London provides some indication of the potential difficulties for Travellers because there is a large number of boroughs that have received designation (21 out of 32 by July 1984) and many of these are contiguous (Fig. 3.1). Legally, large areas of the capital are now 'no-go' areas for families not accommodated on official sites.

The Greater London Council had a non-harassment policy, regardless of designation, but this policy was not implemented by many of the

Figure 3.1 London boroughs that by 1984 had been granted designation under the Caravan Sites Act 1968.

boroughs. A report by the Principal Race Relations Adviser (Greater London Council 1984) maintained that 'many of the boroughs have been critical of the GLC's policy (although they have no constructive alternatives) and have no sympathy with the problems faced by Travellers', even though over one-third of Travellers in London have nowhere legal to stop. There was a familiar conflict between local authorities responding to the hostility of the settled population to Gypsies and the Greater London Council, attempting to implement an anti-racist policy.

An indication of conflicting interests is contained in replies to a letter (September 1984) asking the boroughs for comments on designation, specifically, whether they had found it an effective means of controlling the number of families in their area. Only one, Greenwich, suggested that it was complying with the GLC request to avoid harassing families and was taking positive steps to improve conditions for families camped illegally. The response of three other authorities, Bexley, Havering and Tower Hamlets (an undesignated borough), appeared more characteristic. All these had fluctuating numbers of illegally camped Travellers between 1980 and 1984 (Fig. 3.2). The responsible officers in Bexley and

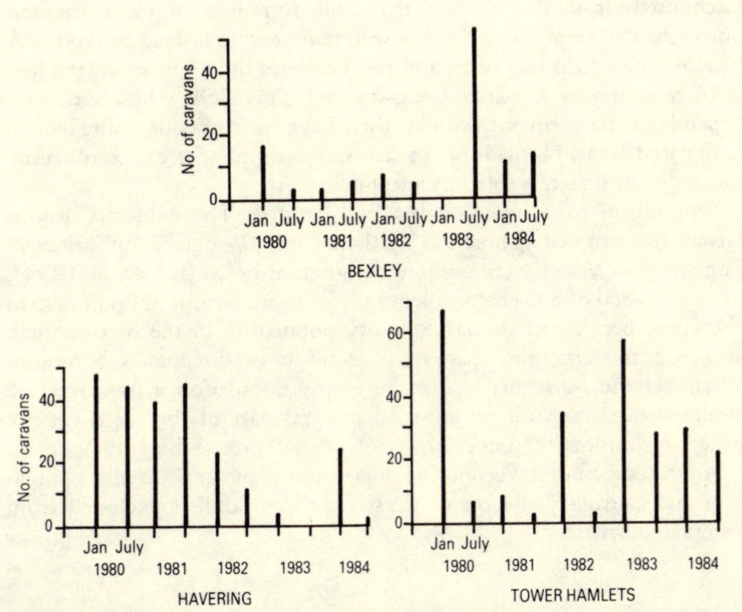

Figure 3.2 Variation in the number of Gypsy caravans in selected London boroughs, 1980–4.

Havering claimed that designation was effective and admitted pursuing a policy of eviction. The Borough Planning Officer for Havering said that 'shortly after receiving designation, there were a large number of prosecutions'. Tower Hamlets, which was then negotiating with the Department of the Environment for designation, was intending to provide 21 pitches but did not feel that it had a responsibility to the rest – ' "the floating" traveller population who by nature of their nomadic life style, are generally excluded from such allocations.' An officer in the Environmental Health department went on to say:

> Over the last 10 years, the Borough has applied for and successfully removed some 30 to 40 unofficial traveller encampments. Tower Hamlets is a densely populated Borough and over the last 20 years a substantial amount of redevelopment has occurred. This created vacant land ideal for unofficial encampments; however, commercial/private development has now reduced the problem.

This comment is characteristic. Local authorities often see the presence of Travellers as a land-use conflict rather than as a question of community relations – the minority group is effectively dehumanized and any problems it may have as a result of local authority practices are discounted. It is also evident from the responses of some London boroughs that responsibility is not seen to extend to nomadic Gypsies. A local population is identified and this becomes the population 'resident in or resorting to' the area. Long-distance Travellers, whose economy depends on frequent movement, then have no rights of settlement in designated areas. Nomadism is a necessary component of the romantic image but is unacceptable in the capitalist city.

Opposition to designation has been muted. The Minority Rights Group campaigned against the legislation in 1979 but 22 authorities in England and Wales were granted designation between 1980 and 1984, making a total of 50. The problem may be more serious in London than elsewhere because of the large Gypsy population in the metropolitan area and the emerging pattern of contiguous designated boroughs. Whatever the consequences for the Gypsy population in England and Wales, spatial controls are now an integral part of the social-control process, compounding the effects of official sites (Sibley 1979, 1981, 1986) in restricting the economic opportunities of the Traveller population and causing problems of stress for those families excluded from designated districts.

An alternative view

I have suggested that racist attitudes to Gypsies, particularly the attribution of deviance, which stems from mythical images of racial and cultural characteristics, have shaped the responses of the dominant society. Exclusion from urban society has been justified on the grounds that Gypsies properly occupy a niche in a (non-existent) rural society. The legitimacy of urban Travellers is thus questioned because their appearance and life-style conflicts with the stereotype. An anti-racist strategy that corrects this image should include a reorientation of academic enquiry to produce a more sympathetic account of Gypsy culture. A number of theoretical and methodological emphases suggest themselves.

The idea of change

The supposed transition from a rural to an urban milieu suggests radical changes in the culture of Gypsies and other peripheral minority groups and tends to confirm the romantic/rural and deviant/urban stereotypes. Some ethnographic research, for example by Brody (1981) and Berger (1979), suggests that indigenous peoples, peasants and Gypsies, all of whom have economic systems that enable them to maintain a degree of autonomy within the capitalist mode of production, are able to adapt to external changes. They change in order to stay the same. Urban Gypsy communities, for example, readily adopt new technology – for example, citizen's-band radio and video-recorders – but the essential features of the culture remain unchanged. Long-term research on adaptations would help further to discredit modernization theory, according to which fundamental cultural change is assumed. Two sources of information that are particularly appropriate in this context are historical analyses and participant observation.

Historical analysis

The history of many minority groups is unrecorded or is invisible to the outside observer. Invisibility may result from an inappropriate theoretical orientation which obscures the past, as in the case of black people in Britain before the 1950s. Oral history, however, may provide clues to written evidence. For example, accounts of life in the city given by Gypsies have been supported by extensive but fragmented written evidence of Gypsy communities in the 19th-century city, in Europe and North America (Sibley 1986). Again, this helps to correct the ethnocentric views of change embodied in modernization theory.

Participant observation

While the need for long-term observation may seem self-evident in order to develop an anti-racist perspective, the form of participation is not. Gypsies understandably resent outsiders making direct enquiries, particularly since concealment is part of the strategy of survival. A hidden economy, for example, has to remain hidden if it is to escape regulation by authority. Thus, successful participant observation must involve the observer in the life of a community in a practically useful way, for example, as a teacher or legal adviser. Frankness about academic intentions is also essential in order to gain the support of the community but it must be conceded that the greater the success of the project, the greater the danger. Jackson (1983, p. 42) recognizes that the researcher 'is often unable to gauge the potential damage which his study may effect' while Gronfors (1982), with particular reference to research on Gypsies, argues that knowledge about such a minority group is likely to damage the interests of the group, if the State is intent on control – any information may be used against the minority. Gypsies have developed strategies for avoiding domination by the larger society and may be able to do without the help of researchers, whatever their commitment to the minority's interests. There is, therefore, a real dilemma. Possibly a reduction in the options available to the Gypsy population as a result of increasing state regulation tilts the balance in favour of involvement and the production of more critical research.

Environmental change

Apart from racism, it is evident that Gypsies in an industrialized society are constrained by the regulation of the environment, particularly the maintenance of boundaries that define single land uses and create nonconformity. The improvizations and adaptations of the built environment by Gypsies appear more deviant in societies where planning regulations emphasize spatial order. The deviance of British Gypsies and the invisibility of Gypsies in the United States (Sutherland 1975) point to this difference, and historically the emergence of a deviant urban minority appears also to be connected with the increasing regulation of the environment. Thus, anarchist critiques, for example by Turner (1976), Ward (1978), Sennett (1970) and Nozick (1974) may be instructive. Multiple uses, weaker boundaries and less state regulation would reduce the visibility of the Traveller community and perhaps increase their economic opportunities. Possibly, racism would also be less of a problem in a more accommodating environment.

Conclusion

In this essay I have tried to identify the sources of the exceptional treatment of Gypsies in British society. While there are affinities with other minority groups, particularly in regard to racist stereotyping, there is an important difference in the state's response in that the Traveller community is the only one subject to spatial controls as an explicit feature of social policy. The fact that there has been very little opposition to this policy suggests a fundamental conflict between a semi-nomadic population and the dominant society. It should be recognized that the uncritical acceptance of social theory, particularly modernization theory, and an academic interest in spatial control, has serious consequences for the minority group. It could be argued that the imperatives of the capitalist economy are of fundamental importance in an explanation of the problem, but an anti-racist perspective is an essential complement to an economic argument in developing a critical response to the marginalization of the Traveller community by the state.

Acknowledgements

I am grateful to Nichola Dennis, graduate in Geography and Mathematical Statistics, Hull University, for providing information on Peterborough; to Janet Smith for typing this chapter; to the late Andrew Bolton for the diagrams; and to Peter Jackson for his useful comments.

References

Berger, J. 1979. *Pig earth*. London: Writers and Readers Co-operative.

Biddiss, M. D. (ed.) 1979. *Images of race*. Leicester: Leicester University Press.

Billig, M. 1982. *Ideology and social psychology*. Oxford: Basil Blackwell.

Blacker, C. P. 1952. *Eugenics: Galton and after*. London: Duckworth.

Boal, F. W. 1976. Ethnic residential segregation. In *Spatial processes and form*, D. T. Herbert & R. J. Johnston (eds), pp. 47–79. London: Wiley.

Brody, H. 1971. *Indians on skid row*. Ottawa: Department of Indian Affairs and Northern Development, Northern Science Research Group.

Brody, H. 1981. *Maps and dreams*. London: Jill Norman.

Central Housing Advisory Committee 1969. *Council housing: purposes, procedures and priorities*. London: HMSO.

Cohen, S. 1985. *Visions of social control*. Cambridge: Polity Press.

Collier, D. 1976. *Squatters and oligarchs*. Baltimore: Johns Hopkins University Press.

Commission for Racial Equality 1985. *Swann: a response from the Commission for Racial Equality*. London: CRE.

Committee of Inquiry 1985. *Education for all* (The Swann report). London: HMSO.

Community Relations Commission 1977. *Housing choice and ethnic concentration*. London: CRC, Reference and Technical Services Division.

Cullingworth, J. B. 1979. *Essays on housing policy: the British scene*. London: Allen & Unwin.

Deakin, N. & B. G. Cohen 1970. Dispersal and choice: towards a strategy for ethnic minorities in Britain. *Environment and Planning* **2**, 193–201.

Douglas, M. 1975. *Implicit meanings*. London: Routledge & Kegan Paul.

Douglas, M. 1978. *Purity and danger*. London: Routledge & Kegan Paul.

Greater London Council 1984. *Review of GLC policy on Travellers* (mimeo.). London: GLC.

Gronfors, M. 1982. From scientific social science to responsible research: the lesson of Finnish Gypsies. *Acta Sociologica* **25**(3), 249–57.

Hardy, D. & C. Ward 1984. *Arcadia for all: the legacy of a makeshift landscape*. London: Mansell.

Jackson, P. 1983. Principles and problems of participant observation. *Geografiska Annaler* **65**(B), 39–46.

Kantrowitz, N. 1981. Ethnic segregation, social reality and academic myth. In *Ethnic segregation in cities*, C. Peach, V. Robinson & S. Smith (eds). London: Croom Helm.

Livingstone, D. 1984. Science and society: Nathaniel S. Shaler and racial ideology. *Transactions of the Institute of British Geographers* N.S. **9**(2), 181–210.

Lorimer, D. A. 1978. *Colour, class and the Victorians*. London: Academic Press.

Marshall, E., no date. *Houses on wheels*. London: James Nisbet.

Mason, D. 1982. Race relations, group formation and power, a framework for analysis. *Ethnic and Racial Studies* **5**(4).

Mik, G. 1983. Residential segregation in Rotterdam: background and policy. *Tijdschrift voor Economische en Sociale Geografie* **74**(2), 74–86.

Mooney, K. A. 1976. Urban and reserve Coast Salish employment. *Journal of Anthropological Research* **32**, 390–400.

Nozick, R. 1974. *Anarchy, state and utopia*. Oxford: Basil Blackwell.

Parekh, B. 1974. *Colour, culture and consciousness*. London: Allen & Unwin.

Perry, J. 1973. *The fair housing experiment: Community Relations Councils and the housing of minority groups*. London: Political and Economic Planning.

Rapoport, A. 1978. Nomadism as a man–environment system. *Environment and Behaviour* **10**, 215–46.

Sennett, R. 1970. *The uses of disorder*. Harmondsworth: Penguin.

Sibley, D. 1979. Classification and control in local government: a case study of Gypsies in Hull. *Town Planning Review* **49**, 319–28.

Sibley, D. 1981. *Outsiders in urban societies*. Oxford: Basil Blackwell.

Sibley, D. 1986. Persistence or change? Conflicting interpretations of peripheral minorities. *Environmental and Planning D: Society and Space* **4**(1), 57–70.

Sijes, B. A. 1979. *Vervolging van Zigeuners in Nederland, 1940–1945*. 's-Gravenhage: Martinus Nijhoff.

Solomos, J. 1984. *Problems but whose problems?* Southampton: Political Studies Association Annual Conference.

Sutherland, A. 1975. *Gypsies: the hidden Americans*. London: Tavistock.

Turner, J. F. C. 1976. *Housing by people*. London: Marion Boyars.

Wallerstein, I. 1983. *Historical capitalism*. London: Verso Editions.

Ward, C. 1978. *The child in the city*. London: Architectural Press.

Western, J. 1982. The geography of urban social control: Group Areas and the 1976 and 1980 civil unrest in Cape Town. In *Living under apartheid*, D. M. Smith (ed.), London: Allen & Unwin.

Williams, J. 1985. Redefining institutional racism. *Ethnic and Racial Studies* **8**(3), 323–48.

Williams, P. 1982. The invisibility of the Kalderash of Paris. *Urban Anthropology* **11**(3–4), 315–46.

PART II

Racism in Britain

If 'race' is to be understood as a social construction, then racism must always be approached historically and in terms of its specific local context. The four essays in this part each make this point in one way or another.

In the first essay, **Paul Rich** criticizes contemporary 'race relations' research for its lack of historical sensitivity. He describes the development of a sociological interest in 'race' in Britain during the 19th and early 20th centuries, following the rise of social anthropology and culminating in a separate subdiscipline of 'race relations' after World War II. Much of this work, he argues, has been guilty of naïve and ahistorical generalizations based on a limited understanding of British colonial history, narrowly focused on situations of black/white conflict. In its place he provides an interpretation of the development of Western ideas about 'race', firmly rooted in the social and political context of their time and based on detailed historical research.

Robert Miles and **Anne Dunlop** also adopt an historical approach in their analysis of the Scottish dimension of British racism. They criticize the view that Scotland has no 'race relations' problem, denying the alleged 'natural tolerance' of the Scots and repudiating the view that the lack of a substantial 'immigrant' population explains the apparent absence of Scottish racism. They ask why the political process has been 'racialized' in England to a much more significant degree than in Scotland, a question which they answer with respect to Scotland's distinctive history, its pattern of class relations and the development of political nationalism.

The idea of 'race' is clearly apparent in the anti-Irish sentiment that Miles and Dunlop identify as a specific aspect of Scottish racism, associated with a long-standing hostility to Catholicism. Anti-Irish racism is common also in England, as **Judy Chance** describes in her discussion of the Irish in London. Her research in Kilburn has focused on the linguistic basis of ethnic identity and explores the distinctive use of English in the maintenance of ethnic boundaries in everyday social interaction. She draws on literature from social psychology and socio-linguistics to make a number of general inferences about ethnic identity, whether the group in question is 'visible' in terms of physical appearance or whether, like the Irish, they are socially 'invisible'.

In the final essay in this part, **Vaughan Robinson** investigates the extent of spatial variability in expressed attitudes towards 'race' as reflected in answers to opinion polls and other social-survey data. From statistical analysis of these data he argues that neither structural nor psychological explanations of prejudice are adequate to account for these regional and local differences, indicating the unique way in which macro- and micro-level forces are played out in specific localities.

4 *The politics of 'race relations' in Britain and the West*

PAUL RICH

In recent years doubts have been expressed on the degree to which 'race relations' can be developed as an intellectual sub-discipline of sociology. Some observers have noted the general failure of 'race relations' sociology to develop any distinctly sharp or novel theoretical insights, while even some of its practitioners have confessed to the area's general academic marginality (Cohen 1972, Fenton 1980, Phillips 1983). This weakness within the 'race relations' paradigm can be perhaps partly explained by its general peripheralization from public policy-making since the split in the Institute of Race Relations in the early 1970s. However, a more crucial intellectual shortcoming has been the disconnection of 'race relations' studies from the historical analysis of racism and racial thought. The study of 'race relations' sociology indeed has all too often been pursued in an historical vacuum in which vast generalizations have been developed with little or no detailed scholarly historical evidence. This disdain for history is itself, though, a legacy of the colonial origins of 'race relations', suggesting the need for major theoretical innovations in the subdiscipline if it is to survive.

This essay thus proposes to discuss the development of 'race relations' sociology in Britain in the context of a wider debate on 'race' in the West since the end of the last century. It will first focus on British conceptions on 'race' in anthropological research in the years up to World War II. This will be followed by an analysis of the British school of 'race relations' that emerged in the postwar years during the phase of decolonization. Finally, the essay will briefly discuss the crisis in 'race relations' research that developed in the 1970s. The approach will therefore relate sociological thinking on 'race' and 'race relations' to a wider set of intellectual and theoretical landscapes. This is born out of a recognition that sociological theory does not stand apart from wider movements in European and Western thought and is to a considerable degree conditioned by them. Some of the best writing in sociology indeed has been impelled by a more general intellectual imagination that transcends the arbitrary distinction of 'science' and 'art' (Nisbet 1977).

'Race', Western expansion and the anthropological tradition

The historical understanding of 'race relations' theory depends initially upon a recognition of the growth of the 'race' idea in Western societies. Before there could be the possibility of race *relations*, there had first of all to be the social recognition of *races*. Thus *ab initio* 'race relations', defined by Banton as 'a general body of knowledge which tries to bring together in a common framework studies of group relations in different countries and in different periods of history' (Banton 1977, p. 2), was concerned with the *subjective* perception of social differences and the manner in which these shaped the resulting pattern of social behaviour. The field thus depends implicitly on an historical understanding of the circumstances that give rise to racial thought.

The growth of racial awareness in the West has been linked by some observers to the rise of European capitalism in the 15th and 16th centuries. Certainly by the time Shakespeare came to write *Othello* and *The Tempest* there had emerged a fairly high degree of racial consciousness linked to the growth of overseas trading and exploration in the period of the Elizabethan renaissance (Cowhig 1985). The symbolic significance of *blackness* and *whiteness*, though, also went far back into medieval Christian religious ritual and the 'race idea' depended to some degree upon a process of secularization within Western Christian thought in which the brotherhood of men in Christ became redefined in terms of a collective brotherhood of the race (Voegelin 1940).

This racial awareness in Europe became markedly intensified in the era of the slave trade in the 17th and 18th centuries and the more formal doctrines of racism in England began to emerge in the latter part of the 18th century as a result of connections with the slave plantocracy in the colonies of the West Indies, especially Barbados (Fryer 1984, pp. 146–65). Black people in 18th-century London though were already a sufficiently cohesive community to be linked to the criminal underworld, and, as David Dabydeen has shown in an important study of the artist William Hogarth, blacks were used to satirize the corruption of an urbanizing metropolis that owed much of its wealth to the overseas slaving connection. This subjective perception of blacks was thus one that linked them to images of criminality, degradation, moral corruption and the upsetting of the traditional moral order (Dabydeen 1985).

A continuing legacy of slave-based racism was carried into later social thought, especially in the case of British imperial expansion in the 19th century. The 'golden age' of this high-tide of empire from the 1830s to the period of imperial decline in the 1940s was accompanied by a variety of 'scientific' racisms based on social Darwinist and eugenic notions of white racial fitness, though by the 1890s there was a growth in anxious social comment on the 'threat' to white racial supremacy from African and Asian 'races'. This fascination with 'race' owed much to the emergence by the mid-Victorian period in Britain of a growing professionalization of interest in the study of observably different 'races', marking an intellectual elaboration of a more deep-rooted European concern with exotic cultures, such as those of the Middle and Far East (Lorimer 1978; Said 1978, 1985; Porter 1983; Stepan 1982). Unlike the 'Orientalism' stretching back to travellers' accounts, such as those of Marco Polo in the 13th century, the interest in 'races' in other areas of British imperial control became linked to an anthropology that increasingly stressed the functional cohesion of tribal societies and the political integration of the governmental system. Thus, as Talal Asad has argued, the British imperial concern with non-European 'races' became bound up with two different images of non-Western government and social structure in India and Africa, which stressed force and hierarchy, and consent and legitimacy respectively (Asad 1973).

The origins of 'race relations' tended to be more of a product of the anthropological concern to preserve and protect African societies in the colonial empire than of a theory of forceful conquest by despotic rulers in Asia. By the 1890s the period of military conquest in Africa was substantially over and the writings of the amateur ethnologist and traveller in West Africa, Mary Kingsley, did much to shift anthropological and colonial interest away from the Indian to the African terrain by the time of the Anglo–Boer War (Rich 1986a). The revival of liberal concern with the rights of nationalities in the Edwardian period led to a brief attempt to have the whole issue of 'races' and nationalities discussed on a global basis at the 1911 Universal Races Congress at the University of London. However, the immensity of the task and the division of the Congress into rival factions inhibited the objective of linking the issues across continents and the experiment was not repeated before the advent of World War I (Biddiss 1971, Rich 1984). India remained in many ways an inscrutable continent of both impenetrable mystery and abject poverty, which few British observers outside the narrow coterie of Orientalist specialists felt qualified to comment upon, and by the 1920s the rise of a more professional school of anthropologists felt more readily drawn to the terrain of Africa where it was possible to observe what seemed to be complete African societies at first hand.

The upsurge of interest in African societies was in part a consequence of the consolidation of colonial authority in the 1920s under the doctrine of 'indirect rule' whereby the powers of traditional chiefs and elders were employed to buttress the power of the white colonial administration. At the same time there was an underlying anxiety that the nationalism that had surfaced in European politics at the time of the Peace of Versailles in 1919 would extend unchecked to areas of the 'backward' colonial empire. Black nationalism in the 1920s was viewed with a considerable degree of alarm in colonial circles for it was linked with movements such as Garveyism and with Pan African ideas of 'Africa for the Africans' and the ending of white colonial rule. 'When that great idea of Ethiopia, a nation and Africa for the Africans, is fully held by a large proportion of the 400 million negroes of the world', proclaimed the journal *Review of Reviews* in 1921, 'it is obvious that we shall be provided with a movement on which the Powers will have to take serious and sympathetic notice' (*Review of Reviews*, June 1921, p. 469).

The rise of anthropological interest in African societies was thus part of a resurgence in interest in developing newer and more modern forms of social control to contain change in a continent which was still seen as being able to avoid the problems of industrialization and class conflict that had emerged in the European setting. Like the Orientalist tradition, the anthropological image of functioning and cohesive African tribal entities was essentially timeless and ahistorical (Asad 1973, pp. 114–16). Therefore if there was by the 1920s a general movement of non-Western peoples into world history, it was conducted to a considerable degree through Western interpretative agencies. In the case of the Orient after World War I it was the Orientalist school who filled the vital intermediary rôle of translating the process of change in the Middle and Far East into recognizable symbols of a world historical process intelligible to informed Western audiences (Asad 1973, p. 240). So, too, in the sphere of 'race relations' it was the emergent school of professional social anthropology that sought to define the cognitive maps of Western, and more particularly official and governmental, understanding of non-Western and non-white societies. This anthropological school in Britain took on, in some respects, the mantle of an earlier colonizing quest except that the object pursued was now one of a holy grail of scientific knowledge rather than quick profits. Unlike the more sophisticated and genteel Orientalist school, however, the new generation of professional anthropologists could not fall back on a large body of written knowledge and had instead to go and find it through organized fieldwork. It was thus the genius of Malinowski to specify in detail the nature and disciplines of this fieldwork quest and, in time, to gather together an able group of professional anthropological fieldworkers at the London School of Economics interested in the phenomenon of 'culture contact'. The ethos was a somewhat predatory one centred on the acquisition of knowledge of non-Western societies as part of the longer-term goals for the emergent anthropological discipline itself of filling in the vacuum of knowledge between West and non-West during a period of critical political and cultural change: '. . . the ethnographer has not only to spread his [sic] nets in the right place', Malinowski wrote in *Argonauts of the Western Pacific*, 'and wait for what will fall into them. He must be an active huntsman, and drive his quarry into them and follow it up to its most inaccessible lair' (Malinowski 1972, p. 8). Furthermore, paralleling the drive towards system and order in colonial rule after an earlier phase of active conquest and pacification, anthropology represented the assertion of 'law and order into what seemed chaotic and freakish. It has transformed for us the sensational, wild and inaccessible world of "savages" into a number of well-ordered communities, governed by law, behaving and thinking according to consistent principles' (Malinowski 1972, pp. 9–10; see also Richards 1957, Wax 1972, Thornton 1985).

There was much in this though that represented a major intellectual advance on the former spurious orthodoxies in the infant discipline of sociology in Britain, such as Wilfred Trotter's ethological concepts of 'herd instincts' or William McDougall's racist theories of group behaviour (Richards 1957; Jones 1980, pp. 124–33). The Malinowskian school of structural functional anthropology was able to gain an important place for itself in British academic and political discussion on colonial policy and 'race relations' by the late 1920s, especially after the establishment of the International Institute of African Languages and Culture in London in 1926. Malinowski himself hoped that anthropology could act as the vehicle for testing and experimenting with more general sociological theories. As he wrote to Graham Wallas as early as 1916:

> My greatest ambition would be to draw the ethnological conclusions from your work, to apply these general Sociological principles to the special problems of Ethnology. After all, Ethnology is really the proper field for a science of Comparative Sociology. And the present prevalent tendency in Ethnology to look for 'origins', 'survivals' and other 'curios' by no means covers the field of Ethnological possibilities. (Malinowski 1916.)

In actual practice, the anthropological enterprise as defined by Malinowski in Britain was to receive little direct support from sociology, though there had been a more general importation by anthropologists such as A. R. Radcliffe-Brown of Durkheim's stress on social cohesion understood through historically variable but interdependent social *functions*. This had liberated French ethnology from its tight intellectual confines and raised it to a level of international debate by World War I (Karady 1981). Sociology in Britain before World War II however still remained a subject largely dominated by amateurs with less formal connection with universities and the academic establishment than those involved in anthropological research. Earlier debates between eugenicists and a school of Oxbridge moral idealists left the aims of the subject ill-defined and the developments in British anthropology in the 1920s tended to be under its own steam and in increasing involvement with the apparatus of colonial policy making (Rich 1986a). For Malinowski the attraction of 'indirect rule' lay in the recognition that it was impossible to 'transform Africans into semi-civilized pseudo-Europeans within a few years' (Malinowski 1929, p. 23). The point was to develop a policy of guided social change based on existing African social institutions, and to extend the indirect-rule doctrine from the political sphere into all aspects of African culture.

Beyond the more immediate concerns of 'practical anthropology', however, there were for Malinowski some universal implications for the anthropological enterprise. Following some lectures by Jan Christiaan Smuts in 1929 entitled *Africa and some world problems* in which a strong plea was made for the scientific study of African colonial societies, Malinowski suggested in a memorandum in 1930 – 'Report on the State of Anthropology and the Possibility of its Development in England' – the need for anthropology to link up with other social sciences in a more comparative approach to social issues 'whether of forecast or reconstruction'. 'We are witnessing', he continued, 'the biggest historical phenomenon in the whole development of human culture, the formation of a universal civilization in which the various elements are rapidly fusing. The dominant influence comes, of course, from our mechanical Western civilization, but this civilization is being modified itself by Oriental, African, Asiatic and Oceanic elements. It would be an inestimable loss to future social science if the early stage of this process were not studied' (Malinowski 1930). This ambitious vision reflected the fact that British colonial policy by the 1930s had become increasingly receptive to anthropological and social-science concepts as the Colonial Development and Welfare programme began to get under way. The growing association of colonial development schemes with imperial aims in the British Commonwealth became especially apparent by the time of World War II. The extension of ideas of state planning and social engineering to the colonies appeared to some Fabian improvers such as W. M. Macmillan and Julian Huxley (who had been involved in the work behind the Hailey *African Survey* in the 1930s) fertile ground for extensive schemes of state planning on the lines of the Tennessee Valley Authority in the United States. Such schemes furthermore offered a moral basis for resisting American claims for the internationalization of British colonial possessions once hostilities had terminated (Lee & Petter 1982, p. 135). The objective thus became in Huxley's words 'development . . . primarily for the benefit of the native peoples, not primarily for the advantage of the colonial power, or of national or international big business, remote in the capitals of white men's countries' (Huxley 1941, p. 93; see also Huxley 1944).

The general climate of opinion, therefore, by the late 1930s and 1940s was towards large-scale international projects in which problems associated with bigness or remoteness were outweighed by the need to pursue goals towards 'one world' in the light of the manifest failure of the League of Nations. In the pursuit of this ideal it was also hoped that the spectre of racism would be finally banished from a new and more advanced form of Colonialism based now on ideas of 'development' and racial 'partnership'. Huxley himself, together with the Cambridge anthropologist Alfred Court Haddon, had sought to undermine the scientific pretensions of this older racism from the mid-1930s, especially in response to the emergence of Nazi racial ideology. The appearance of the important book *We Europeans* in 1935 marked a dismissal of the concept of 'race' in favour of 'ethnic', 'genetic' or 'physical' types which at this time seemed to represent less politicized concepts than that of 'race' (Huxley & Haddon 1935, p. 138). This view, though, clashed to some extent with the anthropology school in the United States and in 1942 Ruth Benedict published a seminal study *Race and racism* which argued that, even in the absence of any systematic scientific evidence to support the 'race' concept, racism itself still remained in societies all over the world as a doctrine or creed asserting the superiority of one social group over another. 'Racism', wrote Benedict, 'is the new Calvinism which asserts that one group has the stigmata of superiority and the other has those of inferiority. According to racism we know our enemies, not by their aggressions against us, not by their creed or language, not even by their possessing wealth we want to take, but by noting their anatomy' (Benedict 1983, p. 2). This argument for the centrality of *racism* as opposed to *ethnic* group distinctions did not immediately make itself apparent in British social thought, but in the American context it represented a more pessimistic view than that of the sociological school of 'race relations', centred on Chicago, which had argued since the early 1920s for an ecological theory of racial contact based on a cycle of contact, competition, adaptation and assimilation. Benedict argued that racial prejudice could only be overcome if a political attack was made on the social conditions that fomented racial prejudice, though she did not feel all American blacks were yet 'ready' for full citizenship (Benedict 1982, pp. 155–62). By the end of World War II, some American sociologists such as Franklin Frazier were also arguing for a more political stand in contrast to what he termed the 'fatalism' of the race relations cycle and began suggesting the need for a more dynamic social theory to account for the rapid social change occurring in American society (Frazier 1947).

Britain tended to remain insulated from these wider debates on the sociology of 'race' until well into the postwar years. Some liberals though had been shocked by the revelation of the full nature and scale of National Socialism in 1945–6 at the time of the Nuremberg trials and lost a considerable degree of faith in the power of reasoned argument to transform the apparently irrational phenomenon of racial prejudice. The significance of Nazi ideology in the 1930s was that it had not just occurred on the basis of a small group of fanatical cranks but with the full connivance of a considerable section of the German academic and scholarly establishment. The former German Nobel Prize winner for Physics, for example, Philipp Lenard, had collaborated with the Nazis as early as 1924 and at the University of Freiburg there was established a Lenard Institute in December 1935. Here the physicist Johannes Stack declared against 'Jewish Physics' (Weinrich 1946, p. 13). More significantly, some scientists and biologists collaborated in the Nazi development of 'political biology' employing eugenic ideas of *extinction* (*Ausmerze*) of the physically or mentally 'unfit' and the active *selection* (*Auslese*) of those deemed suitable for a new Nordic German race. The century-long European debate on 'scientific' racial typologies had come to full fruition (Weinrich 1946, pp. 27–30; Mosse 1979, Field 1977). Some opinion in Foreign Office circles though was that the question of how to combat the doctrines of National Socialism remained one substantially for each individual country itself to decide. There was a feeling that the creed would be 'totally extinct' by the end of the year and that no organized international action was necessary (Cowell 1945). However, Kenneth Grubb, then working in the Ministry of Information, warned of the need for a resurgence of Christian ethics if Nazism was to be met and defeated on the ethical plane:

> When a doctrine has incurred such popular odium as the Nazi race doctrine it may be sufficient to condemn its obvious absurdities and to transfer to the doctrine itself the hatred which it would seek to inculcate against its victims. But when it extends into a more subtle phase I do not know if this is sufficient. I do not know whether racial perversity can, in fact, be finally cured except by counter-balancing a

> more satisfying and more critical doctrine of man; and, for myself, I doubt whether such a doctrine can be properly stated unless it is derived from a correspondingly adequate doctrine of God. (Grubb 1945.)

But Grubb's Christian view did not make much headway in the emerging postwar discussion on 'race', until at least the establishment of Unesco. Following the illness of the first Director-General, Alfred Zimmern, Julian Huxley took over the planning of Unesco's purposes and aims in 1946. Though his ideals of evolutionary humanism were not accepted by all the organization's members, they did in some degree reflect the intellectual climate in which Unesco grew up in its early years (Cowell 1966). Huxley was an ardent exponent of international action transcending political ideology and in some respects this resembled the contemporary international-relations approach based on functionalist co-operation between rival powers. At his farewell address as Director in Beirut in December 1948 he declared that Unesco's programme was geared towards a more 'civilized' and 'unified' world order, though there was 'no common ideology involved in this' for it was a 'common practical idea'. This was a reaffirmation of the liberal creed of individual minds meeting together as 'citizens of the one world of the human mind' (Huxley 1948), and it thus eschewed the question of group cultures, ethnicities and primordial nationalisms that were to overtake the organization (and indeed Beirut itself) over the following decades.

Huxley's ethical and anti-ideological 'one-world' approach accorded with the first two Unesco Statements on Race in 1950 and 1951, which were partly stimulated by the 1948 Universal Declaration of Human Rights. Both these statements represented more detailed scientific support for universalist ideals on 'race' which stretched back to the 1911 Universal Races Congress; both reaffirmed without hesitation that man was one single species, though the First Statement in 1950 saw 'races' as merely 'one of the group of populations constituting the species *Homo sapiens*' representing 'variations ... on a common theme' (Montagu 1972, p. 8). The Second Statement in 1951 marked a more cautious view, impressed by physical anthropology and genetics, and left open the question of how human groups diverged from the common stock. 'Races' for the Second Statement had a greater independent significance than for its predecessor, for a 'race' it saw as being 'reserved for groups of mankind possessing well developed and primarily heritable physical differences from other groups'[1] (Montagu 1972, p. 142; Dunn 1951, pp. 10–11).

While the First Statement argued that biological study lent support to the 'ethic of universal brotherhood' (Montagu 1972, p. 12), the Second

Statement concentrated upon the moral potentials of individuals as shaped by their physical and social environment and emphasized that 'equality of opportunity and equality in law in no way depend, as ethical principles, upon the assertion that human beings are in fact equal in endowment' (Montagu 1972, pp. 145–6). Despite these significant differences of emphasis, both Statements were important milestones in the reassessment of biological theories of racial differences and reflected a growing climate of sociological interest in 'race relations' during a period of European decolonization and decline of belief in the imperial 'trusteeship' over 'backward races'.[2]

The emergent debate on 'race' in Britain

The Unesco Statements on Race in the early 1950s occurred at a critical time politically in Britain as the decline of imperial power ended the earlier hopes of Malinowski and some of the British anthropology school of using the British Empire and Commonwealth as an effective laboratory in a worldwide process of 'culture contact'. The emergence of colonial nationalism was leading to official opinion becoming increasingly involved in the question of 'race relations' in the colonial context as it became apparent that sectionalist processes of 'communalism' were undermining earlier ideals of building larger political entities. The débâcles in the Indian sub-continent in 1946–7, with the fissuring of the former Raj into India and Pakistan, and the humiliating withdrawal from Palestine and the partitioning of the territory after the advent of the state of Israel, indicated the nature of the forces at work in the postwar international order. There was a fear too with the onset of the Cold War that communal tensions could be manipulated by the Soviet Union in the interests of an expanding communist aggression against the West.

British interest in 'race relations' as a field of social research as well as political debate thus, for a period in the 1950s, coincided with rising American interest in the politics of decolonization from the standpoint of what D. Cameron Watt has termed a 'naïve populism' or belief that former colonial territories should in manner be allowed to produce governments in accordance with their people's wishes (Watt 1984, p. 250). The shift in thought during this period was symptomatic of a deeper political crisis in the West stemming from the impossibility of insulating trends in colonial politics from the Western moral standpoint in the war against Nazism and fascism between 1939 and 1945. Both colonial rule based upon conquest by European colonizers and the experience of fascist totalitarianism in Central Europe were rooted in a similar set of political processes, for both organized 'race' thinking in pursuit of rationalized ideologies of control (Ridley 1973). But if the internal racial struggles in America in the 1930s and 1940s allied to the definition of War Aims in the 1941 Atlantic Charter helped remoralize the concept of what Gunnar Myrdal termed the 'American Creed' based on Enlightenment humanism, it was difficult to discern quite the same process in liberal thought in Britain as mechanisms were sought for the handing over of political power in colonial territories to nationalist leaderships (Rich 1986a, pp. 169–71). Here the political handover of power by the late 1950s and early 1960s owed more to the perceptions of *Realpolitik*.

Thus despite the hopes of a number of liberals in the late 1940s and early 1950s for a progressive pattern of decolonization in African and Caribbean colonies on lines similar to a progressive Whig ideal of the extension of parliamentary liberties, the reality proved somewhat different. The débâcle of Suez in 1956–7 ensured an early liquidation of empire which bore little relationship to coherent political principles beyond the simple fact that colonies appeared an increasing anachronism in the bipolar political order of the late 1950s and that they did not really pay for their upkeep either (Holland 1985, p. 127).

The hopes of the sociological school of 'race relations' were thus somewhat prematurely dashed. In 1950 H. V. Hodson had been able to capture a temporary mood of political interest in 'race relations' by suggesting the establishment of an institute to study the area on a Commonwealth-wide basis (Hodson 1950, pp. 313–15). This proposal led to the eventual establishment of a unit attached to the Royal Institute of International Affairs at Chatham House in 1953 under Philip Mason, which finally became an autonomous Institute of Race Relations in 1958. The Institute took a considerable interest in the politics of the Central African Federation which had been established in 1953 on the basis of the doctrine of racial 'partnership', though in the wake of riots in Nyasaland (Malawi) in 1959 and the report of the Monckton Commission on the Constitution of the Federation the following year it appeared that the concept was increasingly less likely to meet the demands of African nationalist leaderships bent on the early termination of white settler rule backed up by the authority of Whitehall (Rich 1986b).

'Race relations' were also beginning to be taught in sociological courses in Britain, especially at the University of Edinburgh where Kenneth Little gathered a group of scholars together in the early 1950s including Michael Banton, Sheila Patterson, Violaine Junod and Sydney Collins. This was a relatively fluid period when the numbers of black immigrants in Britain were small, and the bulk of the research was on communities like Cardiff, Liverpool and Stepney before the emergence of a 'national issue' of 'race relations' which began to be constructed by the press and media in the wake of the 1958 Notting Dale and Nottingham disturbances (Banton 1983, Miles 1984). There was little formal support in government circles for legislation to restrict black immigration following the decision of a Cabinet Sub-Committee in 1951 to avoid this on the grounds that the numbers involved were not sufficiently large to warrant such a measure (Cab 130/6/25215, see Rich 1985). Furthermore, in 1954 the Conservative Home Secretary, R. A. Butler, considered in a Cabinet Memorandum the question of excluding blacks from the civil service but rejected the idea on the grounds of its illiberality as well as the likely protest in the UN and the Commonwealth, both of which still commanded some political respect in Tory circles (Butler 1954).

The international climate in which 'race relations' was being discussed during the 1950s did have increasing political impact as numbers of social scientists, in the United States especially, sought an 'educational offensive' against racial segregation in the South both on the conclusions of Myrdal's *An American dilemma* and the Unesco statements of 1950–51 (Klineberg 1954, Bibby 1956). This optimism in social-science circles was partly a reflection of the particular status of the US liberal intelligentsia, which still retained a considerable element of the moral ethos of the Roosevelt 'New Deal' years (from which the initial commitments against racism such as that of Ruth Benedict had sprung) as well as powerful links with the Eastern Seaboard Jewish intellectual establishment centred around *Partisan Review* and *Commentary*, which had succeeded in making a powerful impact on postwar American political discussion (Novack 1972, pp. 163–6).

In Britain, however, the response from anthropological circles in the 1940s to the emergence of 'race relations' had been sceptical and cautious, while even some sociologists were hesitant about making 'race' a central issue in social-science research. The liberal intelligentsia in Britain was generally less willing to become committed on the 'race' issue than its American counterpart. In 1954, for example, Maurice Freedman, who lectured on 'race relations' at the London School of Economics, wrote in a review of some of the Unesco publications on 'race' that 'Racialism' could be 'only one criterion to be used for marking out a field of study, and it is by no means an altogether satisfactory one. The extent to which racialist ideas enter into group conflict and discrimination varies not only between situations, not only within different situations over time, but also within particular situations at one time'. There was indeed, a certain logic, he concluded, to changing the term 'race relations' to a more general title such as 'inter-group relations', though such a term was so general that it effectively denoted the field of sociology itself (Freedman 1954, p. 343). Even as 'race' started to become a growing political issue in Britain, its academic study remained marginal and in 1961, with the Commonwealth Immigrants Bill indicating a new era of growing state control in the field, Donald Macrae wrote that the majority of sociologists in Britain still considered the whole subject as 'both irrelevant and distasteful'. Given that serious sociology such as that developed by Durkheim, Hobhouse and Weber earlier in the century had rejected 'race' as a useful category of analysis on the grounds that it explained too much too easily and could never be susceptible to proof, the importance of 'race' in social analysis, argued Macrae, lay in the realm of ideology. In this respect, though a racial ideology was by no means inevitable in a society such as Britain, there was a danger of the society repeating in a similar manner the experience of the United States (Macrae 1961, pp. 106–15).

This argument suggested that the framework for the study of 'race' should be anchored in the historical analysis of the social conditions likely to give rise to both racial ideology and racial conflict. A similar view had been expressed by Herbert Blumer in 1955 at an important international conference on 'race relations' at Honolulu.[3] Given that 'race' had meaning for social analysts as a social concept, the problem was to isolate the features accompanying the 'relations' between 'races' that made them markedly different from other kinds of relationships. Here Blumer found it hard to identify the uniqueness of such relationships, for 'aside from the single unique feature that comes from races regarding each other as biologically distinct, all of the relations found between races are also found between groups that are not races. Similarly, the range of relations found generally in human group life may be recovered in the relations between racial groups' (Blumer 1955, p. 6). Blumer's approach to defining the nature of 'race relations' was remarkably open-ended, in the light of subsequent attempts at theoretical development. The nature of 'race' relationships depended upon the 'given historical setting' for the 'organized racial conception' that a social group brought to bear upon a situation was itself an historical product (Blumer 1955, p. 8). Such relations need not even lead to 'tension', though Blumer recognized that one important base for theoretical development lay in the area of European colonial expansion in which various sorts of relationships were established between racial groups on the basis of colonial and imperial rule. In general, an hierarchical relationship was established on the basis of either the peaceful or warlike invasion of the colonized land and the establishment of an alien colonial order; the importation of outside groups into the colonial society; and the 'more or less voluntary migration of people to an occupied area which they enter on a subordinate level and in which they compete economically with members of a dominant racial group' (Blumer 1955, pp. 12–13; see also Mason 1954, Rich 1981).

Blumer's analysis was important for its recognition of the global scale on which these processes were occurring whereby 'the insularity which has marked the racial orders of recent times is dissolving within the framework of our modern world' (Blumer 1955, p. 15). This also meant, given that 'race has no determinative bearing on race relations and since the relations are a result of a process of historical development involving complicated and varying factors', that the establishment of 'scientific' theory was likely to be difficult for this 'would depend seemingly on some stability and constancy in a given process of historical formation' (Blumer 1955, p. 21). The approach was perhaps symptomatic of a pronounced difference of style in American racial discourse, for in the international arena at least the conception of 'race' was not tied to colonial responsibilities as it still was in Britain. As the United States developed to super-power status, its doctrine of containment was increasingly less likely to be applicable to all parts of the globe than it had been in Europe from 1947 onwards, despite the mistaken assumptions regarding the capacities of American power in South-East Asia and Vietnam in the 1960s and early 1970s. Thus from the late 1950s onwards it was becoming clear that the United States was not going to replace former British imperial hegemony with the same manner of control and was anxious both to recognize and to accommodate the claims of nationalist leaderships in Africa and the Caribbean.

By 1960, therefore, as Michael Banton has pointed out, there was a turning point in the direction of 'race relations' research – at least on the international level – as many of the former deterministic assumptions about their trajectory gave way to a more open-ended approach given that 'the behaviour of the subordinated is not completely determined by the social structure, for they have some power to choose the sort of group they will be' (Banton 1974, p. 37). This was the period in which interdisciplinary and area studies became fashionable in academic teaching and research and it did seem that a break with the past had been achieved in terms of a break with the older paradigm of 'race relations' linked to colonial rule over 'subject' races. The field had, in fact, little to say about the politics or history of these former 'subject' peoples, and these topics became increasingly the preserve of emergent African, Indian and Caribbean historical schools.

The ferment that accompanied the emergence of a number of African, Asian and Caribbean nations in the early 1960s became reflected, however, in more specialist and expert thinking on 'race'. The two further Unesco statements in Moscow in August 1964 and Paris in September 1967 indicated a growing division of labour between scientists and sociologists. The Moscow statement reflected the developments in archaeology and genetics over the previous decade, which cast further doubt on the idea of separate biological 'races' even if they were part of one human species (Hiernaux 1965, p. 74). To this extent the consensus amongst most scientists by 1964 reinforced the first Unesco Statement in 1950 and emphasized that human progress was based mainly, if not exlusively, on cultural achievements rather than genetic endowment. It thus concluded that peoples everywhere 'appear to possess equal potentialities for attaining any civilizational level. Differences in the achievements of different people must be attributed solely to their cultural history' (Montagu 1972, p. 153).

This concern with the 'cultural history' of peoples and nations implied the fortification of educational concern with human cultures and began to indicate a growing crisis in 'Western' values in an age of cultural and ethnic pluralism. For some social scientists this was a period in which the social sciences could, in some degree, be used effectively to substitute for the older notion of a Western *mission civilisatrice*. This was now seen as outmoded in an age which was rejecting ideals of grand synthesis of cultures in the form Malinowski had imagined three decades previously. It was now becoming fashionable to champion the smaller ethnic and cultural entity which could as far as possible preserve its own identity in a world order of growing uniformity. Social science, therefore, needed to establish for itself a new mediating rôle, given that its earlier tasks in terms of tackling segregation and racist ideology in Western and metropolitan states seemed to have been substantially achieved by the mid-1960s, especially in the United States where the civil rights struggle had led to the passing by the Johnson Administration of the Civil Rights Act of 1965. 'It is now considered "backward" and "uneducated" in most countries and societies to hold beliefs of racial superiority and inferiority', Alva Myrdal wrote in 1966, exemplifying this social-science optimism; 'where racism is on the wane, it is because insight into the fundamental equality of all human beings is being taught and accepted' (Myrdal 1966, p. 48).

Some social scientists, such as Nathan Glazer and Robert Nisbet in the United States, responded to this situation in the 1960s by emphasizing the saliency of ethnic groups and 'ethnicity' in contemporary 'plural societies'. Others however sought a more radical definition, especially in the light of the Fourth Unesco Statement in 1967 which welcomed 'the anti-colonial revolution of the twentieth century' as opening up 'new possibilities for eliminating the scourge of racism' (Montagu 1972, p. 159). The statement provided an opportunity for a more politicized set of commitments by social scientists to combat the phenomenon of 'racism', which was now seen as a cumulative doctrine which could survive on the basis of structural and social inequalities despite the demise of formal racist ideology. This was seen as necessitating a more comprehensive political and educational attack on institutions that perpetuated racial discrimination than the liberal efforts at combatting racial prejudice at the attitudinal level in the 1950s. 'The major techniques for coping with racism', the Fourth Statement declared, 'involve changing those social situations which give rise to prejudice, preventing the prejudiced from acting in accordance with their beliefs, and combatting the false beliefs themselves' (Montagu 1972, pp. 160–1).

The more militant standpoint of the Fourth Statement reflected a changing political climate in the West as the earlier universalist faith in human rights started to be questioned. The anti-colonial revolution that occurred in the early 1960s took place at the same time as an intensification of racial segregation occurred in the white ruled states in the south of the continent, especially South Africa under its doctrine of apartheid. The continued institutionalization of apartheid seemed manifestly to threaten the hopes held by some in the emergence of independent black African states. The emergence of this black nationalism seemed to Philip Mason to represent a revolt against Western values, as 'associations with color and racism smear the positive aspects of Western ideals' (Mason 1967, p. 350). Some observers went further by suggesting that race was *the* salient social variable. As John Rex wrote in the *Unesco Courier* in 1968:

> Increasingly in some at least of these countries the traditional political issues pale into insignificance beside the problem of racial inequality and man's attempts to fight against it.

The effect, though, was to liken apartheid in South Africa with the 'segregation of coloured people in squalid lodging houses' in some British cities (Rex 1968). 'Race' and 'racism' thus became catch-all terms describing a plethora of group relationships. In so far as they were defined through the Unesco Statements they continued the process of imposing external classificatory systems upon a variety of situations, without necessarily taking account of the popular definition of particular social relationships. As Julian Pitt-Rivers pointed out, there was a need for social-science observers to recognize the different *popular* conceptions of 'race', as in the case of the Latin American term *raza*, which could be employed in at least two different ways: by qualifying individual standing in Hispanic life on the basis of phenotype, and so serving as only one of a number of indicators of social status; and when referring to ethnic status irrespective of phenotype but on the basis of membership of an Indian community (Pitt-Rivers 1973). A difference of focus tended to emerge in the employment of the term 'social race' by fieldworkers at the local level and theorists of 'race relations' who were seeking a sociological 'paradigm' in which to develop a general theory. This difference, moreover, has tended to reproduce earlier divisions within anthropology between attempts at grand theory in the metropolis and more localized efforts at theorization on the basis of more detailed fieldwork at the periphery (Kuper 1980).

The failure of the 'race relations' paradigm

The divisions of opinion among sociologists of 'race relations' in the 1960s can be seen as in part an example of a more general tendency towards splits and cleavages in the international social-science community than is found in the natural sciences (Shils 1967). Despite the professed attempts by some sociologists to develop a degree of scholastic rigour on the basis of the opinions of 'experts' in Unesco (Rex 1970, p. 2), their efforts continued to be hamstrung by the fact that in many ways this kind of work seemed an increasingly outmoded attempt by Western intellectuals to continue in imperial fashion to impose their own terms of academic and intellectual debate on a fast-changing political situation. The fallacy of this approach was revealed by a temporary alliance in the British Institute of Race Relations between a radical group of white social theorists and a group of black militants who, when they had finally overthrown the old order in 1972, proceeded to impose their own tight political orthodoxy through the journal *Race and Class* (Sivanandan 1974, Mullard 1980). The issue illustrated the continuing problems in distinguishing a scholarly from a popular definition of 'race' and by the middle 1970s many sociologists felt a loss of scholarly legitimacy either to speak or to write about the subject with any confidence. There was the added problem, too, that the growing politicization of 'race' in a country such as Britain occurred at a time of growing decline in political deference generally and, as Lord Radcliffe warned in 1969, it was increasingly unrealistic to expect Parliament to 'give a lead' on the issues of 'race' and immigration, and he suggested that the issue of 'strangeness' would decline with the passage of time (Radcliffe 1969).

Despite the absence of any social-science consensus on the term 'racism' – which in the course of the 1970s came to be employed increasingly loosely, especially when linked to the term 'institutional racism' – there was a reluctance to shift the terms of the debate to the alternative concept, originally formulated by the American sociologist William Graham Sumner, of *ethnocentrism*. This suggestion, made by Michael Banton, partly harked back to the anthropological discussion in

Britain in the 1930s and the plea by Huxley and Haddon for an alternative word to 'race', given that 'pure races' could not exist (Banton 1969). Some sociological scholars objected to this mode of argument on the grounds that Banton was merely attacking a straw man, for the debunking of the notion of 'race' behind the term 'racism' still failed to ask the 'sociologically relevant questions about racism or the belief systems that can substitute for racist beliefs'. Banton was thus to be criticized for dealing with the issue 'as a problem in the history of ideas in western Europe' (Moore 1971, pp. 100–1).

'Race' and racism, however, as this essay has sought to show, *are* part of the history of European and Western thought and the study of the changing and evolving set of meanings behind 'race' in European cultures is vital for the subject's intellectual development. In one respect, Weber's early approach to the area of 'race' was also conditioned by the consideration of the circumstances that give rise to the belief in 'ethnic' factors in human communities dictated by the notion of a common 'blood' relationship. Whenever the terms 'nation', 'race' or 'people' were used, he argued, 'there is some implication of an existing political community, however informal, or of memories, preserved in the form of a common epic tradition, of a political community which used to exist at an earlier period; or else there may be some implication of a community of language or at least dialect, or finally of a common cult'. The course of history thus suggested for Weber 'with what extraordinary ease common political activity in particular leads to the idea of "common blood" – as long as there are no differences of anthropological type, or *too* extreme a character, to stand in the way'. This approach emphasizes the nature of ideas of "blood" relationships as well as the social situations that give rise to them and indicates the importance of cultural traditions in order to explain the phenomenon of racial prejudice and racism (Weber 1978).

The radical sociological school of 'race relations', furthermore, has tended to ignore the historical dynamics of racial thought, except as mere 'background' to the study of structurally defined 'race relations situations'. Despite the attempt by Philip Mason in the 1960s to encourage 'race relations' sociologists to investigate historical 'patterns of dominance' based on differing modes of imperial conquest and colonial rule in India, Southern Africa, Spanish America, the Caribbean and Brazil (Mason 1970), more recent work has increasingly veered into ahistorical generalizations based on Western imperial expansion and the phenomenon of 'colonialism'. The attempt by John Rex, for example, to produce a general 'paradigm' of 'race relations' has dwelt almost exlusively on the theme of 'colonial exploitation'. Rex has chosen to read into a passage of Weber on the notion of 'booty capitalism' a general theory on the predatory nature of European capitalism in the colonial context and offers this as an explanation for the emergence of 'race relations' situations based upon the economic exploitation by the colonizer of the colonized (Rex 1981, p. 2). None of this very general analysis, though, explains *why* 'race' came to be employed and extended in the colonial setting and neither does it explain the emergence of National Socialism in Europe itself by the 1930s and 1940s. While colonialism clearly has had an important impact in terms of the universalizing of concepts of 'race', it cannot in all circumstances be considered the sole reason for this. 'Race' concepts need to be investigated by both social scientists and historians in a transcultural mode and compared in both Western and capitalist societies with varying forms of non-capitalist and socialist bloc societies. It is evident from the work of Norman Davies on Polish nationalism, for example, that notions of Slavic identity in comparison to 'Teutonism' have survived in Polish thought throughout the long period without nationhood from 1815 to 1918 (Davies 1981, pp. 10–11, 24–5).

It is thus the essential partiality of much contemporary 'race relations' work and its concentration on both the colonial experience and black/white situations that ensure that many of its conclusions are contained in its underlying premises. As the phase of European guilt about colonialism starts to pass, a considerable quantity of 'race relations' literature appears parochial in nature in so far as it insists on dwelling merely on capitalist and Western societies to the exclusion of other modes of inter-ethnic conflict. The obsession of colonial modes of interracial contact and exploitation on the peripheries of Western imperial expansion is undoubtedly an important dimension of any understanding of the dynamics of contemporary racial hostility and 'racism'. However, key problems remain in explaining how the image of black 'colonial' inferiority became transferred from the colonial periphery to the metropolitan heartland; and the links have tended to be asserted, for example in Rex's work on the black immigrant communities in Sparkbrook and Handsworth, without being concretely demonstrated (Rex & Moore 1967, Rex & Tomlinson 1979, Lyon 1985). There is a persuasive historical case to be made for seeing contemporary British society as in many ways a relic of its imperial past and burdened by past perceptions and stereotypes which have been reshaped to fit a changing social situation in which West Indian and Asian communities have been marginalized into inner-city ghettoes (Joshi & Carter 1984, Rich 1986a, b). However, British society was not exclusively shaped by its imperial past for it was also moulded to a considerable degree by the dynamics of Central European history in which the fascist experience in the 20th century was to have an important part (Holmes 1979, Griffiths 1980).

Understanding the relationship between these two separate strands of imperialism and racism is of crucial importance for historians of 'race relations', and this indicates that many of the issues that have been raised in 'race relations' sociology need to be taken further by more detailed historical research. 'Race relations' studies in essence are at a theoretical and methodological crossroads, and the subject area, if it is to grow, urgently requires a greater cross-disciplinary imagination incorporating research from history, literature and anthropology, as well as from sociology.

Notes

1 Philip Tobias, the South African anthropologist, considered the first Unesco Statement went 'too far' in undermining race as a 'valid concept' of human classification and continued to argue for the conventional racial classifications of human groups into Negroids, Caucasoids, Mongoloids and Australoids (Tobias 1962).

2 Even anthropologists such as Carlton Coon, who insisted on using the notion of 'race' and 'racial systems' in order to understand the evolution of *Homo sapiens* admitted that racial discrimination was a 'holdover from the time when it served a purpose, when a racial division of labour carried with it a certain material and social efficiency. This purpose no longer exists, and now race is a nuisance. It can cease to be so only if we make people understand it' (Coon 1955, p. 182).

3 Hawaii was seen by many social scientists in the 1950s as a model of harmonious 'race relations' (Shapiro 1954). By the 1960s, though, doubts began to emerge on its supposed inter-ethnic tolerance (Samuels 1969).

References

Asad, T. 1973. Two European images of non-European races. In *Anthropology and the colonial encounter*, T. Asad (ed.), 103–20. London: Ithaca Press.

Banton, M. 1969. What do we mean by racism? *New Society*, 10 April.

Banton, M. 1974. 1960: A turning point in the study of race relations. *Daedalus* **2**.

Banton, M. 1977. *The idea of race*. London: Tavistock.

Banton, M. 1983. The influence of colonial status upon black–white relations in England, 1948–1958. *Sociology* **1**(4), 546–59.

Benedict, R. 1983 (1st edn 1942). *Race and racism*. London: Routledge & Kegan Paul.

Bibby, C. 1956. The power in words. *Unesco Courier*, February.

Biddiss, M. D. 1971. The Universal Races Congress of 1911. *Race* **13**(1), 37–56.

Blumer, H. 1955. Reflections on theory of race relations. In *Race relations in world perspective: papers presented at the Conference on Race Relations in World Perspective*, A. W. Lind (ed.). Westport, Conn.: Greenwood Press.

Butler, R. A. B. 1954. Memorandum on 'Recruitment of coloured persons to the civil service'. CAB file 129/65, 2 February.

Cohen, P. S. 1972. Need there be a sociology of race relations? *Sociology* **6**(1), 101–8.

Coon, C. 1955. *The history of man*. London: Jonathan Cape.

Cowell, F. R. 1945. Letter to K. Crubb, 25 May, FO 924/147.

Cowell, F. R. 1966. Planning the organisation of Unesco, 1942–1946: a personal record. *Journal of World History* **10**(1), 210–36.

Cowhig, R. 1985. Blacks in English Renaissance literature and the role of Shakespeare's 'Othello'. In *The Black presence in English literature*, D. Dabydeen (ed.). Manchester: Manchester University Press.

Dabydeen, D. 1985. *Hogarth's blacks: images of blacks in eighteenth-century English art*. Kingston Upon Thames: Dangaroo Press.

Davies, N. 1981. *God's playground: a history of Poland*, Vol. 2. Oxford: Clarendon Press.

Dunn, L. C. 1951. *Race and biology*. Paris: Unesco.

Fenton, S. 1980. 'Race Relations' in the sociological enterprise. *New Community* **8**(1–2), 162–8.

Field, E. F. 1977. Nordic racism. *Journal of the History of Ideas* **38**(3), 533–80.

Frazier, E. F. 1947. Sociological theory and race relations. *American Sociological Review* **12**(3), 265–71.

Freedman, M. 1954. Some recent work on race relations: a critique. *The British Journal of Sociology* **5**, 343–54.

Fryer, P. 1984. *Staying power*: London & Sydney: Pluto Press.

Griffiths, R. 1980. *Fellow travellers of the Right*, London: Constable.

Grubb, K. 1945. Letter to F. R. Cowell, 13 June, FO 924/147.

Hiernaux, J. 1965. Introduction: the Moscow Expert Meeting (August 1964). *International Social Science Journal* **8**, 1.

Hodson, H. V. 1950. Race relations in the Commonwealth. *International Affairs* **26**, 303–15.

Holland, R. F. 1985. *European decolonisation 1918–1981: an introductory survey*. London & Basingstoke: Macmillan.

Holmes, C. 1979. *Anti-semitism in British society, 1876–1939*. London: Edward Arnold.

Huxley, J. 1941. *Democracy marches*. London: Chatto & Windus.

Huxley, J. 1944. Colonies and freedom. *The New Republic*, 24 January.

Huxley, J. 1948. *This is our power . . . .* Beirut: Unesco.

Huxley, J. & A. C. Haddon 1935. *We Europeans*. London: Cape.

Jones, G. 1980. *Social Darwinism and English thought*. Sussex: Harvester Press.

Joshi, S. & B. Carter 1984. The rôle of Labour in the creation of a racist Britain. *Race and Class* **25**(3), 53–70.

Karady, V. 1981. French ethnology and the Durkeinian breakthrough. *Journal of the Anthropological Society of Oxford* **12**(3), 165–76.

Klineberg, O. 1954. 32 social scientists testify against segregation. *Unesco Courier* **6**, 24.

Kuper, A. 1980. The man in the study and the man in the field: ethnography, theory and compassion in social anthropology. *European Journal of Sociology* **21**(1), 14–39.

Lee, J. M. & M. Petter 1982. *The Colonial Office, war and development policy*. London: Maurice Temple Smith.

Lorimer, D. A. 1978. *Colour, class and the Victorians*. Leicester: Leicester University Press.

Lyon, M. 1985. Banton's contribution to racial studies in Britain: an overview. *Ethnic and Racial Studies* **8**(4), 471–83.

Macrae, D. G. 1961. Race and sociology. In *Ideology and society*, D. G. Macrae (ed.), 106–18. London: Heinemann.

Malinowski, B. 1916. Letter to Graham Wallas 27 June, *Graham Wallas Correspondence*, 1/59. London: British Library of Political and Economic Science.

Malinowski, B. 1929. Practical Anthropology. *Africa* **11**, 22–38.

Malinowski, B. 1930. Report on the State of Anthropology and the possibility of its development in England. In *Lord Lothian Papers*, Scottish Public Record Office, Edinburgh, GD40/17/243, encl. in J. H. Oldham to Lord Lothian 6 February.

Malinowski, B. 1972 (1st edn 1922). *Argonauts of the Western Pacific*. London: Routledge & Kegan Paul.

Mason, P. 1954. *An essay on racial tension*. London & New York: RIIA.

Mason, P. 1967. The revolt against Western values. *Daedalus* **96**(2), 328–52.

Mason, P. 1970. *Patterns of dominance*. London: Oxford University Press.

Miles, R. 1984. The riots of 1958: notes on the ideological construction of 'race relations' as a political issue in Britain. *Immigrants and Minorities* **3**(1), 252–75.

Montagu, A. 1972. *Statement on race*. London: Oxford University Press.

Moore, R. 1971. Race relations and the rediscovery of sociology. *The British Journal of Sociology* **22**, 97–104.

Mosse, G. 1979. *Toward the final solution*. London: Dent & New York: Ferbig.

Mullard, C. P. 1980. *Power, race and resistance: a study in the Institute of Race Relations,1952–1972*. University of Durham: Unpubl. Ph.D. thesis.

Myrdal, A. 1966. Two decades in the world of social science. *Unesco Courier* **48**, July–August.

Nisbet, R. 1977. *Sociology as an art form*. London: Oxford University Press.

Novack, M. 1972. *The rise of the unmeltable ethnics*. New York: Macmillan.

Phillips, M. 1983. Danger! astrologers at work: a critical note on the narrow orthodoxy of race relations research. *Community Development Journal* **18**(3), 265–69.

Pitt-Rivers, J. 1973. Race in Latin America: the concept of Iraza. *European Journal of Sociology* **14**, 3–31.

Porter, D. 1983. Orientalism and its problems. In *The politics of theory*, E. Barker (ed.), 179–93. Colchester: University of Essex Press.

Radcliffe, Lord 1969. Immigration and settlement: some general considerations. *Race* **11**(1), 35–51.

Review of Reviews **63**(1921), 469.

Rex, J. 1968. The ubiquitous shadow of racism. *Unesco Courier*, 25 January.

Rex, J. 1970. *Race relations in sociological theory*. London: Weidenfeld & Nicolson.

Rex, J. 1981. A working paradigm for race relations research. *Ethnic and Racial Studies* **4**(1), 1–25.

Rex, J. & R. Moore 1967. *Race, community and conflict: a study of Sparkbrook*. London: Oxford University Press (for the IRR).

Rex, J. & S. Tomlinson 1979. *Colonial immigrants in a British city: a class analysis*. London: Routledge & Kegan Paul.

Rich, P. 1981. The South African Institute of Race Relations and the debate on 'race relations', 1929–1958. *African Studies* **49**(1), 13–22.

Rich, P. B. 1984. 'The baptism of a new era': The 1911 Universal Races Congress and the liberal ideology of race. *Ethnic and Racial Studies* **7**(4), 534–50.

Rich, P. B. 1985. The politics of 'surplus colonial labour': black immigration to Britain and governmental responses. In *The Caribbean in Europe*, C. Brock (ed.) London: Frank Cass.

Rich, P. B. 1986a. *Race and empire in British politics*. Cambridge: Cambridge University Press.

Rich, P. B. 1986b. The impact of South African segregationist and apartheid ideology on British racial thought 1939–60. *New Community* **13**(1), 1–17.

Richards, A. I. 1957. The concept of culture in Malinowski's work. In *Man and culture*, R. Firth (ed.), 15–31. London: Routledge & Kegan Paul.

Richards, D. 1977. The ideology of European dominance. *Presence Africaine* **111**, 3–18.

Ridley, H. 1973. Colonial society and European totalitarianism. *Journal of European Studies* **3**(1), 147–59.

Said, E. 1978. *Orientalism*. London: Routledge & Kegan Paul.

Said, E. 1985. Orientalism reconsidered. *Race and Class* **27**(2), 1–15.

Samuels, F. 1969. Colour sensitivity among Honolulu's Haoles and Japanese. *Race* **11**(2), 203–12.

Shapiro, H. L. 1954. Hawaii: crossroads of the Pacific. *Unesco Courier* **2**, 24–5.

Shils, E. 1967. Color, the universal intellectual community, and the Afro-Asian intellectual. *Daedalus* **96**, 2.

Sivanandan, A. 1974. *Race and resistance: the IRR story*. London: IRR.

Stepan, N. 1982. *The idea of race in science: Great Britain, 1800–1950*. London: Macmillan.

Thornton, R. J. 1985. Imagine yourself set down ... Marsh, Frazer, Conrad, Malinowski and the rôle of imagination in ethnography. *Anthropology Today* **1**(5), 7–14.

Tobias, P. 1962. *The meaning of race*. Johannesburg: SAIRR.

Voegelin, E. 1940. The growth of the race idea. *Review of Politics* **2**, 283–317.

Watt, D. C. 1984. *Succeeding John Bull: America in Britain's place, 1950–1975*. Cambridge: Cambridge University Press.

Wax, M. L. 1972. Tenting with Malinowski. *American Sociological Review* **37**(1), 1013.

Weber, M. 1978. *Economy and society*. In *Max Weber: selections in translation*, W. E. Runciman (ed.), pp. 368–9. Cambridge: Cambridge University Press.

Weinrich, M. 1946. *Hitler's professors: the part of scholarship in Germany's crimes against the Jewish People*. New York: Yiddish Scientific Institute.

5 *Racism in Britain: the Scottish dimension*

ROBERT MILES AND ANNE DUNLOP

> In any case, no racial problems seem to have occurred as yet, whether because of different attitudes held by the indigenous population from those held by inhabitants of other parts of Britain or because the numbers of coloured immigrants are not yet sufficiently large to create problems of the kind experienced in several English cities. (Budge & Urwin 1966, p. 96.)

> We asked Dilip Deb, an Asian lawyer with a practice in Glasgow, why Scotland had so far escaped any race problems, despite a sizeable immigrant population ... He says the situation in Scotland is different. We don't have big numbers of immigrants in specific areas, crammed into ghettoes. (*Sunday Post*, 15 September 1985.)

Analyses of racism in Britain rarely make any specific reference to Scotland. Yet *within* Scotland a clear differentiation is made between England and Scotland, sustained by the claim that 'race relations' is an English problem, absent north of the border. Various explanations for this alleged difference are offered, ranging from references to the size of the 'immigrant population' to a hypothesized 'natural tolerance' of the Scottish people. These explanations are widely articulated in the Scottish press and in the Scottish political process and have been endorsed by a number of academic commentators (e.g. Harvie 1981, p. 67).

This common-sense interpretation has been challenged elsewhere (Miles & Muirhead 1986, Miles 1982). The objective of this essay is to advance further evidence for a contrary argument, namely that what distinguishes Scotland from England is the absence of a racialization of the political process in the period since 1945 rather than an absence of racism *per se*. This argument hinges on the view that it is not possible to draw conclusions about Scottish political processes simply on the basis of events in England. There is no necessary correspondence between political processes in Scotland and England; first because to a significant extent the two political units exhibit distinct structural features, and secondly because they have distinct historical trajectories. This distinctiveness is far from absolute, but to the extent that it does occûr it has implications for our concerns.

We concentrate on three aspects of this distinctiveness. First, for both economic and political reasons, the history of migration both into and out of Scotland is distinct from that of England. Second, and as a consequence, the political and ideological composition and reproduction of class relations have distinct features and this has implications for the content of the political agenda. Third, Scottish political history has been shaped by political nationalism in a distinct manner, reflecting the fact that the Act of Union of 1707 led to political unification while retaining a distinct state structure. We shall discuss these three aspects of difference following an account of the argument that the political process has not been racialized in Scotland in the period since 1945.

Racism and racialization: the colonial dimension

The absence of the process of racialization of politics in Scotland since 1945 stands in stark contrast to the history of Scottish involvement in colonialism and a related reproduction of racism (Miles & Muirhead 1986). This history is an important dimension of the development of Scotland's economic supremacy in the world economy of the 19th century, the roots of which were put down in the West–Central region, and specifically in Glasgow. Between 1830 and 1870 the capitalist economy expanded very fast, the expansion being based on the production first of textiles, and secondly of heavy industrial products such as pig iron and ships. The central factor in this rapid industrialization was that the Scottish economy became an integral and central component of the world economic system during the 19th century, its dependence on the English economy being obscured by the rapid economic and social transformation that was under way (Dickson 1980, p. 183). This position was exemplified in events such as the organization of the International Exhibition in Glasgow in 1888.

The 19th-century development of industrial capitalism in Scotland was rooted, in part, in the international activities of Scottish merchant capital. The Act of Union of 1707 had allowed an emergent merchant class access to English domestic and colonial markets on terms equal to those of English merchant capital. Scottish merchants generated considerable profits from trade with North America and the Caribbean during the 18th century. The tobacco and linen trades were of particular significance, a proportion of the profits being used as a source of investment in the nascent manufacturing sector of the developing Scottish economy. The growth of Glasgow as a commercial and trading centre was therefore accompanied by an increasing intercourse with various parts of the world economy and an increasing involvement in various facets of colonialism.

At the level of economic relations, two dimensions are worthy of mention here. First, Scottish involvement in the British Caribbean, dominated by the slave mode of production, was considerable (Sheridan 1977). The Scottish presence in the Caribbean was not entirely voluntary because the early development was dependent partly upon the use of indentured labour from Scotland (Rinn 1980), one source of which included some of those arrested after the Jacobite revolts of 1715 and 1745. But the Caribbean was also considered to be a place of opportunity for the ambitious, and many Scots voluntarily indentured themselves to take on lowly administrative positions in the plantation economy with the hope of economic advancement. Some were successful, making the transition from book-keeper to attorney and then to plantation owner, subsequently returning to Scotland as absentee landlords where they left their social marks on Scotland in the forms of conspicuous consumption and as the foundation of manufacturing activity. Additionally, Glasgow became a sugar port and, on a few occasions, Glasgow ships participated directly in the slave trade (Fryer 1984, pp. 51–2).

There was significant Scottish economic involvement in the British colonies in India and Africa. Scots were involved in the East India Company and were also represented in the ranks of the free merchants who participated in local trade in India. During the 19th century, Scots became managing agents, gaining control of the production of coffee, tea, jute and indigo (Parker 1985). These commodities were important in sections of the Scottish rentier bourgeoisie with investments in a wide range of activities throughout the world (Dickson 1980, p. 193). There was also some Scottish participation in the development of trade and production in Africa. A firm called Miller Brothers was heavily involved as part of the Royal Niger Company in establishing colonial domination in Nigeria, while the British East Africa Company and the African Lakes Company (the latter promoted by the Church of Scotland) were active in British-controlled East Africa (Dickson 1980, pp. 250–1).

Thus there was an interdependence of Scottish colonial activity and the rise of an industrial bourgeoisie in Scotland during the 18th and 19th centuries. As a result of this process, a section of the Scottish population was in direct contact with the colonized populations whose labour power was integral to British colonialism. This economic interdependence was accompanied by other forms of colonial activity and the production of an imagery of these colonized populations. Scots were heavily involved in missionary activity, through English missionary societies until the Scottish churches established their own, and this spread of the Christian gospel was often linked with both exploration and trade. The writing of Scottish explorers and missionaries was published and circulated in Scotland, often providing a racist imagery of those who received the 'benefits' of 'Scottish civilization'. This imagery was reproduced and legitimated by a specifically Scottish contribution to scientific racism (e.g. Kames 1774, Combe 1825). Finally, the Scottish press in the 19th century reported on various major episodes in the history of British colonialism (Cowan 1946), sustaining a consciousness of Scotland's participation in the British Empire.

In conclusion, we would suggest that Scottish colonial experiences have had important consequences for the contemporary reproduction and expression of racism in Scotland and that there is good reason to expect that racism is an element of Scottish political consciousness. Scottish participation in a wide range of colonial activities has ensured that since the late 18th century there has been an awareness in Scotland of populations in different parts of the world and therefore an imagery of those peoples. No doubt, that imagery was multi-faceted, but one part of it has been racist in character (Miles & Muirhead 1986). However, by arguing that racism is a legacy of Scotland's colonial involvement, we are forced to consider why it has been the case that, in comparison with England, this racism has been less explicitly articulated as a sustained and organized campaign against the arrival of migrants from India and Pakistan in the post-war period.

Scotland and migration

However, the post-1945 Asian migration is neither the first nor the largest migration to Scotland. The development of capitalism in Scotland has been associated with a number of substantial movements of population within and both into and out of the country. The process of dispossession and proletarianization of the agararian labouring class, following by a migration to the growing industrial centres from the late 18th century, is well known, especially in the case of the Highlands (see Dickson 1980, pp. 147–52; Lenman 1981, pp. 122–4). Further, the same process contributed significantly to emigration from Scotland to various parts of the world during the same period (e.g. Cowan 1961, pp. 18–26, 118–9, 209–12; Donaldson 1966, pp. 57–80). The latter migration was additionally assisted by the mechanization of manufacture which led to the impoverishment of independent petty-commodity producers such as handloom weavers (Dickson 1980, pp. 164–7). A tradition of emigration to avoid poverty and unemployment has continued into the 20th century, stimulated particularly by material conditions and frustrated expectations between 1918 and 1931 and again between 1951 and the mid-1970s (Donaldson 1966, pp. 194–200; Harvie 1981, p. 66). It has been estimated that between 1861 and 1939 Scotland lost 43.7 per cent of the natural increase in the population through net emigration overseas, a percentage that increases to around 54 per cent if one includes emigration to other parts of the United Kingdom. Scotland is therefore second only to Ireland in the list of European countries losing population by emigration during this period (Flinn 1977, p. 448).

But the picture of Scotland as a country where emigration is a conscious tradition has to be tempered by the evidence of migration flows *into* Scotland. Capitalist development is necessarily uneven, as is well illustrated by the movement of Irish migrants into Scotland at the same time as financial assistance was being provided for dispossessed tenants in the Highlands to migrate to Canada (Cowan 1961, pp. 212–18). The Irish economy was equally affected by the structural transformations associated with the rise of industrial capital, but in a rather different way (see Miles 1982, pp. 124–6). The development of capitalist farming in the east and north led to dispossession, while in the south and west, subdivision of peasant plots made the production of subsistence increasingly precarious – processes that created a latent reserve army of labour, if not a rural 'lumpenproletariat'. The problem of material reproduction was accentuated by famine on a number of occasions during the first half of the 19th century, an event which intensified the pressure to migrate. But, unlike Scotland, capital investment in industrialized production in Ireland was very limited, and concentrated in the Belfast region, so that there were fewer opportunities for proletarianization within the country. For large numbers of small tenants and peasant labourers, emigration was therefore the main means of escape from rural poverty, unemployment and starvation, and Scotland became one of their main destinations.

An economic and political connection between Ireland and Scotland had been established in the 14th century and consolidated in the 17th by the policy of plantation development in Ulster (de Paor 1970, pp. 1–32). This connection was accompanied by migration. By the end of the 18th century there was a tradition of seasonal migration from Ireland to Scotland as Irish peasant producers sought a cash income from their labour power during the summer months, and this constituted the foundation for the more permanent settlement that accompanied the developments of the 19th century. These Irish migrants were concentrated in the south-west of Scotland and in the Glasgow region, and by 1831 they amounted to around 35 500 persons out of a total population of 202 400 in Glasgow (Dickson 1980, p. 188; also Flinn 1977, p. 456). In 1851, there were 207 367 Irish-born people living in Scotland, a figure which accounted for 7.2 per cent of the Scottish population, the proportion for England and Wales being 2.9 per cent (Flinn 1977, p. 457). They occupied mainly semi- and unskilled positions, notably in construction, cotton production and mining. Their arrival further expanded the growing pool of labour, and this had beneficial consequences for the capitalist class by reducing the tendency for wages to rise and by providing a potential strike-breaking force (Dickson 1980, pp. 194, 196, 198). Competition within the labour market, combined with struggles for housing and other resources, constituted a motive for the identification of a criterion of differentiation by which to establish a hierarchy of legitimacy to determine access to these resources. In a society where the Reformation had occurred under the influence of John Knox, the Irish migrant's Catholicism was pregnant with potential.

Irish migration to Scotland has continued into the 20th century, although on a very limited scale. In the late 19th century there was an influx of migrants from southern and eastern Europe, the 1911 Census recording a foreign (mainly European) population in Scotland of 24 739. Italians constituted about a quarter of these, while most of the remainder were probably political refugees from Russia and Poland (Flinn 1977, p. 458). Since 1945, there has been a migration from the Indian subcontinent. The history of this migration remains to be written and so reconstructing its origin and development is difficult. There is evidence of an Asian presence in Scotland in the late 19th century. These migrants were possibly seamen or itinerant salesmen or both (Salter 1873, pp. 88, 170, 220–3), and they may have retained a presence in Scotland through the first half of the 20th century. To this extent there is a comparison with England, in so far as there too there has been a long-established Asian (and Caribbean) presence (Fryer 1984) which provided a foundation for the economically stimulated migration of the 1950s (Peach 1968).

However, the Scottish economy of the 1950s and 1960s did not undergo the period of growth and transformation that characterized the English economy and that led to significant labour shortages in certain sectors. The growth rate was half that of England and income per head was 13 per cent lower than the average for the United Kingdom (Harvie 1981, p. 62). There was a short boom in the shipbuilding industry after the end of the war, but a slump began in 1957 which affected all heavy industry in Scotland, much of which was undercapitalized and using outdated technology, thus making it uncompetitive internationally.

Additionally, many of the new industries (cars, electrical goods, chemicals) that emerged between 1922 and 1938 and which played a key rôle, along with services, in the reconstruction of capitalism after 1945 were barely evident in Scotland (Harvie 1981, pp. 35–8, 54–60). The weakness of the dependent and specialized character of Scottish capitalism became increasingly evident from the mid-1950s (Dickson 1980, p. 294). One measure of that weakness is found in the increasing level of net migration out of Scotland: in the decade 1951–61, net migration was 282 000 and in the following decade it was 326 000 (Harvie 1981, p. 66).

There is, therefore, little reason to expect a migration to Scotland from the New Commonwealth on the scale that occurred in England (Peach 1968, p. 70). Nevertheless, there has been a migration, mainly from India and Pakistan (with a smaller migration from Hong Kong; see Lothian Community Relations Council 1983). The 1961 Census showed a total population of New Commonwealth origin in Scotland of 4437 (compared, for example, with 126 726 in London and the South-East of England alone: see Peach 1968, p. 67). A recent study has estimated a total 'ethnic minority' population in Scotland in the 1980s of 38 000, of which about 65 per cent are of Indian and Pakistani origin (Scottish Office 1983), although these figures have been revised upwards by other researchers (Dalton 1983/4). Most of these people live in the cities of Glasgow, Edinburgh and Dundee. This presence is due to a number of different migration flows. Many of those who arrived during the 1950s came directly from the Indian subcontinent, whereas those who arrived in the 1960s seem to have migrated via England, including East African refugees of the late 1960s and early 1970s (see, for example, Jones & Davenport 1972, Kearsley & Srivastava 1974). There is evidence of an early penetration of certain forms of petit-bourgeois activity, notably warehousing and restaurant ownership, and although there is no statistical evidence to sustain casual observation, a large proportion of the current Asian population in Scotland appears to be concentrated in the provision of services as small entrepreneurs (especially shopkeeping). If this is so, the absence of a large Asian wage-earning class would be consistent with the evidence of the condition of the Scottish economy in the 1950s. If there were no significant labour shortages, then the migration that did occur was not directly related to economic need. The most likely exception to this generalization concerns the employment of Asians in the transport industry in Glasgow, a feature that seems to originate in the 1950s. In other words, the Asian migration to Scotland was not as centrally related to the demands of the capitalist economy as in the instance of New Commonwealth migration to England in the same period and in the case of the Irish migration to Scotland in the 19th century. In so far as this is true, then there would have been only limited competition between migrant and indigenous labour in the labour market in the 1950s.

Migration and class relations

In this section, we outline some of the political and ideological consequences of the reactions to 19th-century Irish migration to Scotland. We will argue that these consequences amount to the institutionalization of a cultural signification within the process of the reproduction of class relations in Scotland, constituting an important part of the context within which classes are formed and struggle. Anti-Catholic resentment, especially amongst the working class, was intensified by a sense of economic insecurity which had some material foundation during the 19th century (e.g. Handley n.d., Campbell 1979). But what is called, in common-sense terms, the 'sectarian problem' is not a purely religious differentiation. Religious affiliation may be the principal factor but the idea of 'race' has also been articulated consistently by various factors. Underlying this religious signification is therefore a secondary process of racialization (Miles 1982, p. 135–48). In other words, material insecurity within the emergent working class articulated with a specific cultural content to produce a particular fractionalization, the 'sectarian problem'. This was not, however, the product of an instrumental divisive strategy implemented by the bourgeoisie but was grounded in the necessary differentiation within the labour market.

The process of signification and racialization continues to be sustained by various institutional mechanisms, three of which will be discussed below. For the purpose of the argument in this essay the significance of these processes lies in that the Asian migrants of the post-1945 period arrived in a context in which class relations were already significantly fractionalized. Anti-Catholicism predated the arrival of Irish migrants, previously focusing on the small and isolated Catholic communities which were located mainly in the Highlands. The persecution of these communities after the Reformation was commonplace in a society which was self-consciously Protestant in religious observance and which retained a distinctly Scottish church as a result of the Act of Union (Cooney 1982, p. 7–21). The arrival of Irish migrants, the majority of whom were Catholics, was therefore interpreted as a potential basis for a future challenge to Protestant domination, and so the Irish migration became widely interpreted as a 'Catholic invasion'. But in reality the ideological consequences of the migration were more complex because among the migrants were a minority of Protestants, mainly from Ulster, who brought with them the tradition of the Orange Lodges. This 19th-century import to Scotland sustained and intensified anti-Catholic agitation, institutionalizing it in a specific organizational form (Campbell 1979, p. 182; Murray 1984, p. 78; Bruce 1985). During the 19th century anti-Irish conflict became increasingly entrenched and was expressed frequently in physical attacks on both persons and property. But the indigenous working class did not have a monopoly of anti-Irish agitation; the 'respectable' classes were also involved (Murray 1984, p. 96). Thus outbreaks of violence were led and encouraged by verbal expressions of hostility and denigration, particularly in newspapers, pamphlets, reports to Royal Commissions and the pulpit (Handley n.d., pp. 73, 138, 141, 239–48).

This process of fractionalization has been reproduced by a number of mechanisms in the course of the 20th century (Miles & Dunlop 1986). First, the political and ideological consequences of the Irish migration became institutionalized in the establishment of a segregated, denominational education system which reproduces religious and cultural differentiation up to the present day. Again, the structural potential for this distinct process was grounded in the terms of the Act of Union which allowed the retention of a separate education system in Scotland. For our purposes, the first important development occurred in 1872 when the Presbyterian Churches transferred their schools to the State. However, the Catholic and Episcopalian Churches refused and continued to run their schools largely on a voluntary basis but with some irregular state assistance. The latter schools were brought within the state-controlled sector under the terms of the Education (Scotland) Act 1918, which established a national denominational education system in Scotland. Under the Act, Catholic and Episcopalian schools were incorporated into the State system while retaining the right to teach their own religious beliefs. However, although the State had, in effect, to buy the schools, it conceded considerable autonomy in teacher selection and course content. These concessions were the price paid for Catholic and Episcopalian support, but they offended Presbyterian interests who claimed that they had handed over their schools free – hence the slogan 'Rome on the Rates'.

The Act became the subject of a report to the 1923 General Assembly of the Church of Scotland, along with what was perceived as a further Irish 'invasion' in the context of the Irish struggle for independence. That the Church of Scotland was wanting to make a specifically political intervention was made evident when it decided to publish the report as a pamphlet in the same year under the title *The menace of the Irish race to our Scottish nationality*. The idea of 'race' played a prominent rôle in the report (Miles & Muirhead 1986) and was used to sustain nationalist sentiment. It interpreted the Act as a lever whereby the Catholic Church in Scotland could begin 'the destruction of the unity and homogeneity of the Scottish people' (Reports on the Schemes of the Church of Scotland 1923, p. 751).

Hence, for more than a decade after it became law the Act was interpreted by anti-Catholic elements of various persuasions as a threat to the Protestant hegemony while the 1923 Church of Scotland report is a prime example of the expression of racism in the Scottish political process. There were other examples of racism in this period. In 1930, Andrew Dewar Gibb's *Scotland in eclipse* set out a case for Scottish nationalism by arguing, amongst other things, that independence would allow the Scottish people to prevent their degeneration as a result of the presence of the inferior Irish 'race'. And, in *Scotland's dilemma*, a nationalist pamphlet by John Torrence (1938), a similar anti-Irish theme appears (Handley n.d., p. 352). In addition, the Act established a segregated education system which reproduces religious and cultural differentiation and so shapes and reinforces distinct religious identities amongst the Scottish population.

A second mechanism for the reproduction of this ideological and political division in Scotland, and particularly in West-Central Scotland, is the 'Old Firm' rivalry of Celtic and Rangers football teams, grounded in the place of football in male, working-class culture. The history of Celtic and Rangers football teams bears witness to the institutionalization of religious segregation. Rangers (originating as a boys' football club) and Celtic (originally a charitable organization for poor-relief) were from the outset committed to upholding the Protestant and Catholic religions respectively (for a detailed history, see Murray 1984). Over the years, team rivalry has led to intense battles beween club fans, and so football has become an everyday expression of sectarian differentiation and a basis for its constant reproduction.

A third mechanism for the reproduction of ideological and political division in Scotland is the political activism of militant Protestantism, especially in the 1920s and 1930s. During this period, Protestant extremism became electorally significant for the first time in Scotland, and thereby anti-Catholicism became associated with the struggle for political representation. Moreover, during this period the British Union of Fascists was prominent in English politics. It is therefore significant that explicitly fascist parties had very little impact in Scotland at a time when they achieved various levels of success throughout much of Europe. Hence, the historical legacy of sectarian division effectively blocked any significant fascist advance in Scotland and so, in an economic and political context conducive to right-wing political extremism, militant Protestantism became an effective substitute in the context of a distinct political process in Scotland. The Scottish political arena already had a pre-ordered agenda based upon ideólogical and political divisions. Fascist intrusion into Scottish politics, as a result of its failure or inability to play the 'sectarian card', was therefore unable to articulate with the particular cultural expression of right-wing extremism.

Militant Protestantism in the 1920s and 1930s took two main institutional forms, the Scottish Protestant League (SPL) and the Protestant Action (PA). In addition, an Orange and Protestant Party was formed in 1922 to support Protestant electoral candidates. Although the Party's existence was brief, one of its candidates was elected for the Motherwell parliamentary constituency for one term before the organization became reabsorbed by the Unionists (Bruce 1985, p. 48). The SPL was founded in 1920 in Glasgow by Alexander Ratcliffe in a climate of virulent anti-Catholicism surrounding the Education (Scotland) Act and fears of a large-scale Catholic immigration stimulated by the Anglo-Irish War of 1919–21. The Party was populist and anti-Catholic but had little political impact until Ratcliffe published the *Protestant Advocate* in 1922. The first issue criticized the Labour Party for allegedly being dominated by the Catholic Church; later the publication focused increasingly on the education issue (Bruce 1985, pp. 42–3). In the latter half of the decade, the SPL entered the electoral arena. In April 1927, Ratcliffe had stated (cited in Bruce 1985; p. 48):

> What is wanted today is a real Protestant Party in the House of Commons, a party of Christian Protestant men and women to do battle for the cause of Christ and Protestantism against the forces of political corruption and Roman Catholic plot.

Accordingly, Ratcliffe contested the seat for Falkirk and Stirling Burghs in 1929, describing himself as an 'Independent and Protestant'. His political campaign concentrated on the need for controls over Irish immigration and revision of the Education (Scotland) Act of 1918. During the campaign, Ratcliffe offered to stand down if the Unionist candidate agreed to adopt his demands on immigration and education, an offer which was refused, and Ratcliffe finally polled 21.3 per cent of the vote.

The SPL achieved further success in Glasgow municipal council elections, winning two seats in 1931. By 1932 the SPL was taking 11.7 per cent of the total municipal vote and had won an additional seat. In 1933 the SPL polled 11 per cent of the total votes cast and won four further seats in Glasgow. Gallagher claims (1983) that the SPL's success was based on the combination of social reform and 'no popery', a combination which was well received in certain manual working-class areas and areas of lower white-collar occupation where there was previous support for Conservative candidates. But 1933 proved to be the year of peak electoral success for in the following year the SPL lost all its seats, although the organization remained politically active until 1939.

The demise of the SPL was caused by three factors (Bruce 1985, p. 76); competition with Protestant Action, the outbreak of war, and Ratcliffe's late conversion to fascism and anti-semitism. That Ratcliffe was sympathetic to racist beliefs and objectives was evident in the first issues of the *Protestant Advocate*, which printed articles sympathetic to the Klu Klux Klan (KKK) in the United States. However, the appeal of the KKK for Ratcliffe was intensified by the fact that sections of it were also hostile to Catholicism. Ratcliffe's first known connection with fascism came through an involvement in the short-lived Scottish Fascist Democratic Party (SFDP) founded in Glasgow in 1933. The SFDP's draft constitution included a clause that called for the expulsion of Catholic religious orders from Scotland, the repeal of Section 18 of the Education (Scotland) Act and the prohibition of Irish immigration to Scotland. Ratcliffe denied that he was a fascist. However, his move to open support for Hitler came through his hatred for 'popery'. In the *Protestant Vanguard* in March 1934, Ratcliffe declared that 'we are not in any sense whatsoever defending Hitlerism. But we would prefer Hitlerism to popism' (Bruce 1985, p. 78). In the autumn of 1939, Ratcliffe visited Germany and returned to fill the *Protestant Vanguard* with praise for Hitler and defamatory comments about the Jews. In June 1940, he praised Nazi harassment of Roman Catholics. In the context of the British Government's decision to enter the war against Germany, this position sealed the demise of the SPL.

A similar Protestant extremist organization became active in Edinburgh. John Cormack formed Protestant Action in 1933, one of its aims being, in Cormack's own words, 'applied physical Christianity' (Gallagher 1984). Much of its political propaganda attacked Catholics and Catholicism. Between 1934 and 1936, the PA enjoyed short-lived electoral success. Nine councillors were elected and in 1936 the PA gained 31 per cent of votes cast in Edinburgh polls. Cormack disavowed any connection with fascism in explicit terms,but showed a susceptibility to explicit racist sympathies with the formation in 1937 of a cadre of the PA called 'Kormack's Kaledonian Klan' which adopted official KKK symbols. The subsequent decline of the PA was caused by two factors. First, there was little history of Orangeism and militant Protestantism in Edinburgh, unlike Glasgow. Second, the PA (like the SPL) was never faced with an explicitly Catholic opposition, so avoiding the possibility of a mutally reinforcing cycle of conflict. The Catholic political presence was concentrated in the Labour movement, forcing the militant Protestant organization into attacking the latter and, thereby, socialism, a strategy which tended to result in the dilution and destruction of their original ideological message.

The significance of the rise and decline of militant Protestantism should be evaluated in the context of the evidence concerning fascist activity in Scotland in the 1930s. References to this are scant (see Cross 1961, Fielding 1981, Walker 1977, Murray 1984), but there is no evidence of any sustained presence or impact. The British Union of Fascists attempted to organize in Scotland but made almost no impression. Its Motherwell branch claimed 200 members in early 1933 but, even if true, this was untypical (Cross 1961, p. 78). Cross states that it was only in Aberdeen that fascism had any significant impact, although he provides no detailed documentation (1961, p. 108). Sources also refer to the presence in Scotland of the British Fascists which group was later reorganized under the title of the Scottish Loyalists (Benewick 1972, pp. 27, 32, 35, 36).

The only other evidence of fascist activity relates to allegations of a link between the Orange Order and fascist politics by Murray (1984, p. 138). These are based on the activities of William Fullerton, more notoriously known as 'King Billy of Bridgeton', who was leader of the 'Billy Boys', one of Glasgow's street gangs, and who committed himself to the defence of both the Crown and Protestantism. Fullerton claims to have joined a fascist party – which one is not clear – and to have been a section commander responsible for 200 men and women (Murray 1984, pp. 153, 157). Murray's sources are not always stated, leaving one with reservations about his evidence, although an overlap in political ideology between facism and militant Protestantism could have sustained such practices.

The 1920s and 1930s seem to have been the period during which militant Protestantism was most prominent in Scottish politics. Contemporary material is almost non-existent. It is significant, however, that the system of Orange Lodges still exists and that the 12th of July Orange Walk still takes place in Scotland, most notably in Glasgow (Sutherland 1981). The 12th of July celebrations of King William of Orange's victory over King James at the Battle of the Boyne in 1690 marks, for Protestant Loyalists, the deliverance from 'popery' and Protestant ascendancy. Moreover, the outbreak of open conflict in Northern Ireland in the late 1960s has had effects in Scotland, leading to new allegiances in sectarian lines, with sections of Protestant and Catholic communities allying respectively with the various Unionist and Provisional organizations in Northern Ireland. For example, the Scottish Loyalists, an aggressive youth wing of the Orange Order, was formed in 1980 by young members of the Order who were frustrated by the willingness of the police to allow the 'Troops Out' movement the right to march in Glasgow (Bruce 1985, p. 227). Indeed, a journey round any inner-city or peripheral housing scheme reveals graffiti that bear testimony to the continuing allegiances and historical memories of sections of the Protestant and Catholic communities. Given that the Scottish education system remains segregated, and that football remains the focus of religious differentiation, these various forms of political practice are part of a wider process of political and ideological reproduction which structure class relations in the light of the consequence of the migration of Irish workers to Scotland in the 19th century. One observer of recent events has concluded (Sutherland 1981, p. 50);

> Protestant extremism in Scotland remains a potent force, openly condoned by sections of the respectable establishment.

In sum, it is essential to evaluate the political repercussions of Asian migration to Scotland in the context of previous migration history and its effects on class relations. That history shows both that arguments about 'Scottish tolerance' are ideological and that Asian migrants arrived to become part of an already materially and culturally fractionalized class structure.

Political nationalism

The extent to which the Scottish political agenda is distinct from that in England is a matter of dispute (e.g. Miller 1981), but the fact that it is distinct is made particularly evident by the history of political nationalism in Scotland. The 'national question' has been an increasingly prominent issue on the political agenda for much of this century and has been a major issue since the late 1960s. Its significance is that it counterposes the Scottish 'nation' to that of England, a political and ideological signification which is sustained by the long-term decline of the Scottish economy. Since the 1960s political nationalism has actively drawn upon the debate about 'economic decline', locating its cause or solution or both in the Union with England. This means that, in Scotland, one major strand in the ideological dimension of the crisis of capitalism has externalized the problem, locating it in the realm of political relations. In so far as this ideological signification has had electoral consequences it has weakened the political influence of the Conservative Party in Scotland and exposed the continuing hegemony of the Labour Party.

Within Scotland, the nationalist legacy leads to the articulation of a political solution to crisis in the form of some variant of independence or devolution. The idea of 'nation' is employed to counterpose Scotland to

England as distinct cultural units, so externalizing the solution to economic decline. In England, such a process is historically and structurally problematic because it is the dominant economic and political unit (although British membership of the EEC presents parallel possibilities). One important strand in the political response to economic crisis in England has therefore been to seek an internal cause and solution, and a key constituent of that has been the identification of New Commonwealth migrants and their children as an 'alien wedge' which has weakened 'national homogeneity' and introduced 'new' elements of internal dissolution (e.g. by their alleged propensity to crime and violence). Thus, the idea of 'nation' in England articulates with and reformulates the racism which had a long pedigree grounded in colonial domination. In Scotland, the latter historical experience is overlaid by a historical experience and memory of a loss of nationhood, of being the object of a different form of 'colonial' domination by England, and this has been the dominant element which has shaped the political agenda in the post-1945 period. This is illustrated by the main electoral trends in Scotland since 1945. The most significant of these is the steady decline in electoral support for the Conservative Party and the rise in support for the Scottish National Party (SNP), particularly after 1964. Prior to this date, the Conservative and Labour Parties had contended for the majority of electoral support in Scotland since the early 20th century. However, in only two General Elections since 1945 (1951 and 1955) did the Conservative Party gain a majority of the vote (49 per cent and 50 per cent respectively). What is significant is that in all but these two General Elections, the Labour Party had gained a majority of the votes amongst the Scottish electorate. Thus, up until 1964, party allegiance split roughly three ways amongst Labour, Conservative and Liberal Parties, and 'fringe' parties such as the Scottish National Party remained on the absolute periphery of electoral support, averaging only around 1 per cent of the vote.

The 1964 General Election marks the onset of a dramatic electoral decline of the Conservative Party in Scotland and the strengthening of electoral support for the Scottish National Party. The latter rose from being a 'fringe' party to a serious contender for the majority of the Scottish vote after winning 30 per cent of the vote in October 1974. This General Election signified a considerable political advance for the Scottish National Party which won just 6 per cent less of the vote than the Labour Party and 5 per cent more than the Conservative Party. The Scottish National Party appeared to have drawn some support particularly from Conservative voters, evident in the drop from the February 1974 vote of 33 per cent for the Conservative Party to 25 per cent in October 1974. The electoral balance therefore shifted dramatically and the Scottish National Party appeared to have established itself as the 'second party' in Scotland. However, this trend was reversed in 1979 when the Scottish National Party vote fell to 17 per cent. In that General Election, the Labour party proportion of the vote increased to 42 per cent, and the Conservative Party made a small recovery, winning 31 per cent. The electoral decline of the Scottish National Party was reaffirmed in the 1983 General Election when it polled 11.8 per cent of the Scottish vote (Kellas 1984). This electoral evidence reveals three significant features of the Scottish political process and the parameters of the Scottish political agenda. First, it reveals the potential for a distinct Scottish national identity to become politicized and to transform electoral behaviour, a transformation which signals the emergence of a 'new' issue on the political agenda, that of national independence or 'home rule'. Second, it reveals a declining level of support for the Conservative Party, a trend that is counter to the situation in England (e.g. Kellas & Fotheringham 1976, p. 149). Third, it demonstrates the continuing electoral hegemony of the Labour Party in Scotland, a feature that is also quite distinctive when compared with England.

Concerning the first, the Act of Union 1707 preserved specific Scottish institutions which served not only to reproduce a distinct Scottish national identity but also as a reminder of what had been lost, thereby constituting a basis for a continuing political debate about the 'national question'. For nationalists, the formal existence of these institutions testified to the distinctiveness of Scotland as a national and political unit and gave credibility to the argument for the establishment of Scotland as a separate nation-state. There is therefore considerable irony in the fact that during the 19th century, when nationalist political movements transformed the political map of Europe, Scotland was distinguished by the absence of widespread nationalist agitation (Nairn 1981, pp. 126–32). The first significant evidence of nationalist agitation in the 19th century is the formation of the Scottish Home Rule Association in 1886 (Hanham 1969, p. 83). The issue of home rule entered the formal political arena when the Scottish Liberal Association supported the demand in 1888 (Brand 1978, p. 40–3).

Within the Labour movement, support for home rule oscillated throughout the 19th century and into the early years of the 20th century with an increasingly influential trend towards integration with England and a general downgrading of the home-rule issue. The nascent Labour Party was forced into a reconsideration of its position by the support for home rule in Ireland by the Liberal government, and this opened an important phase in the history of the Scottish Labour movement as a whole during the period 1914–31 (Keating & Bleiman 1979, p. 59):

> It saw the movement first veer in a nationalist direction as the leading element in the new Home Rule coalition but then settle decisively for a strategy of UK political advance and a permanent split with the forces of Scottish nationalism.

In brief, the short-term strength of the Scottish economy, brought about principally by wartime pressures on Scotland's traditional heavy industries, made home rule appear to be a viable proposition. To this end, home rule was supported by all factions of the Scotland Labour movement, intensifying after Labour's electoral breakthrough in 1922 which raised further the confidence of the Scottish Labour movement. But Labour's electoral support declined soon afterwards as the economic recession set in, and with the withering away of support for Scottish home rule, Labour's policy shifted from a broad home-rule base built on regional economic strength to a centralized policy of planning and public ownership.

The formation of the Scottish Nationalist Party in 1934 was not followed immediately by a significant electoral success: that did not come until the 1960s and 1970s. Explanations for this success vary, but what is significant about all those surveyed by Webb (1978, p. 102–38) is that they make some reference to the failure of British-based political parties to reverse economic decline in Scotland. Sections of the Scottish electorate responded to this failure by supporting the political party that offered a nationalist solution to economic decline, in a material context where the discovery of North Sea oil could be presented as offering an economic basis for such a political strategy. The history of political nationalism in Scotland since the late 19th century is therefore extremely uneven. For our purposes, two conclusions are relevant. First, whether or not the nationalist movement can be considered to be successful in any period is perhaps less significant than the fact that it existed and was capable of requiring the formal political parties subsequently to take up a position on the 'national question'. Indeed, during the 1960s and 1970s this question largely dominated Scottish politics. The fact that the 'national question' could be maintained as an item on the political agenda in Scotland clearly distinguished political relations from those in England. Second, increased support for political nationalism has been evident in periods of relative economic strength as well as in periods of absolute and relative decline. What is common to both is that in Scotland the case for a nationalist strategy is sustained by an argument about the economic viability of independence.

Concerning the declining electoral support for the Conservative Party and the continuing electoral hegemony of the Labour Party since

1945, our discussion will be briefer and interrelated. Fundamentally, both trends are related to a significant feature of the class structure in Scotland: compared with England, a much larger proportion of the Scottish population is (or was) located in the manual working class, with the corollary that the proportion in non-manual and petit-bourgeois location is relatively smaller (e.g. Money 1982, p. 51; Kendrick 1986, p. 255). The predominance of the manual working class is particularly evident in Strathclyde region which contains almost half of Scotland's population and half of its Members of Parliament (Money 1982, p. 51). It is also more prominent at constituency rather than national level (Kellas & Fotheringham 1976, p. 153). Consequently, and in so far as electoral behaviour is class-related and the Labour Party is seen as a working-class party, the Scottish Labour Party is advantaged in Scotland, and this goes a long way to explain its electoral hegemony (ibid., pp. 152–9). Thus the Conservative Party faces a more difficult task in appealing for working-class support. In the context of increasing electoral support for political nationalism, this has meant that

> Scottish Labour *votes* are the anvil on which the SNP and the Scottish Conservative party had in turn hammered each other half to death without in either case doing much damage to the anvil. (Money 1982, p. 54.)

There are additional factors which explain the electoral weakness and decline of the Conservative Party in the post-1945 period in Scotland. Its identification with the landed aristocracy in a social formation with a large industrial working class is a longstanding source of weakness. This identification is sustained by the fact that Conservative Party electoral success in Scotland is predominantly located in rural constituencies (Kellas & Fotheringham 1976, p. 156). More specifically, the longstanding identification of the Conservative Party with Loyalism, an identification that ensured a certain degree of Protestant working-class support, was broken consciously and openly after 1959 and was not replaced in such a way as to retain or create new working-class support (Money 1982, p. 59). Finally, the weakness of the 'middle class' and petit-bourgeoisie has minimized the social basis for the emergence of an aggressive, right-wing Toryism which has dominated Conservative Party politics since the mid-1970s in England. Until recently, the only well-known adherent of this trend in Scotland was Teddy Taylor, MP until 1979 for the urban constituency of Cathcart in Glasgow.

Conclusion

Racism constitutes a component part of political consciousness in Scotland and yet Scottish domestic politics have not been racialized in the post-1945 period following a migration from the Indian subcontinent. This means that the Scottish situation is anomalous in the wider European context. We locate the explanation for this primarily in certain features of the political process in Scotland; features that are, in turn, grounded in the distinctive trajectory of capitalist development in Scotland. The crisis of British capitalism since the mid-1960s has given rise to specific forms of political and ideological expression north and south of the border (cf. Nairn 1981).

Thus, the migration to Scotland from the Indian subcontinent in the post-1945 period occurred in a distinct political context in which there was a much lower potential for sections of the indigenous population to signify that the Asian presence was a political issue, and subsequently to racialize that presence. Clear lines of ideological fractionalization within the working class were already drawn and actively reproduced and, as the economic crisis has intensified, political nationalism has, for historical reasons, largely dominated the political agenda until the early 1980s. The absence of any significant fascist tradition in Scotland has meant the absence of one political force with a potential for political intervention aimed at mobilizing racism against this Asian population. When attempts were made by English-based fascist parties to intervene in Scotland in the 1970s, they had very limited political success and tended to merge with the fringe of militant Protestantism. In addition, the growing weakness of the Conservative Party in Scotland has ensured that its active rôle in the racialization of politics in England has not been matched in Scotland. In the absence of these right-wing forces with a particular interest in mobilizing racism, the political hegemony of the Labour Party (and movement) has not been challenged at this particular ideological level and has therefore been able to maintain a largely abstract anti-racist position. The left-wing elements within this hegemony have therefore had greater scope for intervention evident in the formation of the Scottish Immigrant Labour Council in the early 1970s, an organization which had the objective of ensuring class unity in a context where racist legislation and agitation in England was considered to constitute a threat in Scotland.

Events during the 1980s suggest that the situation is changing, that racism is becoming organized in a more systematic form in Scotland. First, fascist organizations such as the British National Party and the National Front have maintained a more sustained and active presence in Scotland, particularly in Glasgow, Edinburgh and Dundee (*Dundee*

Standard, 18 July 1980, July 1983, September 1983; *Glasgow Herald*, 4 July 1985). Fascist newspapers are regularly sold in city centres and outside football grounds, and the British National Party has organized political rallies in Glasgow in November 1984 and 1985 with its leader, John Tyndall, as the main speaker (*Glasgow Herald*, 2 December 1985). The propaganda of these organizations often attempts to link the much longer tradition of anti-Catholicism with a racism which focuses on the Asian presence. Thus, the British National Party leaflet advertising its meeting in November 1985 was headed 'Keep Britain White! Smash the IRA!' and Tyndall claimed during the meeting: 'We won't let Mrs. Thatcher sell out the Loyalists of Ulster the way she has sold out the white people of Handsworth and Tottenham' (*Glasgow Herald*, 2 December 1985).

Secondly, Asian and other minority groups are reporting increased levels of verbal, racist abuse (*Glasgow Herald*, 2 July 1984), and a recent survey has shown that racist attacks are a common experience among people of Asian origin in Glasgow (*Scotsman*, 5 February 1986; *Glasgow Herald*, 7 February 1986). Racist abuse and racist violence are also common in Glasgow schools (*Glasgow Herald*, 16 May 1985). This second development is probably not unconnected with the first because fascist leaflets have been handed out to pupils in schools and fascist slogans regularly appear both inside and outside schools.

Third, there is evidence of new forms of political organization within the Asian communities. In the past, 'traditional' religious and cultural organizations have been the main form of political organization but they have now been joined by more self-consciously political organizations which are seeking action to deal with racism and patterns of exclusion and calling for direct Asian participation in decision-making. Particularly important are the Scottish Asian Action Committee and the Minority Ethnic Teachers Association. Such organizations, although small, have been particularly effective in contributing to public debate and in forcing policy makers to begin to think about the implications of the presence of minority cultures and of racism and exclusionary practice. This influence has been particularly evident in the field of education, and is the fourth dimension of change in Scotland.

Official and common-sense discourse remains largely unchanged so far. Press reporting in Scotland on the 'riots' in England in 1985 queried whether such events were possible in Scotland and concluded that they were not because (*Sunday Post*, 15 September 1985):

> Mixed marriages between Scots and Asians are growing. Every night Indian restaurants throughout Scotland are packed with Scots enjoying a friendly night out. Edinburgh and Glasgow night-spots do

> roaring trade with mixed groups of young Scots, Asians, West Indians, and Africans raving it up.

The explanation for this continues to be identified as Scottish 'tolerance' and a small 'immigrant population', arguments which continue to be articulated within the Asian bourgeoisie and petit-bourgeoisie, representatives of which have claimed (*Glasgow Herald*, 16 May 1986):

> Any prosperity and success that we enjoy today is mainly attributable to the warm and welcoming character of the average Scot.

Yet the available evidence concerning the expression of racism, evidence which directly contradicts this discourse, is leading to emergent contradictions in official statements. For example, in response to the evidence of increasing racist abuse and violence in a Glasgow school, a newspaper reported the following official response (*Scotsman*, 5 February 1986):

> An education authority spokesman said that there was no racial problem within the school and that every time an incident had happened in the school's vicinity the staff had called the police. They were determined to put a stop to the attacks.

If the processes identified above continue, it will be increasingly difficult within official discourse to deny the reality of racism in Scotland.

A second racialization of domestic Scottish politics (the first occurred in response to the Irish migration) is therefore probably now under way, demonstrating that the early absence of a process of racialization in Scotland in response to the Asian migration of the 1950s and 1960s cannot be equated with the absence of racism in Scotland. What, then, has changed? First, electoral support for Scottish nationalism has been in decline since the late 1970s. If a British political frame of reference has become more widespread since then, there is greater political space for the identification of an internal 'cause' of material deprivation. Second, the economic crisis is even more severe in Scotland than in England and this may be testing the effectiveness of the traditional 'scapegoat' and may be initiating a search to find a new one. The Asian communities, commonly identified as having successfully occupied a petit-bourgeois position in the Scottish economy, may be sufficiently prominent to act as a trigger to the articulation of a long-established racism in Scotland.

References

Benewick, R. 1972. *The fascist movement in Britain*. London: Allen Lane.

Brand, J. 1978. *The national movement in Scotland*. London: Routledge & Kegan Paul.

Bruce, S. 1985. *No pope of Rome*. Edinburgh: Mainstream Publishing.
Budge, I. & D. Unwin 1966. *Scottish political behaviour*. London: Longman.

Campbell, A. B. 1979. *The Lanarkshire miners*. Edinburgh: John Donald.
Combe, G. 1825. *System of phrenology*. Edinburgh: John Anderson.
Cooney, J. 1982. *Scotland and the papacy*. Edinburgh: Paul Harris.
Cowan, H. I. 1961. *British emigration to British North America: the first hundred years*. Toronto: Toronto University Press.
Cowan, R. M. W. 1946. *The newspaper in Scotland*. Glasgow: George Outram.
Cross, C. 1961. *The fascists in Britain*. London: Barrie & Fockliff.

Dalton, M. 1983/4. The New Commonwealth and Pakistan population in Scotland, 1981. *Scottish Council for Racial Equality Annual Report (1983–4)*.
de Paor, L. 1970. *Divided Ulster*. Harmondsworth: Penguin.
Dickson, T. (ed.)1980. *Scottish capitalism*. London: Lawrence & Wishart.
Donaldson, G. 1966. *The Scotsman overseas*. London: Hale.

Fielding, N. 1981. *The National Front*. London: Routledge & Kegan Paul.
Flinn, M. (ed.) 1977. *Scottish population history*. Cambridge: Cambridge University Press.
Fryer, P. 1984. *Staying power: the history of black people in Britain*. London: Pluto Press.

Gallagher, T. 1983. The year Red Clydeside turned pink. *Weekend Scotsman*, 5 November.
Gallagher, T. 1984. The clansman's reign of terror. *Weekend Scotsman*, 17 November.
Gibb, A. D. 1930. *Scotland in eclipse*. London: Humphrey Toulmin.

Handley, J. n.d. *The Irish in Scotland*. Glasgow: John S. Burns.
Hanham, H. J. 1969. *Scottish Nationalism*. London: Faber & Faber.
Harvie, C. 1981. *No gods and precious few heroes*. London: Edward Arnold.

Jones, H. R. & M. Davenport 1972. The Pakistani community in Dundee. A study of its growth and demographic structure. *Scottish Geographical Magazine* **38**(2).

Kames, Lord 1774. *Sketches in the history of man*. Edinburgh.
Kearsley, G. W. & S. R. Srivastava 1974. The spatial evolution of Glasgow's Asian community. *Scottish Geographical Magazine* **90**(2).
Keating, M. & D. Bleiman 1979. *Labour and Scottish Nationalism*. London: Macmillan.
Kellas, J. 1984. *The Scottish political system*. Cambridge: Cambridge University Press.
Kellas, J. & P. Fotheringham 1976. The political behaviour of the working class. In *Social class in Scotland: past and present*, A. A. MacLaren (ed.). Edinburgh: John Donald.

Kendrick, S. 1986. Occupational change in modern Scotland. In *Scottish government yearbook 1986*, D. McCrone (ed.). Edinburgh: Edinburgh University Press.

Lenman, B. 1981. *Integration, enlightenment, and industrialisation: Scotland 1746–1832.* London: Edward Arnold.

Lothian Community Relations Council 1983. *Needs of the Chinese community in Lothian.* Edinburgh: Lothian Community Relations Council.

Miles, R. 1982. *Racism and migrant labour.* London: Routledge & Kegan Paul.

Miles, R. & A. Dunlop 1986. The racialisation of politics in Britain: why Scotland is different. *Patterns of Prejudice* **20**(1).

Miles, R. & L. Muirhead 1986. Racism in Scotland: a matter for further investigation? In *Scottish political yearbook 1986*, D. McCrone (ed.). Edinburgh.

Miller, W. L. 1981. *The end of British politics?* Oxford: Clarendon Press.

Money, W. J. 1982. Some causes and consequences of the failure of Scottish conservatism. In *Conservative politics in Western Europe*, Z. Layton-Henry (ed.). London: Macmillan.

Murray, B. 1984. *The Old Firm – sectarianism, sport and society in Scotland.* Edinburgh: John Donald.

Nairn, T. 1981. *The break-up of Britain.* London: Verso.

Parker, J. G. 1985. Scottish Enterprise in India, 1750–1914. In *The Scots abroad: labour, capital and enterprise 1750–1914*, R. A. Cage (ed.). London: Croom Helm.

Peach, C. 1968. *West Indian migration to Britain.* Oxford: Oxford University Press.

Reports on the Schemes of the Church of Scotland with the Legislative Acts Passed by the General Assembly, 29 May 1923. *Report of Committee to Consider Overtures from the Presbytery of Glasgow and from the Synod of Glasgow and Ayr on 'Irish Immigrants' and 'The Education (Scotland) Act 1918'.*

Rinn, J. A. 1980. Scots in bondage: forgotten contributors to colonial society. *History Today* **30**(July), 16–21.

Salter, J. 1873. *The Asiatic in England.* London: Seeley, Jackson & Halliday.

Scottish Office Report 1983. *Ethnic minorities in Scotland.* Edinburgh: Central Research Unit Paper.

Sheridan, R. B. 1977. The role of the Scots in the economy and society of the West Indies. *Annals of the New York Academy of Sciences* **292**, 94–106.

Sutherland, I. 1981. Doing the Orange walk. *New Society*, 9 July, 49–50.

Torrence, J. 1938. *Scotland's dilemma.* Edinburgh: Bellhaven Press.

Walker, M. 1977. *The National Front.* London: Fontana.

Webb, K. 1978. *The growth of nationalism in Scotland.* Harmondsworth: Penguin.

6 *The Irish in London: an exploration of ethnic boundary maintenance*

JUDY CHANCE

Whatever the academic quandary about the validity of racial classification of the species *Homo sapiens*, I would suggest that if as social geographers we are to understand our observations fully, we must be prepared to examine social interaction initially in the terms used by those involved in the interactions. Having achieved this participant understanding we must then, as academics, proceed to a more abstract analysis of the relationships between the perceptions and the actions observed. This is a point that has been convincingly expounded by Mitchell in his discussion of ethnic perception and behaviour in Zambia (Mitchell 1974).

At the same time, given the sensitivity of the issue of racism, it is important that it should be quite clear that our *usage* of what Mitchell refers to as 'common-sense constructions' and 'interpretations' does not imply our *acceptance* of them as our own position.

There have been several studies of the history of anti-Irish racism in Britain, exploring both prejudice and discrimination – for instance the work of L. P. Curtis (1968, 1971), Lebow (1976), Parsons (1983), and E. Curtis *et al.* (1984). These all stress the impact of colonial relations on British treatment of the Irish, and argue that the derogatory stereotype of the Irish as uncivilized and subhuman serves to justify Britain's exploitation of the resources and people of Ireland. For instance, Lebow explores Edmund Spenser's claim that the Irish were so savage that they could only be tamed by a policy of complete occupation of Ireland and forceful subjugation of the native population. Such proposals are backed up by Spenser's description of the Irish:

> Marry those be the most barbaric and loathy conditions of any people (I think) under Heaven. . . .. They do use all the beastly behaviour that may be, they oppress all men, they spoil as well the subject, as the

> enemy; they steal, they are cruel and bloody, full of revenge, and delighting in deadly execution, licentious, swearers and blasphemers, common ravishers of women, and murderers of children. (Spenser 1596; quoted in Lebow, p. 16)

There is, however, a dearth of material on the contemporary position, although within the Irish community, and especially among the second generation, there is growing discussion of the problems caused by anti-Irish racism. A number of Irish community groups are collecting statistical evidence to back their claims of current discrimination in employment and access to services. For instance, in January 1986 the Lambeth branch of the Irish in Britain Representation Group advertised the vacant position of a full-time outreach worker, part of whose work entailed the compilation of information on specific social, cultural and welfare needs of the Lambeth Irish community.

Being a first-generation Irish immigrant myself, I seek to look from the inside out, and to grasp the Irish experience of interaction with other groups, defined both by ethnic and by class traits. This focus allows a concentration on the ways in which the Irish interpret and negotiate situations. It must be made clear here that such attempted manipulation may be conscious or unconscious – I would suggest that in the great majority of situations it is unconscious, and indeed the phenomenon appears to be a universal human habit, not just another quirk of the devious Irish mind.[1]

Apart from its inside-out approach, this study is also unusual in that it focuses on what, with regard to the core British population, is a largely invisible minority. In spite of the prognathous renditions of the Victorian cartoonists (Curtis 1971) and their counterparts in the work of modern cartoonists such as Jak[2] the Irish are not easily distinguished by visible cues. Indeed the most reliable cues to Irish identity lie in the spoken word – not so much in the use of Irish itself, as the distinctive use of English and the verbal content of the speech. Thus one's attention is necessarily focused on those behavioural cues which in other more highly visible minority populations may well go unnoticed.

This physical 'invisibility' of the Irish is echoed by their invisibility in many official statistics. Thus although they form the largest single immigrant group in Britain, the need to find them by quite literally knocking on front doors has prevented any large-scale number-crunching analysis of their countrywide position, or indeed even of their distribution across the whole of any of the major cities with large Irish populations, for example London, Liverpool and Manchester. Once again their peculiar position helps to force attention onto the detailed local-community study, for example in Bronwen Walter's work

(Walter 1980, 1984, 1986). Attempts have been made to sketch a national picture (Jackson 1963, O'Connor 1974), but even Jackson's book, by far the more useful of the two, is more a general history of Irish immigration than a detailed spatial analysis.

It can be argued that this enforced concentration on small-scale community studies has an important contribution to make to the larger-scale studies of more visible immigrant groups in Britain, such as those from the West Indies and South Asia. At present most of the geographers working on such studies are white and are therefore visibly members of the outgroup from the point of view of the groups they are studying. This distinctiveness may well be affecting interactions. Indeed it is highly improbable that it will have no effect, especially when one considers the general level of racial tension and the influence of stereotypes on perception and behaviour (Hewstone & Jaspars 1982, Tajfel 1982, Apfelbaum 1979). This, however, is not the key problem: the real issue is that many geographers appear to be *unaware* of their own impact on the situations they are observing. This is not simply because relatively few geographers have read much social psychology or sociolinguistics but, more fundamentally, because their attention is drawn to the more obvious ethnic markers. Hence they cannot pay so much attention to the subtler and often more responsive behavioural cues.[3]

For instance, the most obvious cues to ethnicity include skin colour, hair type, facial features, dress and speech. The first three, in addition to other physical variables, have been used as the bases of racial typologies but, as has become clear with the increase in our knowledge of human physiology and genetics, these external differences mask an internal homogeneity across all racial divisions. Nonetheless, in terms of the common-sense construct, it is the relatively unimportant external adaptations to environmental conditions that are the badges of racial difference. In the study of interactions one must not ignore the impact of these perceptions in one's explanation of one's observations.

Dress and speech are also fairly obvious ethnic cues, but are much more amenable to modification than the physical variables discussed above. They are therefore potentially of much greater interest in that they may be used to reflect an individual's perceptions and attitudes. The real interest lies in the changes in speech and dress patterns. In the case of the Irish, dress is no longer a useful cue but for other groups it may be extremely important, most obviously perhaps in the case of women from the Indian subcontinent. Deborah Phillips touches on this issue in her discussion of the characteristics of the dispersed group of Asians in Leicester (Phillips 1981).[4]

Speech is of especial interest in that it can provide direct and indirect cues, in content and in paralinguistic cues – accent, tone, rhythm and pitch. Argyle gives a useful introduction to this area and also to the broader field of non-verbal communication, much of which appears to be shared across cultures (Argyle 1967).

Irish migration to Britain

The Irish have been migrating to Britain for over 1600 years, and form the largest single immigrant group in Britain today,[5] yet they are largely ignored by social geographers except in the provision of statistical comparisons in studies of Asian and Afro-Caribbean migrants, and as a rather dubious but not infrequently cited model of the West Indian future in Britain. The irony of their latter rôle is that nobody really understands in any detail the Irish position in Britain in the 1980s.

A review of the social-scientific literature on the Irish in Britain shows that there is a host of studies of the Irish in 19th-century cities, both contemporary accounts and modern analyses (e.g. Papworth 1982, Lees 1979), with very little work on the Irish in Britain in the 20th century.

A major problem for any modern study is the paucity of census data and the absence of migration data. The International Passenger Survey specifically exludes any coverage of movement between Britain and Ireland.[6] The very absence of such data illustrates the way in which the history of the British Isles, as well as the propinquity of the Irish, has led to their not being seen as foreign, while nonetheless remaining a distinct and not altogether acceptable group. Without entering into a heavy-handed sociological analysis of the Irish joke, even a sporadic perusal of the letters column of the *Irish Post*, a national weekly aimed at the Irish in Britain, will indicate the extent to which the iniquity of these jokes exercises the minds of its correspondents.[7] I do not want to imply that the Irish are subject to the same levels of popular prejudice and discrimination as other immigrant groups in Britain. My aim is to demonstrate how, in the absence of obvious physical cues, the Irish can be recognized, and how the cues used are manipulated in a range of social interactions to strengthen or to weaken ethnic boundaries.

What does need to be pointed out is the most important error in the stereotypical image of the Paddy in Britain. The generally accepted view of Irish immigrants forming one homogeneous group is false. Apart from the highly mobile population of Irish travelling people there have in the 20th century been four main groups of Irish immigrants (i.e. those from the Irish Republic). Of these, one is Protestant; the other three are Catholic. Between 1916 and 1939 50 per cent of Irish Protestants emigrated (Kennedy 1973), but I have been unable to find any data concerning their destinations and their subsequent history. The second group consists of the Irish professional class, many of whose members were educated in Britain and are thus well prepared for rapid assimilation.[8] The third group is the Irish middle class, a relatively new social group; and the fourth and largest category is that of the rural migrant, the great majority coming from the tiny farms of the West of Ireland. As recently as 1955, 24 per cent of all Irish farms were of less than 5 acres, most of these being unviable without some external source of income. In 1956 60 per cent of all Irish-born children could expect to emigrate, most going to Britain or North America, and most coming from the poorest areas of the West.[9]

Of the general material available on the Irish in Britain, the most valuable is Jackson's work, dating from the 1960s. Of particular interest is his concept of the Irish as colonists in Britain, exploiting British resources but with their life still focused on Ireland and on their ambition to return home (Jackson 1963, 1964). In other words, among the first-generation Irish there is an absence of identification with the British core population, one of Gordon's seven aspects of assimilation (Gordon 1964).

Jackson also highlights the importance of World War II in opening up new avenues of employment for the Irish (Jackson 1963). Before the war they were heavily concentrated in the following occupations: seasonal farmwork, unskilled labour, the Catholic priesthood and orders, domestic service and laundering. At the start of the war Irish movement to Britain was stopped, but it was soon realized that seasonal labour was vital if the harvests were not to be wasted, and subsequently Irish immigrants were encouraged to provide labour in weapon and ammunition factories, clerical jobs, medical services and other skilled and white-collar positions previously denied them. Jackson sees this development as the key to improved social mobility, although the part played by advancement in the Irish education system should not be ignored. Nor should it be assumed that all Irish migrants benefit equally from the improved chances of social mobility. Indeed O Brennían's discussion of the 1981 Labour Force Survey presents rather depressing statistics about Irish employment in London (O Brennían 1985).

Since Jackson's work in the 1960s the position has changed in at least three major ways. The Common Market (and the present economic depression) have effectively evened out a good deal of the gap between the living standards in Ireland and Britain; the escalation of violence in Ulster since the late 1960s has created new tensions for the Irish in Britain; and the 1968 alteration of the American immigration regulations sharply reduced the flow of Irish people to America, leaving emigration to Britain as the main alternative to staying at home. Among many Irish families for all the children to stay at home would be a major departure from established norms.

The Irish in Kilburn

Judging from the position in Kilburn, the early and mid-1970s saw a net flow of young people back to Ireland, seeking to take advantage of the short-lived 'economic miracle' there. But now the flow is once again into Britain, and consists largely of young single people who cannot find work in Ireland.[10] The other major component of the Irish population in Kilburn is the elderly. Indeed, in 1985 there were approximately 1.5 funerals for every 1 baptism in Kilburn's parish church, but it is not absolutely clear how much this reflects a drift among the young away from the church rather than simply a decline in population numbers.

In Kilburn the Catholic church is run by the Oblate Fathers, an Irish missionary order, who are also very active in explicitly Irish activities – for instance they have a ceilidh in the church hall every Wednesday, they were involved in the founding of the Irish Centre in Camden, they run a hostel for Irish boys and they organize a range of social services for their parishioners. This close association of Irishness and Catholicism has been seen as an area of potential confusion and at least one attempt has been made, in a Liverpool community, to separate out the effects of the two. I am a little wary of such a divorce, believing that Irish Catholicism is quite distinct from English Catholicism, and that therefore, at least within a group such as one finds in Kilburn, it should be seen as an important element of Irish identity. Certainly in the course of interviews several respondents seemed to see the terms 'Irish' and 'Catholic' as interchangeable when describing their friends and neighbours. This was more common among the older residents.

To date, the fieldwork in Kilburn has involved a questionnaire survey, meetings with the local priests, teachers and workers in Irish organizations, and general observation of the area. The impression on first arrival is that one might well be in Ireland – Irish papers, both local and national, are on sale, the main Irish banks have branches in the High Road, Irish names appear above shop fronts and Irish accents abound. In fact, there is also a large Asian population, especially to the north and west, while to the east lie the middle-class white areas of Hampstead and St. John's Wood, and to the south there is a concentration of West Indians.

This multiplicity of ethnic and social groups was an important bonus in the choice of Kilburn as a study area since the aim is to look at the variation occurring in interactions with a wide range of groups.

Unfortunately, the analysis of these encounters must be in terms of the *Irish* perception of the situation rather than embracing each group's interpretations – to achieve that would require an immense amount of fieldwork.

This study of the Irish in Kilburn is intended to explore Irish attitudes to the other ethnic and social groups that they identify and encounter, and to elucidate the means by which they maintain their own ethnic boundaries and identity. The Kilburn area is divided between two local authorities, Brent and Camden. Of these, Brent is particularly well supplied with a range of Irish associations, and the number of such institutions is increasing in Camden. The Irish therefore have a selection of institutions that provide information, support, and a focus for specifically Irish interests and activities. While many of these have a relatively long history – for example a branch of the Gaelic Athletics Association, and some of the county associations – many others are new (less than ten years old), and of these several have received local authority or GLC grants, or both.

The Irish are not the only group to have their own ethnic and cultural associations. There is also a range of institutions serving the various Asian and Afro-Caribbean populations. My own observations suggest that these services are not ethnically exclusive – in particular the Irish and West Indian music groups attract a mixed group of participants. This mixing, however, is not without its own problems. In particular there have been complaints by young blacks of harassment when they seek entrance to one dance hall which is a major venue for Irish bands. It is also noticeable that in the immediate aftermath of events such as the Harrods bombing individuals are much more prone to express their hostility to the groups seen as responsible. It would be of considerable interest to chart the way in which both local and national news coverage affect local interactions in Kilburn.

The methodology currently being used in the study of a relatively invisible immigrant population lays much stress on the behavioural cues to identification, and on the ways in which these cues can be and indeed are modified in response to varying perceptions of an interaction. In the study of more immediately visible populations their very visibility distracts the observer's attention from the less obvious behavioural cues which may illustrate attempts to modify the boundary imposed by the observer's own perceptions of physical differences. This modification may of course tend to change the position in two ways, either increasing or decreasing the hardness of the boundary between the groups concerned. If the study of a relatively invisible group can clarify and categorize these behavioural cues the approach may then be carried over and applied to the study of more immediately visible ethnic groups.

Methodology

What follows is a brief outline of the projected stages of the work, to be supplemented by a more detailed discussion of various aspects.

The first of three phases is the questionnaire survey. Questionnaires are administered in individual interviews with each respondent, touching on general socio-economic and demographic facts, the strength of their feeling of Irishness, and their previous history in Britain. Having started the pilot studies with a series of direct questions it soon became clear that it would be more profitable to conduct a more general conversation along those lines, filling in the questionnaire immediately after the interview. The interviewees are also presented with two maps, one being a street map of Kilburn and its surroundings on which they are asked to delimit their own neighbourhood and also to mark any areas that they associate with any particular groups. In many cases the concept of neighbourhood proves problematic. After trying a number of alternative approaches the best solution turned out to be a dot-to-dot system, with the dots provided by mapping the location of various places visited at least once a month and within reasonable walking distance of the home. Interestingly, the respondents who seemed best able to define a neighbourhood as initially requested were those who owned dogs, in which cases the boundaries often reflected the routes taken during walks.

The second map presented is of the whole of Greater London, with the names of well-known districts or centres marked. Once again respondents are asked to label those that they identify with any particular groups, and also any areas with which they are familiar.

Finally there is a Bogardus-style survey of their attitudes to a range of social and ethnic groups, the list reflecting the groupings that were generated in the pilot questionnaires, and the rankings based on their acceptability as neighbours.

The second phase will involve a smaller sample, drawn from the respondents in the first phase. (These have been selected from the electoral register with the help of a dictionary of Irish surnames, and also from the records of local schools.) It entails an activity analysis, based on Boal's work in Belfast (Boal 1969), looking at shopping patterns, routes to work, leisure activities, premarital addresses, and graffiti. It is hoped to visit each member of this sub-sample again, to ask them to map the route they would use to visit a number of points east and west off the High Road, which forms a north–south line bisecting the area. It would be interesting to see whether they follow the High Road, turning off it for each separate call (as one might expect from the local mapping exercise in the first phase), or whether they use side streets parallel to the

High Road in a more direct route. While this may appear a peripheral exercise, a foray into the realm of mental mapping, it is important that before the final phase of the fieldwork the form of each respondent's cognitive image of the area is clearly understood. As a geographer it is tempting to assume that everyone *shares* a spatial view of their environment, whereas in fact many of the respondents in this study appear to see Kilburn as a series of discrete points between which journeys are made.

The third and final phase will involve a subgroup of the participants in the previous phase. It will be the most time-consuming, and each participant must be prepared to accept my spending a fair amount of time observing their daily interactions and taping their conversations, both at work and in their leisure time. This is the phase in which I hope to observe their behaviour in interactions with people from a wide range of backgrounds. My main interest is in linguistic variation but attention will also be paid to other cues which may help to provide the context in which the language should be interpreted – for instance the physical setting, the social setting, and the individual's attitudes as revealed in the initial interview and survey.

The first two phases will allow the construction of a model of each respondent's image of his or her own social environment, plus their own level of identification as Irish, and to observe how much correspondence exists between boundary perceptions and behaviour patterns. This is to act as the background for the last phase which will be an attempt to see whether the Irish really do manipulate their linguistic behaviour in any consistent way in order to maintain or to relax ethnic boundaries, according to what is seen as appropriate in an encounter.

Obviously I am drawing heavily on Barth's work on ethnic boundaries, details of which are lucidly laid out both in his own work (Barth 1969) and in Jackson's discussion of its relevance in social geography (Jackson 1980). Given the contraints of space and the growing familiarity of his ideas among geographers, this essay will concentrate on the linguistic work on which I have drawn, which is perhaps less well known than it deserves to be in disciplines other than sociolinguistics and social psychology.

The starting point for much sociolinguistics lies in the work of Labov, especially his monumental studies of Black English Vernacular (BEV) as used by urban blacks in the United States of America (Labov 1972a, b). Apart from detailed phonetic analysis he devotes some of his time to documenting the association between gang membership and the use of BEV, which is most marked among adolescents. Labov argues that the use of BEV is a means of group identification, and this concept is further developed in a tantalizingly brief report of a study he made in Martha's Vineyard, a small island off the Massachussetts coast, settled since the

18th century, and formerly the base for a thriving whaling fleet (Labov 1972a).

The islanders' survival now depends on their letting their houses to summer tourists, with the result that a formerly isolated community is now exposed to large annual tourist invasions. Its speech patterns had been well documented prior to Labov's work, being of interest because of their retention of many pre-1800 traits. His study has shown that among the more conservative islanders these traits are now being increasingly accentuated. In addition, among young people returning to the island from college on the mainland there is an even greater degree of this hypercorrection, whereas those who intend to leave the island discard the island speech patterns in favour of more standard mainland ones.

Labov's explanation centres on the use of distinctive language to express the difference between the two populations. The initial contrast is seized upon and exaggerated to show social identity in the face of external pressures. Such an explanation has rather obvious echoes in the work of social psychologists such as Tajfel, who has written widely on the dynamics of group membership and in/out group relations in terms of the individual's search for a positive self-identity (Tajfel 1970, 1982).

There have also been sociolinguistic studies in Britain, notably by Trudgill who has written on class variation in speech as well as on local dialects (Trudgill 1974, 1983), and by Milroy, who has looked at the relationship between network multiplexity and levels of vernacular usage in three areas of Belfast (Milroy 1981); but in neither case has there been as clear a focus on language modification in relation to inter-group interaction as in Labov's study of Martha's Vineyard.

However, if one moves from sociolinguistics to social psychology in Britain one comes across the work of Giles which is very firmly focused on exactly that issue. He has developed an approach which he has labelled 'accommodation theory', and which is a logical extension of a well-established set of observations of various channels of non-verbal communication – for instance the mirroring of physical actions, which reflects consensus between the (albeit subconscious) copier and the initiator (Giles 1977; 1979a,b).

Giles draws on Tajfel's work on inter-group behaviour which suggests that the members of any one group will seek to differentiate themselves from all non-members by means of distinctions which the ingroup members see as positive, and which can therefore contribute to each individual's positive self-identity (Tajfel 1970). Giles's development has been to apply this idea to language in order to develop a model that can predict how the language of one group will differ during interaction with another group. In particular he has been interested in the part played by psycholinguistic distinctiveness in what he sees as the revival of ethnicity, evinced in the 1970s in many parts of the world (Giles 1979a).

He seeks to explain the use of speech markers with reference to five variables. The first is the relative status of the groups involved – in general he predicts that the members of the lower status group will show some convergence towards the higher-status speech markers except when these are specifically reserved for the higher-status group, for example the Brahmin dialect in India. This is termed 'upward convergence'; but one may also come across 'downward convergence' – for example if a high-status individual seeks a service from someone of lower status. In any interaction one may of course find change or stability in the speech of more than one individual, the net result of which may increase, decrease or maintain the distinctiveness of speech patterns. For instance, upward convergence may be met with upward divergence (i.e. further accentuation of the higher-status markers), thus maintaining the linguistic expression of status differentials. Divergence is seen as a means of asserting ingroup distinctiveness, as in Martha's Vineyard, and as such is of obvious relevance in any attempt to understand group identification and isolation.

The second variable discussed by Giles is compounded of the strength of the individual's desire for a positive social identity, and the importance attached to group membership. The effect will depend on whether group membership is seen as conferring a positive social identity. He has argued that if members of a subgroup believe themselves to be of low status, and regard their poor position simply as a reflection of their low intrinsic worth, one might expect to find upward convergence; but if group membership is seen as *positive* one is more likely to find maintenance or divergence.

The third of Giles's concepts is that of the recognition of cognitive alternatives to the group's social situation. This relates to the group's self-perceived identity: if a low-status group recognizes that its status is not an inescapable adjunct of its low inherent value, and if other higher-status groups had followed different courses then the low-status group could by now have been equal or superior to them, this knowledge can enhance their self-esteem. Perhaps the classic example of this has been the development of Black consciousness in America, which effected a redefinition of values. Such action was proposed by Tajfel as one means whereby members of a low-status group could reconcile group membership with their need for a positive self-identity (Tajfel 1970).

This particular variable is of some interest in Kilburn as the area is on the whole rather run-down, with a bad record of burglary and street violence. It is possible, judging from conversations with respondents, to see a divergence between the younger and older sections of the Irish population. Some of the older people spontaneously mentioned the local decline, blaming the new arrivals from Ireland. For instance, one man claimed that the only young lads coming over now were the ones who wouldn't have made it in Ireland, let alone over here. The younger people, on the other hand, blame rising housing costs and unemployment for their plight.

Giles has termed his fourth variable 'perceived ethnolinguistic vitality'. It refers to a group's image of itself and of its own ability to survive as a distinct ethnic and linguistic entity. Important components include demographic characteristics, group status in a range of spheres, and the group's level of representation in formal and informal institutions. In general, a group that sees itself as having high ethnolinguistic vitality will be more likely to behave as a distinctive unit in intergroup situations than a group that sees its vitality as low. However, one might posit a U-shaped curve: if a particular language is seen as in danger of being lost this may act to generate an active core group seeking to recover group vitality and to safeguard the language. It is possible that a study of the revival of Welsh nationalism, with its strong emphasis on bilingualism in all institutions, could help to elucidate this question.

This study of the Irish in Kilburn focuses on their use of English rather than Irish. It might seem that an accent is less distinctive than a different language, but Giles claims that:

> intralingual ethnic speech markers in an outgroup tongue can be as important symbolizers of ingroup identity as a distinctive language itself. (Giles 1979a, 280–1.)

Thus the fact that the Irish in Kilburn use English as their first language need not be seen as a problem. Indeed it allows for readier inter-group communication and therefore facilitates a study of the maintenance of ethnic boundaries by interaction. It is perhaps of special interest to note that there is a growing pressure for the provision of Irish language classes in Kilburn, as in many other Irish communities in Britain, catering for both adults and children. In Kilburn some of the proposals make specific reference to the provision of mother-tongue teaching for Asian groups in Brent. The *Kilburn Times* of 21 August 1981 contained in its regular 'Irish Scene' column an article demanding Irish lessons in local schools on the grounds that these would facilitate the integration of children if they should go back to live in Ireland – this in spite of the fact that fewer than 7 per cent of the Irish population use Irish as a vernacular language, and that the requirement of a qualification in Irish for all Civil Service posts in Ireland was abolished in 1973. It has become clear, from both conversations in Kilburn and discussions at a number of Irish-interest conferences I have attended over the past 18 months, that much of the demand for Irish-language teaching is coming from second-generation immigrants and that the issue is not simply one of cultural identity – it also carries strong political connotations. The Development of Irish Youth Association (DIY), set up in London in 1985, captures the prevailing second-generation attitude accurately in its magazine *Irish Dissent*:

> We are a different breed from that of our parents. Of course we know and enjoy Ireland, but London is our home, our city. We can't try to recreate a lost Ireland in the midst of '80s London. Neither are we prepared to put up with the shabby treatment once meted out to our parents . . . we believe that the only way the Irish community in London will ever be treated on a par with the home nation is through asserting its Irish identity. (O'Brien 1985, pp. 1, 4.)

Giles' final variable is derived from Banton's work on perceived ethnic boundaries (Banton 1978). Banton argues for a continuum from hard to soft boundaries; Giles proposes a modified version with two orthogonal axes (Fig. 6.1). The shading represents relative levels of accentuation of ethnic markers to be expected from individuals in each quadrant. Giles argues that total hardness is the product of the two dimensions he has isolated:

$$\text{total hardness} = \text{linguistic hardness} \times \text{non-linguistic hardness}$$

and that these factors can be played off against each other to maintain a steady total hardness in a dynamic situation. The degree of ethnic-

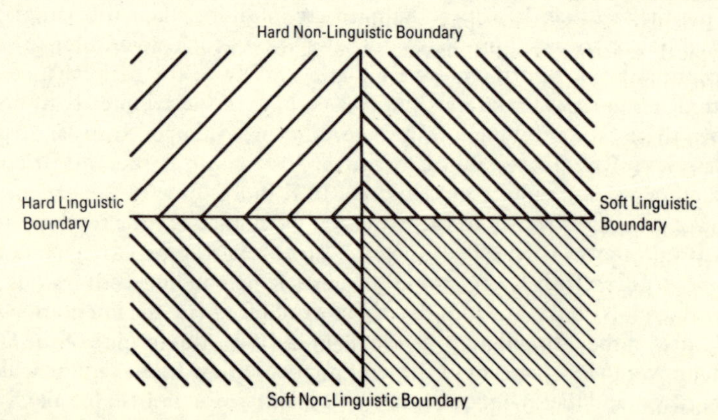

Figure 6.1 Hard and soft linguistic and non-linguistic boundaries (after Giles 1979a).

marker accentuation will be greatest where boundaries are *seen* to be softest – this perceptual emphasis allows members of one group to fall into more than one quadrant and can thus account for the variations observed within groups. Giles argues that the flexibility of language makes it especially suitable for manipulation in the maintenance of identity.

It is clear that Giles' work is well suited to a study of inter-group interactions in the field rather than just in the confines of strict experimental conditions. As Ley has shown (Ley 1974), the use of psychological models in social geography can give rise to valuable new approaches in our understanding of an essentially subjective phenomenon – a group's understanding of and relation with its physical and social world. Furthermore, Giles' work is particularly suited to the study of a group like the Irish, who, in the absence of any immediately striking non-linguistic markers such as skin colour or dress, might be expected to place relatively great emphasis upon the use of speech markers to maintain group identity.

These speech markers need not, of course, be limited simply to specific features of a dialect or accent, but may well include the expression of opinions on certain key issues. In the case of the Irish in Britain perhaps the most obvious such area is the situation in Northern Ireland, especially in the context of an Irish/British interaction. However, in their encounters with other immigrant groups it may well be that other issues, particularly the expression of racist sentiment, would be more important. While interviewing respondents in the questionnaire, and especially when talking about their preferred neighbours, it became clear that many of them held racist attitudes, although the women in particular seemed unhappy about expressing their views openly. A typical sentence would begin: 'I wouldn't like you to think I'm prejudiced but . . .' The commonest stereotype for West Indians was violent, and I was offered many reports of local incidents to support this claim, and also of counter-measures of like kind taken by some Irish men. The most frequent complaint about South Asians was that their food smelt – a point discussed by Deborah Phillips in her study of housing policy in Tower Hamlets (see Ch. 9, below).

It should be mentioned that the study in Kilburn is using only first-generation Irish immigrants, on the assumption that their ethnic speech markers and their Irish identity will be stronger than those of individuals born in Britain. In the course of fieldwork, however, as mentioned above, it has become clear that much of the interest in the Irish cultural associations, and to a lesser extent in other institutions, comes from second-generation immigrants. In the context of Giles' work, this can be seen as the stressing of non-linguistic cues in the maintenance of ethnic boundaries, given the fact that for this generation, brought up in London, the traditional speech markers have become much less obvious, especially with regard to accent rather than content.

Kilburn has had Irish connections for over a hundred years, with the earliest arrivals being labourers on the roads and railways, and on local suburban developments. There is a growing interest in local history, with the Irish Centre in Camden acting as a meeting place for discussions and some exhibitions. Other cultural expressions of Irish ethnicity include ceilidhs, concerts by Irish bands, feises (the Irish equivalent of eisteddfodau), Gaelic Athletics Association events, and boiled bacon and cabbage suppers (the Irish equivalent of brown rice meals), which are generally organized as fund-raising events by the very active county associations.

The desire to return home eventually is common among immigrants from many countries in Britain, but actual return has been relatively common among Irish migrants, especially since the 19th century, when many young men came over as seasonal labourers, returning home each autumn with their wages, which were used to pay the home farm rent as many of the tenant farmers were really working at subsistence level. The very cheap cost of travel across the Irish Sea, particularly with the advent of the steam packets in 1818, facilitated this annual migration. For instance, in 1824 the cost was 5d on deck, 10d in steerage – this was the result of fierce competition between a number of businesses running boats, mainly in and out of Liverpool and Glasgow (Jackson 1963).

Fares are still within the reach of many, and there are a lot of organized visits quite apart from private ones. Two very important groups involved in this traffic are the county associations, which also hold regular social events, raise funds for local causes and provide help for people on first arrival or in cases of special need, and the Gaelic Athletics Association. The latter was founded in Ireland in 1884 by Michael Cusack, who was also associated with the Irish Republican Brotherhood (Finnegan 1983), and it embraced the whole concept of an independent Gaelic state rather than just the revival of traditional sports such as hurley. Over time the sporting element became more important, not least because of the achievement of independence for the 26 counties of the Republic, but the Republican element lingers on in some branches.

Given my interest in boundaries it is worth recording that the traditional St. Patrick's Day parade, which had been allowed to lapse in Kilburn, was revived in 1982; but in 1986 it once again lapsed. However, in February 1986 there was a march to commemorate the anniversary of Bloody Sunday, the day in 1972 when fourteen Civil Rights demonstrators were killed in the North of Ireland. Such parades provide an opportunity for a highly visual demarcation of a group's territory, as is graphically demonstrated by, for instance, the Apprentice Boys' marches in the North of Ireland, the Chinese New Year celebrations in several American and European cities, and the old custom of beating the bounds in several British parishes.

In conclusion, quite apart from the value of looking at the Irish, who have been largely ignored by social geographers, in order to gather basic data on their spatial behaviour, their very invisibility forces one to look for cues which in more visible groups may be overlooked. One possible problem in the concentration on the more visible cues is that interpersonal attitudes are markedly influenced by minor behavioural variables of which we are often unaware. There is a very large body of psychological literature on non-verbal communication which can be used to facilitate interpretation of encounters in the field.[11] It seems that many social geographers are unaware of the impressions that they both give and receive via channels such as eye contact, paralinguistics and proxemics.[12] Obviously, for those people working largely from census data or other statistical sources this is of less importance; but for those using interviews and participant observation it is important to have at least some appreciation of the psychological elements of an interaction if one is to interpret it with accuracy.

Notes

1 For further discussion of the ways in which we all seek to influence situations see two collections of papers which contain useful introductory papers on the various approaches of social psychologists (Argyle 1973) and social linguists (Giglioli 1972).

2 One particular cartoon by Jak, published in the *Standard*, London's evening paper, on 29 October 1982, has been the cause of complaints. Initially the Commission for Racial Equality asked the Attorney-General to prosecute the *Standard* but it later withdrew this request on the grounds that the prosecution would not have been successful. For a discussion of anti-Irish racism in contemporary cartoons see Curtis (1984).

3 The point at issue here is the one made by Milgram in his stimulus-overload theory – there is a limit to the amount of incoming information or stimulation that we can process at any one time. Thus, when presented with high levels of stimulation we must select what we process and thus what we respond to. The more obvious the stimulus, the more likely we are to notice it (Milgram 1970).

4 This point is also discussed by Suttles in his study of the Addams area of Chicago (Suttles 1968). He argues that males display higher levels of cultural distinctiveness than females in their dress, general personal display and also dance. However, he does not explore situational variation in individual ethnic expression.

5 The 1983 Labour Force Survey (OPCS 1984) gives a figure of 556 000 for the Irish-born population in the UK. This is the largest figure for immigrants from a single country. The second largest group is the 417 000 Indian-born population.
6 There are some Irish estimates of the levels of migration from Ireland to Britain. In December 1985 the Irish Minister for Labour, Ruairi Quinn, said that the net outflow over the past few years had been 8000 a year. However, the *Irish Post* of 15 March 1986 quotes from a report published by the Economic and Social Research Institute in Dublin which estimated net annual emigration to Britain of 15 000 for the years 1986–90 inclusive, assuming that economic conditions are relatively good in Ireland. The actual figure may well be much higher.
7 For an excellent albeit brief analysis of the Irish joke, see Leach (1979).
8 For a discussion of class divisions in Ireland, and their differences from the British pattern, see Peillon (1982).
9 It is this fourth group that provides the popular image of the Irish immigrant and that is also most likely to be found in areas with a high concentration of Irish immigrants, such as Kilburn. This study is therefore not representative of all Irish migrants; but, given sufficient time and patience, using resources such as records of former school pupils and university graduates, and also using the 'snowball' interview technique, it should be possible to trace and study samples of the other more elusive groups and thus to arrive at a more balanced picture of the Irish in Britain.
10 A survey in 1985 of young Irish people aged 15–24, using a range of hostels and advisory agencies, showed that 72 per cent had been unemployed for at least some of the time between leaving school and emigrating. Seventy per cent had less than £100 on arrival and 34 per cent had less than £30. Fifty-six per cent had found no employment since arriving in Britain, and 27 per cent had at some stage slept rough in London (Connor 1985).
11 See Argyle (1967, 1973) and Deaux & Wrightsman (1984) for useful introductory accounts and further references.
12 'Proxemics' is a useful blanket term for body posture, direction and spacing relative to the posture and position of other people.

References

Apfelbaum, E. 1979. Relations of domination and movements for liberation. In *The social psychology of intergroup relations*, W. Austen & S. Worchel (eds). Monterey, California: Brooks/Cole.

Argyle, M. 1967. *The psychology of interpersonal behaviour*. Harmondsworth: Penguin.

Argyle, M. (ed.) 1973. *Social encounters*. Harmondsworth: Penguin.

Banton, M. 1978. *Race relations*. London: Tavistock.

Barth, F. 1969. Introduction. In *Ethnic groups and boundaries*, F. Barth (ed.). Boston: Little, Brown.

Boal, F. 1969. Territoriality on the Shankhill/Falls divide, Belfast. *Irish Geography* **6**, 30–50.

Connor, T. 1985. *Irish youth in London (1985)*. London: London Irish Centre.

Curtis, E. *et al.* 1984. *Nothing but the same old story*. London: Information on Ireland.

Curtis, L. 1968. *Anglo-Saxons and Celts*. Connecticut: University of Bridgport.

Curtis, L. 1971. *Apes and angels*. Newton Abbot: David & Charles.

Deaux, K. & L. Wrightsman 1984. *Social psychology in the '80s*. Monterey, California: Brooks/Cole.

Finnegan, R. 1983. *Ireland: the challenge of conflict and change*. Boulder, Colorado: Westview.

Giglioli, P. (ed.) 1972. *Language and social context*. Harmondsworth: Penguin.

Giles, H. (ed.) 1977. *Language, ethnicity and intergroup relations*. London: Academic Press.

Giles, H. 1979a. Ethnicity markers in speech. In *Social markers in speech*, H. Giles & K. Scherer (eds). London: Cambridge University Press.

Giles, H. 1979b. Sociolinguistics and social psychology: an introductory essay. In *Language and social psychology*, H. Giles & R. St. Clair (eds). Oxford: Blackwell.

Gordon, M. 1964. *Assimilation in American life*. New York: Oxford University Press.

Hewstone, M. & J. Jaspars 1982. Explanations for racial discrimination. *European Journal of Social Psychology* **12**, 1–16.

Jackson, J. 1963. *The Irish in Britain*. London: Routledge & Kegan Paul.

Jackson, J. 1964. *The Irish in London: aspects of change*. Centre for Urban Studies Report no. 3, University College London.

Jackson, P. 1980. *Ethnic groups and boundaries*. Research paper no. 26, School of Geography, Oxford University.

Kennedy, R. 1973. *The Irish*. Berkeley: University of California Press.

Labov, W. 1972a. *Sociolinguistic patterns*. Philadelphia: Pennsylvania University Press.

Labov, W. 1972b. *Language in the inner city*. Philadelphia: Pennsylvania University Press.

Leach, E. 1979. The official Irish jokesters. *New Society* **50**, 7–9.

Lebow, R. 1976. *White Britain, black Ireland*. Philadelphia: Institute for Study of Human Issues.

Lees, L. 1979. *Exiles from Erin*. Manchester: Manchester University Press.

Ley, D. 1974. *The black inner city as frontier outpost*. Monograph no. 7, Association of American Geographers, Washington D.C.

Milgram, S. 1970. The experience of living in cities. *Science* **167**, 1461–8.
Milroy, L. 1981. *Language and social networks*. Oxford: Blackwell.
Mitchell, J. 1974. Perceptions of ethnicity. In *Urban ethnicity*, A. Cohen (ed.). London: Tavistock.

O Brennían, S. 1985. The invisible Irish. *Irish Studies in Britain* **7**, 9–10.
O'Brien, D. 1985. Editorial and DIY cultural management. *Irish Dissent* **1**, 1–4.
O'Connor, K. 1974. *The Irish in Britain*. Dublin: Torc Books.
OPCS 1984. *OPCS Monitor*, Ref. L.F.S. 84/2 & P.P.I. 84/5. London: OPCS.

Papworth, J. 1982. *The Irish in Liverpool 1835–1871*. Unpublished Ph.D thesis, Liverpool University.
Parsons, B. 1983. Mad Micks and Englishmen. *Irish Studies in Britain* **5**, 11–15.
Peillon, M. 1982. *Contemporary Irish society*. Dublin: Gill & MacMillan.
Phillips, D. 1981. Social and spatial segregation of Asians in Leicester. In *Social interaction and ethnic segregation*, P. Jackson & S. J. Smith (eds). London: Academic Press.

Suttles, G. 1968. *The social order of the slum*. Chicago: University of Chicago Press.

Tajfel, H. 1970. Experiments in intergroup discrimination. *Scientific American* **223**(5), 96–102.
Tajfel, H. 1982. The social psychology of intergroup relations. In *Annual Review of Psychology* **33**, M. Rosenzweig & L. Porter (eds). Palo Alto, California: Annual Reviews.
Trudgill, P. 1974. *Sociolinguistics: an introduction*. Harmondsworth: Penguin.
Trudgill, P. 1983. *On dialect: social and geographical perspectives*. Oxford: Blackwell.

Walter, B. 1980. Time–space patterns of second-wave Irish immigrants into British towns. *Transactions of the Institute of British Geographers*, New Series **5**, 297–317.
Walter, B. 1984. Tradition and ethnic interaction: second-wave Irish settlement in Luton and Bolton. In *Geography and ethnic pluralism*, C. G. Clarke, D. Ley & C. Peach (eds). London: Allen & Unwin.
Walter, B. 1986. Ethnicity and Irish residential distribution. *Transactions of the Institute of British Geographers* **11**(2), 131–46.

7 *Spatial variability in attitudes towards 'race' in the UK*

VAUGHAN ROBINSON

Within geography, the 1970s and 1980s have witnessed a forceful reaction against what most practitioners now regard as the arid and excessively abstract strain that permeated the discipline during the quantitative revolution. This reaction has taken the form of a search not only for social relevance (see Johnston 1979) and a consequent rebirth of applied geography (see Briggs 1981, Sant 1982), but also for new techniques and new ways of approaching traditional or well-established topics.

The latter trend has manifested itself in two contrasting forms. Some geographers have turned away from the generalization of spatial science and looked instead to micro-level explanations of how people actually behave in the real or perceived world. They have been concerned with how people make sense of the world around them and how they use their perception and past experiences to shape their use of space and territory. But above all in their search for explanation, they have focused upon small groups of people or even individuals. In contrast, other geographers have looked to the macro-scale for their new approach and methods. They have focused upon the social and political structure of society and argued that the unequal distribution of scarce resources between different sections of the population creates a pattern of social relations which is the underlying explanation for the behaviour of any given group within society.

Within the literature on the geography of ethnic relations, it is clearly this latter trend that has gained the greater number of adherents. Practitioners have looked beyond their maps and indices of spatial mixing, and have moved from the description of pattern to the explanation of macro-level forces and processes. This has inevitably weakened the bond between geographers in this field and those working in other areas of geographical endeavour; but, in compensation, it has strengthened existing alliances between social geographers and their counterparts in sociology and political science. Some may say that, not a moment too soon, geographers of ethnic relations have abandoned a course that looked set to drive them onto the rocks of spatial fetishism: the belief that problems could be explained spatially, in isolation from the social and structural context within which they existed; the belief that process could, with certainty, be inferred from spatial form. As Hamnett (1978, p. 257) put it in relation to the study of deprivation:

> A concentration on area-based explanations of deprivation is likely to obscure the fundamentally structural rather than spatial or pathological origins of deprivation. Attention is likely to be diverted away from the existence of a socially structured opportunity set which entails that given the existence of poor jobs, poor housing, poor schools, and the like, some people are going to be filling those jobs, living in those houses, and attending those schools wherever they may be. Area effects may intensify or compound such deprivations but they should not blind us to their origins.

However, the accelerated development of ethnic geography in the UK, from being a descriptive backwater within the discipline in the early 1970s to its current position as part of a revived social geography which is in the vanguard of the discipline's development and growth, has not been without problems. Three particular issues have relevance here. First, whereas geography as a whole remained in its descriptive 'capes and bays' era for some 2200 years (Abler, Adams & Gould 1971) until all the rows of Berry's geographical data matrix were full and until all the major continents were located and mapped, ethnic geography in the UK was swept on from the descriptive phase after less than 30 years. As a consequence, there are still simple issues and problems that have never been addressed even descriptively, and these may be of some significance. The process of 'ghetto morphogenesis' (Deskins 1981), for instance, has never really been accurately or methodically described for any British city in the way that Deskins and others have studied it in leading American cities. In Britain we must rely upon a handful of undergraduate dissertations. Similarly, the actual process of ethnic residential succession has never really been described in detail for British streets and neighbourhoods except for the work of Woods (1977) and Robinson (1981). A glance across the Atlantic reveals a rich and varied literature on this topic encompassing description initially, but latterly explanation too.

Secondly, the accelerated development of ethnic geography from simple description to generalized explanation has meant that workers have effectively omitted the intervening intellectual stage of categorization, classification and comparison. Again this means that some research questions have never been asked, let alone answered. Despite the fact that Britain's black population is totally urban in its spatial distribution, only two authors have even described the characteristics of the country's black urban system (Jones 1978, Robinson 1986), and only one typology exists that classifies ethnic settlements by the social and economic characteristics of the town within which they are found. Surely such work is essential if we are to know how our individual case studies interrelate?

Thirdly, humanistic explanations based upon individual perception and experience, and macro-scale analyses relying upon Marxian modes of interpretation, frequently share the same fundamental weakness from a geographer's point of view. They are often blind to the importance of locality: they both regard it merely as the passive and superficial context within which more important dynamics are at work. In doing so they ignore the unique history and local culture that make locality and place such active elements within any explanatory framework, and they also negate the distinctive skills that geographers have always offered by virtue of their traditional and unique concern with spatial differentiation and with the importance of place.

Some, of course, would argue that such a catholic definition of geographical skills is likely to achieve only the relegation of geographers to mere spatial technicians, providing annotated spatial addenda to the work of sociologists or political scientists, and being forced to rely upon these for theory and explanation that make sense of our own work. Perhaps this has always been the case. Others would argue that such a disciplinary division of labour is both unreal and unnecessary and that the existence or operation of geography as a separate and coherent discipline is a matter of little consequence. These arguments should not be allowed to cloud the issue: if present trends continue, locality and place will continue to have a diminishing part to play in the study of ethnic relations, and this will weaken our ability to understand real-world situations.

Prejudice

The literature on prejudice, particularly that relating specifically to racial prejudice, can also be compartmentalized to reveal the contrasting emphases of different research traditions. Cox's arguments presented in *Caste, class and race* (1948), reflect one such tradition. He argues that

> race prejudice is a social attitude propagated among the public by an exploiting class for the purpose of stigmatizing some group as inferior so that the exploitation of either the group itself or its resources may both be justified. (Cox 1948, p. 393.)

In the case of prejudice against blacks in Western Europe, Cox suggests that this resulted from the colonial era when whites attempted to justify and perpetuate colonial expansion and exploitation by stereotyping blacks as inferior, in need of protection and the 'white man's burden'. In his efforts to account for prejudice, he is thus calling into play not only economic relations but also the course of international history. Clearly, such macro-level explanations are many steps removed from individual white residents who find themselves living in a multi-racial neighbourhood of Birmingham. Nevertheless, the prejudices of such individuals probably cannot be explained without reference to such apparently distant factors.

Another important body of literature takes the opposing view and looks for explanations of prejudice within the individual rather than within his or her social and historical context. Adorno's (1964) classic work on the authoritarian personality exemplifies this trend well. In it, Adorno and others argue that prejudice is more frequently found within individuals who have a certain personality type, this often being linked to the way in which they were brought up by their parents. At a similar level of analysis, other writers claim that in order to 'explain' prejudice it is necessary to understand how individuals develop love-prejudices towards their own life-styles and groups, and how they also use cues and associations to exclude those who are thought to be a threat.

At a superficial level, these different approaches to prejudice could easily be regarded as contradictory. Allport (1954), however, in his seminal work *The nature of prejudice*, showed that these apparent alternatives need not be mutually exclusive, but are instead elements within a nested hierarchy of interacting causes ranging from broad historical forces at one point to individual perception at another. Allport showed how the broader macro-level forces are a necessary prerequisite, but how these are mediated through a series of progressively more localized, more particular and even more personal circumstances down to the individual level. His argument reinforces the point made above about the importance of local circumstances ('locational explanations'); and indicates that, contrary to trends within ethnic geography, such factors are still regarded as significant within the study of prejudice. Since all those factors listed under the heading of 'locational explanations' are likely to vary spatially, it suggests that there is a valid rôle here for geographers, who are well equipped to occupy this middle ground.

However, although local circumstances may well mediate personal and macro-level forces, it would be a mistake to accord to them greater authority than they in fact possess. So although this chapter focuses predominantly upon such locational 'explanations' of prejudice, it should not be seen as an attempt to distract attention from the endemic and institutional racism which exists within British society. This is taken as given, as is the significance of psychological factors. What this chapter seeks to do, then, is to consider at a different level of analysis whether local circumstances encourage racism to become even more acute in certain parts of the country than it is in others. It is therefore concerned with the description of a spatial pattern which has not previously been researched in any detail, and it also presents an analysis that illustrates the value of meso-scale research alongside the increasingly popular macro- and micro-level studies.

Caveat emptor

A final, and perhaps central, issue that needs to be aired before the presentation of any analysis, is whether the study of people's expressed attitudes actually tells us anything about their likely behaviour.

Many social psychologists have assumed that there is a direct link between attitudes and behaviour, and that the former can therefore be studied as a surrogate for the latter. Vernon (1938) provides an example of such thinking. He wrote:

> Words are actions in miniature. Hence by the use of questions and answers we can obtain information about a vast number of actions in a short space of time, the actual observation and measurement of which would be impracticable.

Milner (1973) sees the persistence of the social psychologist's view that attitudes and behaviour are directly linked more as a product of pragmatism and convenience than of any deeply held belief that has received rigorous empirical support. On the contrary, in fact, the majority of empirical research into the value of attitudes as a predictor of behaviour tends to show a rather poor fit between the two sets of variables. Moreover, this lack of fit was demonstrated as early as 1934 in LaPiere's pioneering work involving a Chinese couple receiving hotel accommodation and service in restaurants that later professed not to serve 'orientals'. Despite this early lead, LaPiere's message has still not been fully taken up by researchers, many of whom studiously ignore the issue entirely.

However, LaPiere's work and that of other researchers reviewed by Wicker (1969) does not invalidate the importance of studying attitudes. There are, of course, very good reasons why attitudes are not directly translated into behaviour. Wicker suggests a whole series of intervening factors which modify or constrain how individuals decide to react or are able to react to a given attitude. He divides these into 'personal factors' and 'situational factors'. The former include the intervention of other competing or overriding attitudes; the presence of competing motives; the verbal, intellectual and social ability of an individual to articulate accurately his or her views and then to translate these effectively into desired behaviour; and a person's levels of activity or drive. 'Situational factors' include the actual or considered presence of certain people; normative prescriptions of proper behaviour; the availability of alternative behaviours; the expected or actual consequences of a given act; and the intervention of unforeseen extraneous events. In each case, these factors put a filter or barrier between expressed attitudes and the eventual reaction in a particular situation. Fishbein (1967) has provided one of the most cogent summaries of how these factors interrelate and therefore of how attitudes influence behaviour. He suggests that individuals hold views on how they feel they *ought* to react to particular circumstances but that these personal views are, in practice, nested within perceptions about how they feel society *expects* them to react. In turn, these normative beliefs are influenced by the individual's motivation to comply, beliefs about the likely consequences of behaviour, and evaluation of the consequences.

What one can conclude from this is that attitudes are important and worthwhile objects of study in their own right, that they do provide an input to those factors which determine behaviour, but that it would be very naïve indeed to expect a direct and mechanistic relationship between attitude and behaviour. Milner (1975, p. 92) captured the essence of the real relationship between these two phenomena when he wrote: 'to say that "it depends on the person and the situation" is not an equivocation or a lame excuse, but the best possible account of what happens in reality'. He went on to argue that it is the very indirectness of the relationship that makes it essential to research both attitudes *and* behaviour, rather than simply regarding one as a surrogate for the other. He illustrated this point as follows:

> To take an extreme example, a South African black *might* have a very positive attitude towards whites and wish to integrate and socialize with them; in reality, *apartheid* proscribes any such behaviour. Ascertaining his attitudes would not, in itself, tell us very much about his everyday behaviour and the legal limitations on it; similarly, in deducing his attitudes simply from his observable actions, we would be misled. (Milner 1975; p. 91.)

Clearly then, to study behaviour without considering attitudes, or vice versa, is to study only part of a larger whole. This chapter seeks to provide a preliminary analysis of attitudes at a relatively coarse geographical scale, both for itself and as a means of providing a complementary context within which one can place the extensive literature that considers overt spatial behaviour. The analysis of variations in attitudes towards 'race' and minorities gives us a clear view of the local climate of opinion within which black people have to pursue their lives; and it gives us an insight into the attitudes, and willingness to express these attitudes, of their potential colleagues, neighbours, friends, and providers of goods and services.

Historical trends in attitudes towards 'race'

It would require a book to discuss how attitudes towards 'race', immigration and minorities have developed and changed within the UK. Fortunately, such a book exists in the form of Fryer's (1984) *Staying power* and it is therefore unnecessary to repeat this material here. What *is* necessary though is to place the 1984 Attitude Survey on which this chapter is based into some kind of limited time-frame and, in particular, to indicate whether attitudes expressed in the mid-1980s are the culmination of a development process which began at the end of the 1950s with large-scale black immigration to this country.

Surprisingly, such a task is by no means as simple as it might at first appear. Data are not especially plentiful and are frequently incomplete, inconsistent, and not comparable either year by year or survey by survey. However, the Gallup organization has collected and retained a sizeable body of information on changes in attitudes towards black people since 1958, and much of this information derives from similar or identical questions asked at different points in time. Even so, it is not possible to guarantee complete comparability or consistency, and the conclusions presented here must thus be regarded as no more than indicative.

Table 7.1 summarizes some aspects of these data and reveals a number of points of significance. First, expressed attitudes towards the immigration of black people to Britain changed markedly in the 1960s but have remained essentially stable since. Only a tiny minority of people now feel that Britain should allow free entry to 'coloured people from the Commonwealth'. Whether it was changing attitudes that stimulated restrictive legislation or the reverse is, however, impossible to say. Secondly, attitudes about whether the immigration of coloured people has been of benefit to the UK, or whether it has generated serious social problems appear to have remained remarkably consistent since 1965. Fewer than one in five respondents feel that Britain benefited from coloured immigration and around half the sample feel that coloured people raise serious social problems. Thirdly, there is broad pessimism about future ethnic relations in the UK: throughout the period 1959–81,

Table 7.1 Changing attitudes to black people in Britain, 1959–82.

Percentage agreeing with statement or question	1959[a]	1961	1964	1965	1966	1967	1968	1971	1972	1973	1975	1976	1977	1978	1979	1981	1982
(1) There should be completely free entry to UK?		21						5	7			4		7			
(2) UK benefited from coloured immigration?				16		9	16		20			19		20			
(3) Coloured people raise serious social problems?				55		55	69	52	65	61	53	63	59	59	58	56	45
(4) Feelings between black and white are improving?	16		24	18	19	13	6		17			12	16	14		15	
(5) Is there a colour problem in your area? Yes.			12	11	7	10			7			11				5	
(6) I strongly dislike coloured people as:																	
neighbours			17													10	
friends			12													5	
schoolfellows/children			11													4	
fellow workers			13													4	
employer			28													13	
son-in-law			44													33	
daughter-in-law			44													33	
(7) Conflict between black and white is a very important cause of increases in crime and violence.										16		14				32	

Source: All data provided by Gallup Ltd.

[a] In some cases questions were asked of more than one sample during a year. The table always relates to the last sample in any given year.

only 15–20 per cent of respondents considered that 'feelings' between black and white were improving. Contrary to this, the 'colour problem' seems to be regarded by most people as a national issue which affects other parts of the country and not their own. Fourthly, the results of two surveys in 1964 and 1981 suggest that, at a personal level, individuals may be becoming more tolerant to inter-ethnic contact in a variety of social contexts. Alternatively, these data could simply indicate the growing unacceptability of expressing prejudice overtly, even in the impersonal circumstances of an interview. Lastly, there is some evidence that the black population is increasingly being perceived as associated with crime and violence.

All of these data are open to conflicting interpretations, but what they seem to suggest is that attitudes towards blacks have not appreciably softened over the past quarter century. White Britons still regard black people as a problem, as a source of conflict, and as a group to be avoided, if possible.

Social attitudes, data, and the structure of the analysis

That the study of variations in social attitudes between different parts of the country is an under-researched field of enquiry in the UK is indicated clearly by the fact that whereas the two most recent social atlases of Britain (*The facts of everyday life* by Tony Osman, 1985, and *The state of the nation* by Fothergill & Vincent 1985) contain between them some 76 maps of how various social or economic phenomena vary over space, neither contains a single map of how social attitudes vary over the same space.

Several reasons can be suggested for this obvious gap in our knowledge: social attitudes are notoriously difficult to measure or scale – 'there is a tendency to see observable everyday behaviour as somehow more "real", and therefore more important, than the invisible world of attitudes' (Milner 1973, p. 93) – and until recently reliable national data on an array of attitudinal issues have simply not been available. In the case of attitudes towards 'race', these general difficulties are further compounded by the fact that in many circumstances it is neither morally nor legally acceptable for respondents to admit to prejudicial or discriminatory attitudes, and that early attempts to analyse prejudice in this country by Abrams (1969a,b) produced a barrage of acrimonious criticism of methodology, terminology and presentation (Rowan 1969a, Lawrence 1969, Rowan 1969b, Lawrence 1974).

As a consequence, systematic work has since been sporadic, and in many cases of only local significance (e.g. Bagley 1970, Robinson 1985), and the main thrust of research into prejudice is now focused upon the individual. Such work correlates the demographic, economic, or political characteristics of respondents with their expressed opinions on, for example, allocating council houses to black people, or further restricting black immigration (e.g. Schaefer 1973a). Although this research direction has undoubtedly highlighted persistent statistical regularities in those groups that are more prone to express prejudice, it has tended to draw attention only to the individual and personality attributes, to the exclusion of how such attitudes vary between different, objectively defined, sociocultural contexts. Moreover, concentration on the characteristics of the individual – 'the individualistic fallacy' as Rasmussen (1973) terms it – also masks the importance of conformity to perceived local social and cultural norms which themselves are the product of the unique history, culture, and character of each region.

This chapter seeks to shift the balance somewhat by considering both the demographic correlates of prejudice at an individual level and the spatial variation in its occurrence at the ecological level. The analysis is divided into four parts:

(1) a description of how expressed attitudes towards black people vary between different parts of Britain;
(2) an analysis at an individual level of how expressed attitudes towards black people vary between different population groups;
(3) an analysis of whether the sociocultural characteristics of a locality influence the expressed attitudes of those individuals who live there;
(4) a consideration of whether hostility towards black people results from the same causes in all localities.

Each of these four parts relies upon the same data set and it is therefore necessary to provide a brief description of this before proceeding to a discussion of findings. The data were collected by Social and Community Planning Research (SCPR) as part of their British Social Attitudes series which began in 1984. The researchers aimed to survey a representative sample of people over 18 years of age from all parts of Britain.

The sampling was multi-stage, beginning with the selection of 114 parliamentary constituencies (out of a total of 552) by probability methods within strata defined by the Centre for Environmental Studies' Planning Applications Research Group (PRAG) cluster analysis. A single polling district was selected within each constituency, and 23 addresses were then drawn from the electoral register by systematic sampling for each of the 114 polling districts. The households living at these addresses were contacted. If the elector who had been selected from the register was sti… random… addresses… of the sa… response ra…

Each resp… variety of so… Respondents… self-completio… the analysis re… excluded, thereb… the actual sample… missing values and…

SCPR kindly ma… 376 variables were e… from existing data. Th… scale. The P-scale soug… against racial discrimin… such laws) with how th… …ical circumstances, i.e. having… …their boss and having a black person… …marriage. The aim here was to produce a… …of indirectly expressed prejudice which combines… …different facets of the overall concept. The responses to the three questions were recoded on to a common scale and then combined in an additive way to produce a Likert-type scale with four categories ranging from 'very prejudiced' to 'no expressed prejudice'. The F-scale followed a very similar format but was designed to measure conservatism and authoritarianism. It is, however, by no means as rigorous or complete as Adorno's original F-scale, which spawned the 'Authoritarian Personality'.

Spatial variations in attitudes towards 'race'

The first task was to ascertain whether attitudes towards 'race' did indeed vary spatially. Previous research is contradictory. The Institute of Race Relations 1966–7 survey of British 'race relations' generated an attitudinal data set which has since been analysed by a number of authors who have produced conflicting conclusions. Abrams, (1969b, p. 624), the team leader for this part of the survey, concluded that 'in parts of the country where there is little or no first-hand knowledge of coloured people as neighbours and fellow-workers, there is the same minority expressing hostility and the same majority of either tolerant or tolerant-

e line, arguing that
dividual personality traits
gional factor.
ce of research based on the post-
kes, reached a not dissimilar conclu-
r back as 1964, opinions on immigration
ed throughout various contextual circum-
ter very little in which region respondents reside'
75). And he concluded that 'at a time when the thesis
l and political homogeneity is under increasing attack,
minder that in some attitudes Englishmen are still very much
here is little evidence here to indicate that the West Midlands and
ndon deserve their reputations of being racist and cosmopolitan
respectively' (Studlar 1977, p. 179). Indeed he suggested that perceived regional differences may well have their roots in how the media report 'race' and the extent to which local politicians respond to the issue.

In direct contrast to this 'homogenist school' is Schaefer's work. Schaefer (1975) also used the 1966–7 IRR data set but found that there were significant regional variations in expressed prejudice such that on a scale from 0 to 2 the South-West scored 0.94 and the South-East scored 0.59. He isolated five regions as being 'high prejudice areas', these being the South-West, North-West, North Midlands, Scotland and Greater London. Schaefer accounted for these deviations in terms of social change, arguing, rather unconvincingly, that the two most prejudiced areas were the ones which had undergone, respectively, the least and most social change. Schaefer also considered whether attitudes varied at an *inter*-urban level using the IRR five-borough data. He concluded again that there were significant differences, with Bradford being the least prejudiced and Wolverhampton the most. No rigorous explanation for this pattern was presented but it was noted that the most prejudiced towns were also those with the highest percentage of their population of New Commonwealth and Pakistani origin and the greatest degree of competition for jobs and housing.

Given this background, two different measures of attitudes were used; first P-scale; and secondly responses to a direct question asking respondents whether they considered themselves 'very prejudiced', 'a little prejudiced' or 'not at all prejudiced'. It is interesting to note that despite the directness of this last question 4.3 per cent of respondents admitted to being 'very prejudiced', while 32.8 per cent described themselves as 'a little prejudiced'.

These two measures were then cross-tabulated by the 11 economic regions of England, Wales and Scotland to produce the results shown in Tables 7.2 and 7.3. These reveal that there *are* statistically significant

Table 7.2 Self-rated prejudice.

Region	Very prejudiced (%)	Prejudiced (%)	Not prejudiced (%)
North	3	26	71
North-West	4	30	65
Yorkshire	5	39	56
West Midlands	6	35	59
East Midlands	8	29	63
East Anglia	3	36	61
South-West	6	33	61
South-East	4	38	58
Greater London	4	37	59
Wales	4	23	73
Scotland	1	23	76
mean	4	33	63

$\chi^2 = 37$, significance = 0.01, $n = 1664$.

Table 7.3 Indirectly expressed prejudice (P-scale).

Region	Prejudiced[a] (%)	Tolerant[b] (%)
North	20	80
North-West	22	78
Yorkshire	21	79
West Midlands	28	72
East Midlands	26	74
East Anglia	11	88
South-West	19	79
South-East	16	84
Greater London	19	81
Wales	14	86
Scotland	19	81
mean	20	80

$\chi^2 = 18$, significance = 0.06, $n = 1675$.

[a] Categories 3 and 4 of the P-scale.

[b] Categories 5 and 6 of the P-scale.

differences in the degree of expressed prejudice found in the regions of Britain, especially in self-rated prejudice. Figures 7.1 and 7.2 look only at the deviation from the national mean of the percentage prejudiced in each region. They demonstrate that certain regions, such as the West Midlands and Yorkshire, have above-average percentages of people who are willing to admit prejudice either in response to direct or indirect questions. Individuals in the East Midlands and the North-West are also more likely to express prejudice but only when asked in an indirect way. Conversely, both Scotland and Wales demonstrate below-average expressed prejudice on both counts, while the South-West, South-East, Greater London and East Anglia record below-average expressed prejudice on indirect questions but above-average willingness to admit prejudice when asked directly.

Similar areal variation is also apparent at the finer spatial scale of the constituency, although such variation only proved to be statistically significant for P-scale (0.0001), not for self-rated prejudice (0.63). Examples of variability in P-scale include Wolverhampton NE and Meriden where approximately 18 per cent of respondents fell into the most prejudiced class (against a sample average of 4.7 per cent); and, at the opposite extreme, Daventry and Carmarthen where around 90 per cent of respondents showed no indirectly expressed prejudice at all (again compared to a sample average of 48 per cent).

P-scale and self-rated prejudice were not, however, the only data collected on attitudes towards ethnic minorities, and a brief consideration of other variables allows a more detailed picture to be produced, even if it is not possible to discuss this in full. Table 7.4 provides a summary of this information and differentiates between regions where individuals were positive or negative on various issues, and whether the presence of these views strongly or weakly deviated from the sample average. The table reveals a number of points. First, it shows that the Celtic fringe of Britain is, on the whole, positive towards ethnic minorities, no doubt due to its physical isolation from major areas of black or Asian settlement, and perhaps to the fact that the Welsh and Scots regard themselves as minority groups. This seems to be particularly the case with the Welsh who favour special teaching of parental culture and mother tongue as a reflection of their own struggle to retain a Welsh identity. Secondly, both the West and East Midlands appear to be negative towards ethnic minorities as indeed does the isolated and depressed Northern region. Thirdly, East Anglia is an area of weakly held attitudes, which might reflect its lack of experience of multi-racial living. And fourthly, individuals in the North-West, South-West and Greater London evince an above-average willingness to adjust to the presence of different cultural and ethnic groups within their populations.

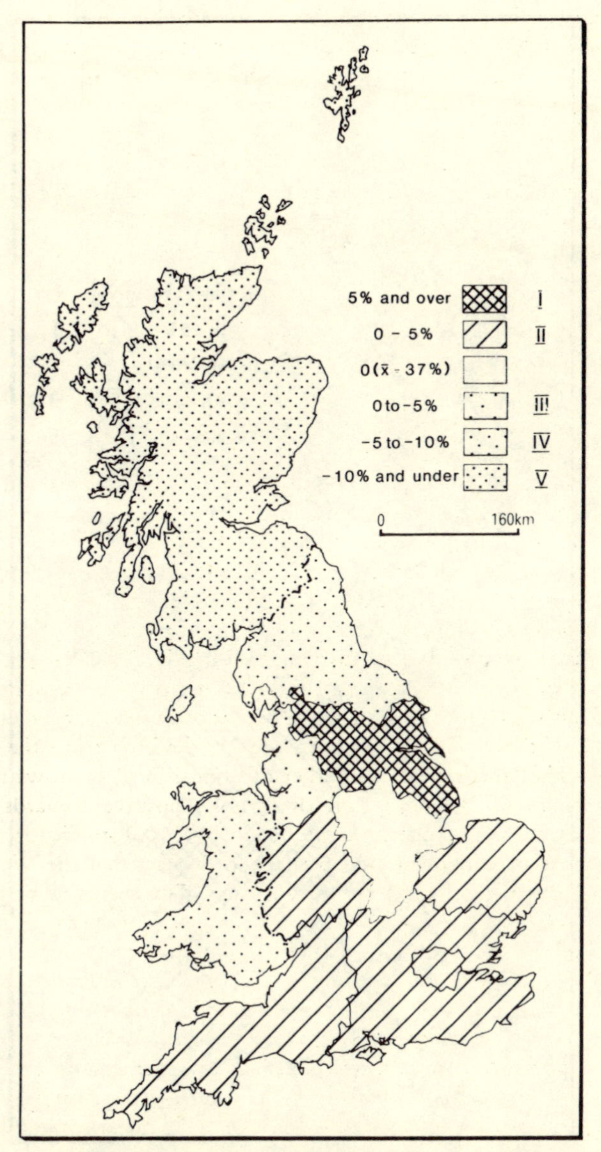

Figure 7.1 Self-rated prejudice in Britain, 1984.

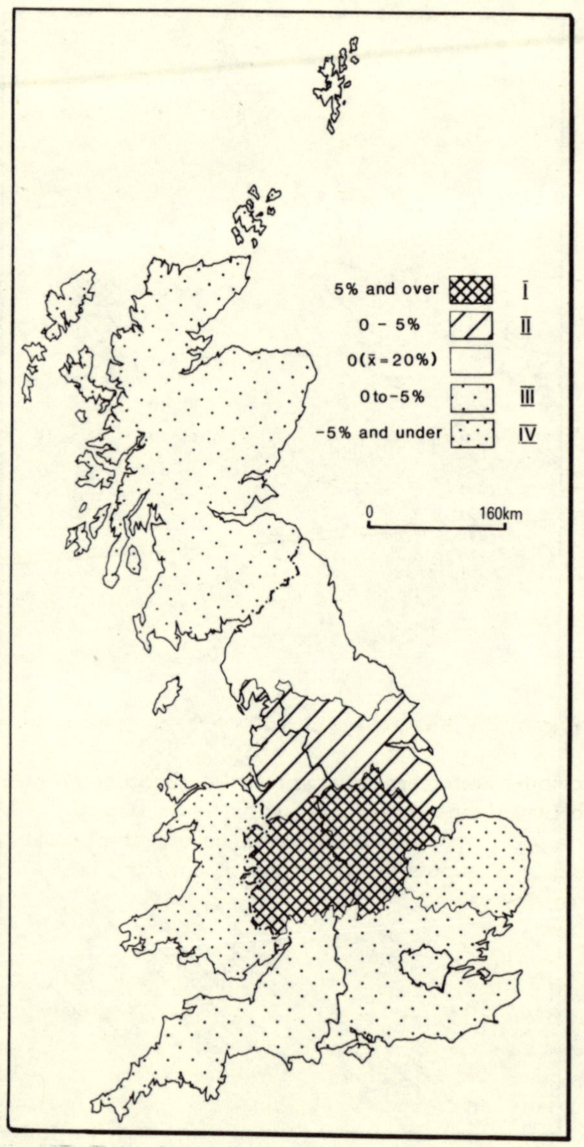

Figure 7.2 Indirectly expressed prejudice in Britain, 1984 (P-scale).

Table 7.4 Summary of regional variations in attitudes towards ethnic minorities.

Region	Variables (see below)										
	A	B	C	D	E	F	G	H	I	J	K
North		■	●	○	■	■	□	□	■	○	○
North-West	●	■	○	□	●	○		○	○	○	□
Yorkshire	■		○	□	■	○	□	○	○	■	■
West Midlands	■	■	□	●		□	●	□	■	■	■
East Midlands	■	●	□	○	■	□	□	■	□	□	□
East Anglia			□	○		□	○	○	○	□	○
South-West	■		■	□	●	○	●	○	●	□	●
South-East		■	○	○	■	■	○	○	■	□	■
Greater London		●	■	○	●	○	●	■	●	●	●
Wales	●	●	○	●	●	□	■	□		●	○
Scotland	●	●	○	●	●	●	□	●	○	○	□

Symbols: ■, strongly negative; ●, strongly positive; □, weakly negative; ○, weakly positive; blank cells represent ambivalence.

Variables: A, self-related prejudice; B, optimism/pessimism about future race relations; C, prejudice against West Indians; D, prejudice against Asians; E, government help to ethnic minorities; F, further immigration; G, provision of English-language classes for minorities; H, allow children to wear traditional dress at school; I, provide teaching on child's parental religion; J, provide teaching of child's parental mother tongue; K, provide teaching on child's parental culture.

Demographic variations in attitudes towards 'race'

As previous research has shown, prejudice varies not just over space but also between different groups within the population. Although a detailed analysis of this issue would be out of place in the present context, it is impossible to exclude it entirely. Table 7.5 therefore summarizes this stage of the analysis, again using the two measures of prejudice.

On the whole, the results confirm the findings of previous research in the field (Abrams 1969a,b; Bagley 1970; Schaefer 1974; CRC 1976; Marsh 1976), although several variables are tested here that have not previously been used. The most likely to express prejudice were the elderly, the unskilled, the unemployed, those with little formal education, Tory voters, and those predisposed to conformity. In common with other research, trade union membership was found not to influence levels of expressed prejudice significantly.

Table 7.5 Demographic and social correlates of prejudice.

	Indirectly expressed prejudice		Self-rated prejudice	
	χ^2	Significance	χ^2	Significance
employment status	59	0.000	36	0.007
social class	26	0.01	29	0.000
self-rated social class	25	0.11	26	0.01
social mobility	2	0.88	9	0.07
housing tenure	31	0.27	29	0.05
age	103	0.000	22	0.01
sex	1	0.69	6	0.06
trade union membership	8	0.06	2	0.80
political party	57	0.002	75	0.000
church attendance	32	0.22	19	0.39
F-scale	100	0.000	17	0.27
self-rated prejudice	575	0.000		
religion	39	0.21	44	0.003
age finished education	87	0.000	19	0.08

Certain findings, however, either contradicted or updated previous research. The data revealed no difference in prejudice between the sexes, whereas women have previously been found to be more tolerant. There was no significant relationship between expressed prejudice and intergenerational social mobility, in contrast to Abrams' (1969) data where upward social mobility was associated with increased prejudice. Again there was no significant variation in expressed prejudice between different housing-tenure groups, although Abrams found those in private rented or council properties to be more prejudiced than owner-occupiers. And lastly, and perhaps most disturbingly, the uncomplicated linear relationship between age and expressed prejudice commented upon by Bagley (1970) has been replaced by a more complex pattern in which the very youngest respondents (aged 18–20) are actually more prejudiced than either the 21–30 year olds or the 31–40 year olds. This is powerful confirmation of a trend noted by Marsh (1976) and Cochrane and Billig (1984) towards a resurgence of prejudice amongst the very young. This expressed prejudice may well be symbolic, but it does suggest that the future no longer holds the steady progression towards tolerance envisaged by early authors.

Other variables included in this analysis for the first time produced mixed results. Self-ascribed social class produced an interesting pattern in which those thinking of themselves as 'upper middle class' were the most prejudiced, followed by the 'poor', the 'middle class', the 'working class' and, lastly, the 'upper working class', who evinced the least prejudice. Differences in expressed prejudice between religions and between different frequencies of attendance at worship proved to be insignificant, although the former did influence the respondents' willingness to admit prejudice in response to direct questions.

Locality as an influence upon attitudes

The third stage of the analysis aims to marry together the previous two phases in an attempt to find those variables that have the greatest power to 'explain' regional variations in prejudice. The demographic and social characteristics discussed above are carried forward into the third phase but they are converted from data relating to individuals into data relating to spatial units. In addition, other data have been introduced to quantify both the structural and attitudinal context within which individuals must operate day by day.

Allport (1954) suggested a number of *structural* circumstances that encourage the development of prejudice. These were (a) relatively rapid in-migration by a visible minority group who then adopt a segregated settlement pattern; (b) the presence of rapid social change, which encourages individuals to feel that they have lost control of their destinies and that the future is uncertain and threatening; and (c) the presence of competition between ethnic groups for scarce resources such as housing or employment, which might provide the potential for conflict to develop.

There are also those variables that tap some aspect of the local climate of opinion and thereby indicate *attitudinal* norms which will underpin social relations and to which individuals will be encouraged to conform. The importance of this attitudinal context has already been demonstrated in a number of pieces of research relating to both political attitudes in the UK and racial attitudes in the US. Rasmussen (1973) for example has shown the importance of local circumstances in determining political allegiance, while in America Pettigrew's (1959) work in the 1950s showed how an individual's attitudes could change if he or she were moved to a different locality with contrasting attitudinal norms. More particularly, he demonstrated how American Southerners are more prejudiced than Northerners, not because of any demographic characteristics but because of their need to conform to the locally imbedded racist culture which has developed because of historical circumstances unique to the region.

British research on the importance of the local sociocultural context for attitudes towards 'race' and minorities has produced results which are by no means clear cut. Schaefer (1973b, 1974, 1975) in a number of related pieces of research argued consistently that the local social context was one of the key determinants of spatial variations in prejudice at both a regional and inter-urban level. His findings also received support from work by Elkin and Panning (1975), who demonstrated how the climate of opinion within a neighbourhood was influential in shaping the attitudes of its residents to the issue of 'race'. An alternative view has been forcefully argued by Studlar (1977). Studlar undertook a detailed analysis of the effect of social context upon attitudes, using bivariate and multivariate techniques on census and survey data. He concluded (1977, p. 178) that 'the low explanatory power and instability of comparative importance of the social context variables lends little credence to any of the hypotheses about the effects of social context upon individual attitudes'. It has to be said, though, that Studlar was actually analysing attitudes towards immigration, not immigrants. Although Studlar was happy to regard the two as synonymous they are, in practice, empirically and conceptually distinct.

In order to test the conflicting claims of Schaefer and Studlar, scores were collected on 35 variables for each of the 114 constituencies sampled by SCPR (see Table 7.6). Some of the variables came from the SCPR data set, others from the 1981 Census or the 1979 General Election results. They covered a wide range of structural, attitudinal, demographic and personality characteristics, although it is, of course, always possible to argue over the selection of such variables. While it would be desirable to provide a justification for the selection of each variable in turn, this is not practicable, although what has been said so far should provide a clear indication of the main issue that each variable is supposed to reflect.

The data were then correlated against the percentage of the SCPR sample in each constituency who had described themselves as 'very prejudiced' or 'a little prejudiced'. The coefficients that resulted from this analysis revealed that, at constituency level, Studlar's conclusions were fully supported. Only two variables produced coefficients above 0.30 and these were not surprisingly those relating to the other measure of expressed prejudice (P-scale). The average value for the other 33 variables was only 0.16, revealing that spatial variations in directly expressed prejudice did not seem to be explained at the constituency level by local demographic, social, structural, or attitudinal characteristics. Even the two personality measures in which workers such as Deakin and Abrams had placed so much faith proved to be insignificant. When the same analysis was undertaken using the 11 regions as spatial units in preference to constituencies, the results were very different. The mean

Table 7.6 Variables used in the analyses, and their sources.

	Variable (%)	Source
1	describing self as 'very' or a 'little' prejudiced	SCPR
2	describing self as 'upper-middle' or 'middle' class	SCPR
3	describing self as 'poor' or 'working-class'	SCPR
4	upwardly socially mobile between generations	SCPR
5	downwardly socially mobile between generations	SCPR
6	attending church more than once per week	SCPR
7	never attend church	SCPR
8	Tory supporters	SCPR
9	Labour supporters	SCPR
10	Alliance supporters	SCPR
11	unemployed 1984	SCPR
12	feeling their neighbourhood in decline	SCPR
13	owner-occupiers 1984	SCPR
14	council tenants 1984	SCPR
15	leaving school at, or before, 15 years of age	SCPR
16	aged 61 years or more	SCPR
17	aged less than 30 years	SCPR
18	population loss 1971–81	1981 Census
19	pensioners	1981 Census
20	foreign-born	1981 Census
21	New Commonwealth or Pakistan born	1981 Census
22	male unemployment 1981	1981 Census
23	owner-occupiers	1981 Census
24	council tenants	1981 Census
25	living at more than one person per room	1981 Census
26	lacking/sharing a bath	1981 Census
27	without a car	1981 Census
28	support for National Front	1979 General Election
29	feeling area unsafe after dark	SCPR
30	worrying about becoming victim of crime	SCPR
31	trade union membership	SCPR
32	liberal on F-scale	Calculated from SCPR
33	authoritarian on F-scale	Calculated from SCPR
34	not manifesting prejudice (P-scale)	Calculated from SCPR
35	manifesting extreme prejudice (P-scale)	Calculated from SCPR

coefficient rose to 0.41 and 13 of the 35 values exceeded 0.5 in strength. The 15 largest of these are shown in Table 7.7. They reveal the strong relationship between expressed prejudice and social characteristics including their behavioural and material correlates. Beyond this, a number of structural variables do appear (unemployment and social change, the latter being represented by population change and mobility), although it is noteworthy that percentage New Commonwealth and Pakistani is not one of them. Similarly, the analysis shows that only one attitudinal variable (social conformity, represented by church attendance) appears in the list.

Clearly, then, at the regional level expressed prejudice does vary, and changes in its magnitude are associated with variability in some demographic, structural and to a lesser extent attitudinal variables. This suggests that different regions do have different sets of local attitudinal norms towards 'race', and that these are grounded in social characteristics.

However, one of the difficulties of such an analysis, as Table 7.7 reveals, is that variables may well be inter-correlated and therefore simply measuring slightly different aspects of the same underlying

Table 7.7 Correlation coefficients between self-rated prejudice and input variables.

Percentages	Coefficient	Source
Labour vote in SCPR sample	−0.76	SCPR
trade union membership	−0.74	SCPR
poor or working-class (self-rated)	−0.73	SCPR
Tory vote in SCPR sample	0.71	SCPR
council tenants	−0.69	1981 Census
male unemployment 1981	−0.65	1981 Census
owner-occupier	0.65	1981 Census
more than 1 person per room	−0.63	1981 Census
Alliance vote in SCPR sample	0.63	SCPR
downwardly mobile	0.63	SCPR
not owning car	−0.60	1981 Census
leaving school at 15	−0.59	SCPR
attending church weekly	−0.56	SCPR
unemployed in SCPR sample	−0.54	SCPR
middle or upper-middle class (self-rated)	0.52	SCPR
National Front vote 1979	0.50	1979 General Election
upwardly mobile	0.48	SCPR
constituency population loss 1971–81	0.46	1981 Census

Table 7.8 Results of stepwise multiple regression on dependent-variable, self-rated prejudice.

Variable (%)	Adjusted R^2	R^2 change	β	Significance
Labour (SCPR)	0.534	0.534	−0.45	0.01
attend church more than once per week	0.709	0.175	−0.63	0.01
liberal (F-scale)	0.754	0.045	−0.50	0.01
constituency population of New Commonwealth or Pakistan birth	0.851	0.097	0.45	0.01
aged less than 30 yrs (SCPR)	0.920	0.069	0.31	0.01
council tenants (SCPR)	0.984	0.064	0.22	0.01
prejudiced (P-scale)	0.998	0.014	0.12	0.01
authoritarian (F-scale)	0.999	0.001	−0.05	0.05
trade union membership	1.000	0.001	0.02	0.05

phenomenon. In this case bivariate analysis reveals only limited information and it is better to use a multivariate technique that can cope with the problem of multicollinearity. Consequently, the same data set was used as the basis of a multiple regression exercise at regional level. Stepwise multiple regression was used in order to avoid the worst excesses of multicollinearity: this solution was selected in preference to using a smaller number of orthogonal components as input variables as it was felt that the latter approach would involve an unacceptable loss of detail, and severe problems of interpretation. Relevance to real-life circumstances was therefore selected in preference to statistical elegance.

The results of this analysis are presented in Table 7.8, where each of the nine variables appearing in the regression model is listed along with its related standardized partial regression coefficient (Beta weight), significance level (calculated by the F test), and the adjusted coefficient of multiple determination (R^2). In contrast to Studlar's findings, Table 7.8 indicates that a sizeable amount of the variance in directly expressed prejudice could be accounted for by the complete regression model, and that a number of individual variables contributed relatively large percentages of the total variance. Social characteristics again appear to be the predominant predictor of variability in prejudice, but regression reveals that this is also underpinned by local attitudinal variables (church attendance and local attitudes to prejudice), structural variables (percentage NCWP and percentage council tenants), demographic variables (age), and psychological variables (percentage liberal or authoritarian on the F-scale).

Clearly then, although spatial variability in expressed prejudice *is* related to the distribution of certain population groups across the regions, it also reflects the unique attitudinal and structural circumstances within each region. Both place and space thus have a rôle to play.

Racial hostility as a function of locality

The fourth and final stage of the analysis represented an attempt to discover whether the relative importance of those variables associated with expressed prejudice differed between the regions. To explore this, stepwise multiple regressions were run for each of the 11 regions using the same 35 predictor variables. The results were then analysed, selecting those variables that 'explained' greater than 10 per cent of the variance and that could produce significance levels of more than 0.01 on the F test. The exception to these criteria was East Anglia, for which both variables explaining in excess of 10 per cent of the variance were only significant at the 0.05 level. The outcome was to produce what are in effect brief profiles of the combination of variables which were associated with directly expressed prejudice in each of the 11 regions.

Of more interest than the specifics of this is that the regions could be grouped together into categories that shared similar associations. There were five such groups. The largest included East Anglia, the West Midlands, the South-West and Greater London. In this group, spatial variability in prejudice seemed to be a direct product of social characteristics, with more prejudiced areas being those with working-class populations who rented council houses, had left school early in their lives (West Midlands), were trade union members (London), and might be unemployed (South-West). In these regions, prejudice seemed to be a response to competition for scarce resources, and it was groups in competition with black people that were most prejudiced.

In the second group, which contained Yorkshire, the North and Wales, expressed prejudice seemed not to be a response to direct competition but to a defensive fear of possible status loss held by the upwardly mobile members of the working class. Prejudice here was thus a response to a perceived rather than a direct threat.

The third group contained the North-West and Scotland, and in these cases prejudice seemed to be a reaction to structural conditions and social change. It appeared to reflect the displacement of anxiety generated by a loss of control over the individual's destiny. In the North-West, the prejudiced were the elderly, working-class, owner-occupiers who were regular church-goers and who feared for their personal safety because of the neighbourhoods in which they lived. These are the people who cannot afford to suburbanize like other whites and are stranded instead in the inner city, where they blame their black neighbours for the perceived deterioration in their neighbourhood. In Scotland, the picture is somewhat different. Directly expressed prejudice again seems to be part of the inner-city problem. It is associated with unemployment and with areas where people do not feel safe after dark. Again, for people in these circumstances, blacks form a convenient scapegoat group who can be blamed for the problems of the British economy.

The last two groups contain only one region each. The South-East seems to provide a good example of situational prejudice. There is sharp spatial differentiation between areas of prejudice and areas of tolerance. This variability does not result from social class or structural variables but rather from the presence or absence of black people. In areas where black people have settled in numbers, prejudice is expressed. In areas where settlement has not occurred, prejudice is not apparent.

Finally there is the East Midlands. The results of the analysis suggest no clear explanations for variations in prejudice within this region. As in the South-East, some areas are prejudiced while others are less so, but this does not seem to be associated with the presence or absence of black people. Rather, it seems to be linked to local attitudinal norms, but what factors were originally responsible for generating these are not revealed by this data set.

Conclusion

The analysis demonstrates that there are significant regional variations in expressed prejudice; that certain groups in society are more prone to express prejudice than others; that the variation in prejudice across the UK results from a combination of factors, some of which are relatively universal in nature whereas others are the product of unique local socio-cultural circumstances; and finally that levels of expressed prejudice within regions arise from differing sets of reasons, and that while in certain regions prejudice may be associated with social class, in others it is linked to status threat, social change, or local situational and attitudinal factors.

What this indicates is that by abandoning the middle ground and looking either to universal features or to those specific to small groups of individuals, geographers of ethnic relations are ignoring important elements of the explanatory mix. The present study indicates that purely structural and purely psychological accounts of prejudice are in themselves incomplete and that it is only by looking at how these macro- and micro-level forces are played out in specific localities with their unique characteristics that a full explanation can be arrived at. Perhaps now that geography has matured in its approach and methodology we could usefully rediscover the importance of the unique, and the sense of place that characterized our work prior to the quantitative revolution.

Acknowledgements

The author is indebted to SCPR for providing the data set on which this research is based and to the ESRC Data Archive for supplying the data on tape. The material on historical trends in attitudes was made available by Gallup Ltd, courtesy of Robert Wybrow.

References

Abler, R. F., J. S. Adams & P. R. Gould 1971. *Spatial organization: the geographer's view of the world*. Englewood Cliffs, NJ: Prentice Hall.

Abrams, M. 1969a. Attitudes of the British public. In *Colour and citizenship*, E. J. B. Rose (ed.). London: Oxford University Press for the Institute of Race Relations.

Abrams, M. 1969b. Attitudes of whites towards blacks. *The Listener*, 6 November, 623–4.

Adorno, T. *et al.* 1964. *The authoritarian personality*. New York: Wiley.

Allport, G. W. 1954. *The nature of prejudice*. Cambridge, Mass.: Addison-Wesley.

Bagley, C. 1970. *Social structure and prejudice in five English boroughs*. London: Institute of Race Relations.

Briggs. D. J. 1981. The principles and practice of applied geography. *Applied Geography* **1**, 1–9.

Cochrane, R. & M. Billig 1984. I'm not National Front myself but. . . . *New Society*, 17 May, 255.

Cox, O. C. 1948. *Caste, class and race*. New York: Doubleday.

CRC 1976. *Some of my best friends*. . . . London: Community Relations Commission.

Deakin, N. 1970. *Colour, citizenship and British society*. London: Panther.

Deskins, D. R. 1981. Morphogenesis of a black ghetto. *Urban Geography* **2**, 95–114.

Elkin, S. L. & W. H. Panning 1975. Structural effects and individual attitudes. *Public Opinion Quarterly* **39**, 159–77.

Fishbein, M. 1967. Attitude and the prediction of behaviour. In *Readings in attitude theory and measurement*, M. Fishbein (ed.). New York: Wiley.

Fothergill, S. & J. Vincent 1985. *The state of the nation*. London: Heinemann Educational.

Fryer, P. 1984. *Staying power: the history of black people in Britain*. London: Pluto.

Hamnett, C. 1978. Area-based explanations: a critical appraisal. In *Social problems and the city*, D. Herbert & D. M. Smith (eds). London: Oxford University Press.

Johnston, R. J. 1979. *Geography and geographers*. London: Edward Arnold.

Jones, P. N. 1978. The distribution and diffusion of the coloured population in England and Wales, 1961–71. *Transactions, Institute of British Geographers* **3**, 515–33.

LaPiere, R. T. 1934. Attitudes versus action. *Social Forces* **13**, 230–37.

Lawrence, D. 1969. Letter to *New Society*, August 21, 300.

Lawrence, D. 1974. *Black migrants, white natives: a study of race relations in Nottingham*. Cambridge: Cambridge University Press.

Marsh, A. 1976. Who hates the blacks? *New Society*, 23 September, 649–52.

Milner, D. 1973. The future of race relations research in Britain. *Race* **15**, 91–99.

Milner, D. 1975. *Children and race*. Harmondsworth: Penguin.

Osman, T. 1985. *The facts of everyday life*. London: Faber & Faber.

Pettigrew, T. F. 1959. Regional differences in anti-Negro prejudice. *Journal of Abnormal and Social Psychology* **59**, 28–36.

Rasmussen, J. 1973. The impact of constituency structural characteristics upon political preferences in Britain. *Comparative Politics* **6**, 123–46.

Robinson, V. 1981. *The dynamics of ethnic succession: a British case study*. Oxford: Oxford University School of Geography, Working Paper No. 2.

Robinson, V. 1985. Racial antipathy in South Wales and its social and demographic correlates. *New Community* **12**, 116–24.

Robinson, V. 1986. *Transients, settlers and refugees: Asians in Britain*. Oxford: Clarendon Press.

Rowan, J. 1969a. Letter to *New Society*, 14 August, 262.

Rowan, J. 1969b. Letter to *New Society*, 11 September, 408.

Sant, M. 1982. *Applied geography: practice, problems and prospects*. London: Longman.

Schaefer, R. T. 1973a. Party affiliation and prejudice in Britain. *New Community* **2**, 296–300.

Schaefer, R. T. 1973b. Contacts between immigrants and Englishmen: road to tolerance or intolerance? *New Community* **2**, 358–71.

Schaefer, R. T. 1974. Correlates of racial prejudice. In *Sociological theory and survey research*, T. Leggatt (ed.). San Francisco: Sage.

Schaefer, R. T. 1975. Regional differences in prejudice. *Regional Studies* **9**, 1–44.

Studlar, D. 1977. Social context and attitudes towards coloured immigrants. *British Journal of Sociology* **28**, 168–84.

Vernon, P. E. 1938. *The assessment of psychological qualities by verbal methods.* London: Medical Research Council, Industrial Health Research Board, Report No. 83.

Wicker, A. W. 1969. Attitudes versus actions: the relationship of verbal and overt behavioural responses to attitude objects. *Journal of Social Issues* **25**, 41–78.

Woods, R. I. 1977. Population turnover, tipping points and Markov Chains. *Transactions, Institute of British Geographers* **2**, 473–89.

PART III

Racism and anti-racism in housing and social policy

Institutional racism is a pervasive and insidious feature of British society that has proven resistant even to the most well-intentioned anti-racist strategies. The four essays in this part of the book all focus on the nature of institutional racism in housing and social welfare provision, and include an analysis of the effectiveness of particular anti-racist policies.

John Cater and **Trevor Jones** concentrate on the effects of institutional racism in the housing market. They explore the paradoxical position of Asians in Britain, simultaneously regarded as the victims of racial disadvantage and as the latest embodiment of an entrepreneurial success story. Taking the example of Bradford, Cater and Jones expose the myth of 'Asian success', regularly portrayed in the popular Press. While British Asians do have above-average levels of owner-occupation, ownership of property does not necessarily confer the rewards of middle-class status, particularly if the property in question is confined to the most depreciating areas of the inner city. The segregated 'ethnic community' offers a poor alternative to participation within mainstream society, and ethnicity frequently serves as little more than an ideological mask to disguise the Asians' exploited position in British society where, in housing as in other sectors, they remain highly marginalized.

There is also clear evidence of institutional racism in public-sector housing allocation. **Deborah Phillips** charts the rise of the anti-racist movement in Britain and describes some of the policies that local authorities such as Tower Hamlets have adopted to counter the effects of institutional racism in council house allocation. Even where a positive anti-racist stance has been adopted, as occurred during the GLC's *London Against Racism* campaign in 1984, local initiatives are bound to fail in the absence of a genuine commitment from central government to challenge racism at all levels. At present, such commitment remains largely rhetorical.

In his review of the changing needs of Britain's ethnic minorities, **Mark Johnson** reports disturbing findings concerning the deteriorating position of black people within the welfare state. Black people's needs are consistently marginalized with respect to mainstream social welfare provision. Decisions that in effect 'blame the victim' are then legitimized by politically convenient myths such as the Asians' alleged 'self-reliance'. Johnson reviews the various demographic, economic and geographical dimensions along which institutional policies and practices are currently working to the detriment of black people's welfare needs.

Finally in this part of the book, **Stanley Waterman** and **Barry Kosmin** discuss the problems of welfare provision that arise from the spatial concentration of London's Jewish population. They begin by asking why the Jews have continued to cluster geographically despite a history of social mobility over several generations. They argue that concentration is a spatial strategy that allows the Jews to maintain a strong institutional base of ethnic organizations while avoiding the stigma of segregation. Concentration without segregation provides the Jews with a range of housing types to suit the needs of a diverse population, and does not isolate them from access to the benefits of mainstream society. Waterman and Kosmin conclude by arguing that the monitoring of Jewish residential patterns at a time when social welfare provision is increasingly being devolved by the state to voluntary agencies is clearly of more than academic interest.

8 *Asian ethnicity, home-ownership and social reproduction*

JOHN CATER AND TREVOR JONES

In a paper in 1983 the present authors and their colleagues noted the apparently contradictory position of Asians[1] in the British class structure.

> A condensation of all the relevant sources – popular and scholarly alike – would produce the dazzling insight that Asians are exclusive yet excluded, ethnically assertive yet racially oppressed, economically deprived yet commercially successful. One may be excused the thought that we are witnessing the birth of a new social category, the bourgeois underclass, a group of highly successful failures. (Aldrich *et al.* 1983, p. 1.)

Although this passage is a deliberate caricature, it nevertheless captures the Janus-like qualities of the British Asian, who appears simultaneously a victim of racism and an upwardly mobile over-achiever destined to follow in the 'rags-to-riches' footsteps of British Jewry (Rex 1973). According to Billig (1978), even the working-class racist finds difficulty in slotting Asians into the customary pigeonhole of racial and cultural inferiority. We suggested that this rather schizophrenic reaction on the part of British observers

> stems from the fact that Asian Britons are simultaneously identified by both *race* and *ethnicity*. . . . Ethnicity is the identity which members of the group place upon themselves, race is a label foisted on to them by non-members . . . while racial identity may be a crippling disability, ethnicity acts as a positive force for the protection and promotion of group interests . . . especially where cultural consciousness and fraternal solidarity are as powerfully developed as they are among the Indian and Pakistani groups of Britain. (Aldrich *et al.* 1983, pp. 2–3.)

Until quite recently British geographical analysis of Asian residential segregation tended to concentrate rather one-sidedly on ethnic choice and to play down racial constraint, presenting residential segregation as a form of self-assertion on the part of minority groups anxious to preserve and promote their ethnic identify by maintaining strict social distance from the white majority (Kearsley & Srivastava 1974; Robinson 1979a,b, 1981; Peach 1983). Such an approach inverts the classic logic of minority-relations analysis. Normally we assume inner-city residential segregation to be a racist institution, the means by which a dominant white majority excludes a subordinate black minority from equal participation in the housing market. Residential segregation is imposed *on* blacks *by* whites. In the case of Asian Britons, however, the laws of social gravity are suspended – or so we are asked to believe.

It is not our purpose here to re-open the old choice-versus-constraint debate. As applied to ethnic residential segregation, this is now regarded as a trite and rather banal framework for discussion (Brown 1981, Jones 1983/4), unresolvable within its own parameters and leading to the unhelpful conclusion that every ethnic group exercises residential options but some more so than others. In seeking a way out of this analytical cul-de-sac this chapter will try to make sense of Asian residential space by placing it within the wider context of class relations in late capitalist society. Such an approach requires that we reverse the conventional arguments relating to Asian ethnicity: instead of asking questions about the degree to which ethnicity confers advantages upon its own adherents, we must now ask about its advantages for the dominant class, for capital, and for the capitalist state.

Before developing this argument, however, it is important to acknowledge that much of the analysis of ethnicity as a positive community resource is entirely valid – though only within its own terms of reference. In its neutral sense, the term 'ethnicity' is generally used to denote cultural identity (usually that of emigré groups) and is often broadened to equate with 'community', the interactive network of mutual loyalty, solidarity and co-operation which centres on that identity. Such ethnic communities are exclusive; as a Mirpuri Muslim resident in a British city, one's first loyalty is to others of that ethnicity. It is also unarguable that this exclusiveness has a territorial dimension:

> a common territory or at least some degree of spatial concentration is the physical prerequisite for interaction within the group. (van den Berghe 1978, p. xviii.)

In the British Asian context Husain (1975, p. 121) lays weight on the

> strong community feeling and distinctive cultural choices of these people, which have influenced their residential location decision.

The importance of this kind of analysis is that it highlights ethnicity as an authentic and legitimate source of group identity in its own right (Lyon 1972/3, Khan 1976, Ballard & Ballard 1977, Anwar 1979). It also reminds us that there is in reality no such thing as an 'Asian' ethnicity, but rather a multiplicity of identities defined by religion, language, nationality and region. Without such an emphasis there is a danger that ethnicity may be dismissed as an archaic pre-capitalist relic or, worse still, labelled as a deviant form responsible for undermining social order. By substituting cultural pluralism for crude assimilation theory we shift the focus away from the immigrant as the problematic element.

All in all the emphasis on choice-based ethnic identity may be deemed laudable in its sensitivity, fair-mindedness and apparent explanatory power. The concept of Asians as a set of self-defined close-knit communities, each pursuing its own autonomous goals, seems to go far in explaining where they live, what they live in and even how they earn a livelihood (Werbner 1980, Jones 1981/2, Aldrich *et al.* 1984). Currently, moreover, the model's appeal is heightened by its relevance to the political mood of the moment. Any disadvantaged community, ethnic or otherwise, that displays a 'preference' for home-ownership (Robinson 1979a) and mobilizes its own resources to promote business enterprise, is clearly the living embodiment of all the sacred values – thrift, self-denial, competitiveness, industry – upon which Britain's economic resurgence must be built. In consequence, the Asian business community now finds itself the darling of the political establishment and the subject of a media hype which attempts to convince us that Asian millionaire tycoons are commonplace (Crace 1978, Forester 1978, Bose 1979, Werbner 1985).

While recognizing the importance of ethnicity as a social force, the present authors nevertheless dissent from the kind of cultural determinism that explains social outcomes primarily by reference to the behaviour of the ethnic actors themselves. On theoretical grounds this approach is unjustifiable in that it pays scant attention to the structures in which these actors are obliged to operate. By spotlighting the economic and cultural advantages that ethnicity allegedly confers upon its adherents we draw a veil over the disadvantages created and maintained by racism. Thus by portraying Asians as exclusive, assertive and achieving, we mystify their position as victims. Because they may opt out of racist job and housing markets, racism does not affect them – or so the implicit argument runs. At best, this perverted logic merely encourages the view that state intervention on behalf of the racially oppressed is unnecessary, since presumably a group made up of successful entrepreneurs and property-owners is well capable of looking after itself. At worst, it condones the fascist caricature of Asians as invaders and usurpers of resources which rightfully belong to British whites.

Empirically, too, the model simply does not square with many of the known facts about the Asian condition in Britain. A low socio-economic profile, high unemployment and high levels of housing deprivation are all indicators which flatly contradict any notion of ethnic autonomy. Our contention is that such contradictions stem from a misinterpretation of the class position of Asians in British society, one that is part of a wider misconception of the class position of petty property owners in late capitalism. In the kind of analysis criticized above, much weight is given to the status of Asians as *property owners*, both in the sphere of production (small-business owners) and in the sphere of consumption (owner-occupiers of housing). We have dealt extensively elsewhere with the role of Asians as business owners (Cater & Jones 1978; Aldrich *et al.* 1981, 1983; Jones 1981/2; Jones & McEvoy 1986), demonstrating the marginality of much petty entrepreneurship. Accordingly for the purposes of the present discussion we focus centrally on Asian home-ownership. Even in the absence of an ethnic dimension, the class position of homeowners is a matter of intense debate. Hence our task is two-fold: first to examine the class location of owner-occupiers in general; and then to establish precisely how Asian owner-occupiers (or the majority of them) fit or fail to fit into this framework. The discussion draws heavily on a thought-provoking paper by Saunders (1984).

Owner-occupation and class location

The widespread owner-occupation of housing in late capitalist Britain poses an acute intellectual dilemma for Marxist commentators, one that did not exist in the lifetime of the founding fathers. In Marx's and Engels' own era there existed a fairly close correspondence between ownership of the means of production and consumption; non-owners of the means of production usually did not own their housing either, in the vast majority of cases being forced to rent it. With the passage of time and the growth of home-ownership from less than 10 per cent of British households in 1919 (Agnew 1981) to over 60 per cent today (Central Statistical Office 1985), labour markets and housing market positions have diverged sharply. Now the bulk of the sellers of labour power have achieved the status of home-owner, a category that includes no less than 56 per cent of skilled, 35 per cent of semi-skilled and 27 per cent of unskilled households (Central Statistical Office 1985). Clearly the ownership or non-ownership of productive capital has little bearing on the ownership of property or 'home capital'.

It is of course possible to resolve this dilemma without abandoning the central tenet of classical Marxism that capitalist society is polarized into the two mutually antagonistic classes of owners and non-owners. One solution, entirely consistent with Marx's insistence on the subordination of consumption to production, is to dismiss home-ownership as of marginal or at best tangential significance in class relations, a position exemplified by Westergaard and Resler's terse statement, '. . . owner-occupation of housing . . . [is] . . . irrelevant . . . to property for power.' (1975, p. 122.) More recent contributors have continued to insist that house-owners and mortgagors should not be regarded as property owners in the true sense, since petty capital in the means of consumption does not confer the power to expropriate surplus value from non-owners (Forrest *et al.* 1986).

Saunders (1984), however, argues that this position is now untenable, since housing is not only a vital determinant of life chances but also provides its owner-occupier with what he terms 'exclusivity' or 'personal autonomy and control' (1984, p. 209; see also Rakoff 1977). Elsewhere he asserts that

> because housing plays such a key role in determining life-chances, in expressing social identity . . . and in modifying patterns of resource distribution and economic inequality, it follows that the question of home-ownership must remain central to the analysis of social divisions and political conflicts. (1984, p. 207.)

This reasoning is by no means a new departure. It derives its authority from the widely cited work of Rex and Moore (1967) and Pahl (1975), among others, who demonstrated that the housing market possesses a certain relative autonomy as an allocating mechanism, so that the housing position of individuals and groups is not necessarily determined by their labour-market position. For Saunders, divisions within the housing market should be seen as analytically distinct from class conflict, a term that is best restricted to the opposition between capital and labour in the sphere of production.

Within the housing market itself, Saunders identifies the key division (or 'sectoral consumption cleavage', Dunleavy 1979) as that between owners and tenants. On the one hand, owner-occupiers enjoy privileged access to a form of tenure that confers numerous material and non-material advantages. On the other hand tenancy is becoming an increasingly disadvantaged form of tenure. The private rented sector is now widely recognized as a residual form, commonly a last resort for those who have few options (Darke & Darke 1979, Taylor & Hadfield 1982). As for council tenancy, this is one of the chief victims of the Thatcherite policy of 'rolling back the State': council housing has become systematically devalued by the privatization of its more desirable stock and by an ideological campaign designed to de-legitimize it as an acceptable housing option. Hence a new social fault-line is being opened up between owner-occupiers and non-owners. The former enjoy their shelter privileges at the expense of the excluded minority of tenants:

> those who by virtue [sic] of race, religion, gender, age or education cannot achieve access to basic consumption resources . . . a relatively large number of people *exploit* an increasingly marginalised minority for whom collective provision remains the only and strictly second best option. (Saunders 1984, p. 215, italics added.)

Asians above the cleavage?

If at this stage we reintroduce Asians into the argument, we find that we are obliged to locate the great majority of them *above* rather than *below* the fault line. Taken at face value, Saunder's framework leads us straight back to what we regard as the heretical position satirized at the start of this chapter. It calls into question the entire battery of terms – minority group, white dominance, black exclusion, racism, oppression, exploitation, enforced segregation, ghetto – that we find addictively indispensable. As a minority group, Asians 'ought' to fall below the cleavage, but, since in practice they are actually over-represented among home-owners,[2] they must be located above it. So, far from being trapped in disadvantaged tenures, they have shown remarkable mobility, spreading out from their original lodging-house concentrations and achieving a considerable penetration of the owner-occupied sector in less than three decades. Far from being 'constrained', they appear to have outpaced the white working class in gaining access to the form of tenure that ranks higher than any other in terms of popular choice. Far from experiencing domination, oppression and exploitation, they are members of an 'exploiting' sector, one that denies housing access to a 'marginalised minority' (Saunders 1984, p. 215). Their tenure status is entirely consistent with the model of ethnic autonomy and perhaps even suggestive of some kind of ethnic supremacy.

Quite predictably, in view of our stated position, we take issue with Saunders, though our disagreement may be more a matter of emphasis than of content. While in perfect accord with the concept of the housing market as a relatively autonomous source of social divisions, we dissent from any suggestion that these can be reduced to tenure divisions. Equally important (indeed vital in the Asian case) are the cleavages that exist *within* rather than *between* forms of tenure. At one point Saunders (1984, p. 204) explicitly recognizes that the privately owned sector is not a uniform monolith but consists of a 'heterogeneity of market situations', which divide its occupants into various levels of material wellbeing and act as a means of regressive transfer of wealth *within* the sector. As he says, '. . . the most substantial (monetary) gains have been secured in the higher echelons . . .' of the owning sector but, since his chief (and in our view laudable) aim is to construct an updated and positive socialist response to private ownership *per se*, he fails subsequently to develop this point. For our purposes, however, this is the very point that must be laboured. Asian residents are an outstanding element among those who might be said to 'lose out' from homeownership.

Living below the norm: Asian owner-occupation

As denizens of the lower echelons of the property market, Asian owner-occupiers fail to gain the benefits that would normally accrue from this form of tenure. It has not escaped observers that the great Asian move into property ownership has generally not taken them into suburbia, 'rurbia', the new towns or even the respectable 'zone of working men's homes' (Smith 1976, Karn 1977/8, Rex *et al.* 1977/8, Rex & Tomlinson 1979). Instead they remain concentrated in parts of the inner city adjacent to their initial 'port-of-entry' zones – the huddled clusters of multi-occupied lodging houses, many of which have since been demolished, which sheltered the first migrants prior to the entry of their wives, children and other relatives. Several studies have documented this spatially restricted outward diffusion of segregated Asian residential neighbourhoods (Jones & McEvoy 1974, Cater & Jones 1979, Woods 1979). Unsurprisingly, the type of housing they have 'voluntarily' elected to occupy is materially inferior to that of the mainstream home-owning sector. It consists largely of residual housing abandoned by suburbanizing white residents, a process of filtering that in the United States would be given the uncompromising label 'ghettoization'. According to some observers, this inner-city filtered housing ranks lower than most council-rented accommodation, so that Asian owner-occupiers constitute 'an inferior housing class in terms of their access to the resources of the housing system' (Rex *et al.* 1977/8, p. 124). Karn (1977/8, p. 50) makes the point even more forcibly: 'the coloured [sic] population is progressing towards a monopoly of the areas of worst housing in the city and for the most part their housing in these areas is owner-occupied'.

All this raises thorny questions about the very concept of ownership, with its connotations of personal autonomy, successful achievement of goals and exclusive access to valued housing resources. Indeed such reasoning is substantially undermined when the resource in question is clearly not valued by many other potential competitors in the housing market. Thus Saunders' (1984) comments about home-ownership as expressing the popular will of all sections of society are not fully applicable to these submerged housing sub-markets. Even his contention that ownership confers 'exclusivity' (personal control over one's immediate living space) appears shaky in this context: what kind of power is it that controls only resources that no one else wants and that confers costs rather than benefits upon its possessors? On the contrary, such a housing situation would appear to contain all the definitive attributes of powerlessness.

What precisely are the 'normal' privileges of owner-occupation that Asians fail to gain? For our purposes, two benefits of home-ownership are particularly germane: first *house-price appreciation* (usually in excess of the general rise in retail prices),[3] which provides owners with a hedge against inflation and an opportunity for personal wealth accumulation; secondly *tax relief*, a major benefit to owners purchasing by means of mortgage credit (the vast majority).

As Ward (1982) points out, however, the age structure and location of dwellings typically owned by ethnic minorities preclude the enjoyment of most or all of these rewards. In the four years to December 1985, nationally the average terraced house increased in value by £9770 (+52.7 per cent), while detached and semi-detached properties appreciated by £19 830 (+57.3 per cent) and £12 430 (+52.9 per cent) respectively (Nationwide Building Society 1986). Asian owner-occupiers are concentrated in the first category. In particular, inner-city terraced housing tends not to be an appreciating asset (in real terms). This is due not solely to low demand for inner-city houses but also to a persistent reluctance on the part of building societies to grant loans on older inner-city property, a reluctance that becomes even more manifest when that housing is in a racially changing area (Duncan 1977). Our own data for Bradford show that over the period 1974 to 1984 the average dwelling in the nine inner-city wards (where Asians constituted 28.2 per cent of the total population in 1981 – 41 106 of 145 728) rose by only 256 per cent (marginally below the Retail Price Index) to £11 516. In contrast suburban house prices increased more than three-fold. A sample of over 200 Asian-bought properties had an average value (based on advertised prices, and converted to December 1984 levels using the Nationwide Building Society's House Price Index) of just £8640. This average value is less than half of the *appreciation* on a typical detached house over the past four years detailed above.

With regard to state mortgage subsidization, it is widely understood that the principal benefactors are not those in greatest need (Boddy 1980). On an average mortgage advance of £25 080 in the fourth quarter of 1985 (Nationwide Building Society 1986) the borrower would receive a subsidy of approximately £875 in the first year of repayments in tax relief at current interest rates (12 per cent in April 1986). In contrast many Asian owners are not (formally) mortgaged, or are mortgaged for only paltry sums, or have insufficient income against which to offset allowances. In the past, several writers have made great play of the Asian population's capacity to bypass the state and to house itself, through mobilizing the collective savings of relatives, friends and the community (Dahya 1973, 1974; Anwar 1979). This view is now generally seen as seriously flawed. Not only does it de-emphasize the possibility that collective self-help is a last resort, a defensive reaction against exclusionary racist practices; it is also exaggerated. As Duncan (1977) demonstrated, Asians in Huddersfield were driven by building society discrimination to seek alternative funds, not from their communal resources but at draconian interest rates from backstreet moneylenders. In this way the neediest homeseekers were obliged to pay more for a given quantity of shelter than any other section of the population.

Even if we ignore this type of evidence and accept the romantic notion that Asian residents choose to house themselves via collective self-help, we would still be describing a highly regressive transfer of real wealth – in this case from a generally low-income population to the state (Jones 1983/4). In a system where mortgagors are statutorily entitled to a tax subsidy, the cash buyer of cheap housing is automatically ineligible for this major item of financial support. Asian homeowners are not simply ill-rewarded relative to their mortgaged white suburban counterparts; they are in effect penalized for demonstrating all those Victorian values of self-reliance that currently bulk so large in official state ideology.

We are well aware that the above conclusions need some degree of qualification. Not all Asians are non-mortgaged owner-occupiers; not all are inner-city dwellers; not all are poor. Moreover, not all whites are affluent mortgaged suburbanites. The fact remains, however, that the majority of Asian homeowners do fall within the lower echelons of that particular housing market sector and they are there, to put it bluntly, because they are not white and because they are not acknowledged by the dominant majority as having a legitimate claim to the kind of privileges that the native white homeowner can assume as of right.

Before concluding this section, it should be noted that the term 'housing subnormality' may be taken to refer not only to regressive wealth loss but also to objective material conditions of shelter. Homeownership cannot simply be equated with superior or (necessarily) even decent housing conditions. In practice many owners, and particularly blacks and Asians, live in substandard accommodation, housing that falls below the minimum acceptable standards of a rich welfare capitalist society. The low value of much inner-city terraced housing noted above helps to illustrate this point, as do physical indicators of housing quality in successive censuses. There is no doubt that, with the focus on improvements, housing conditions for all groups have improved markedly over the past two decades. Indeed by 1981, census statistics on the absence of basic household amenities demonstrated little other than the outdated and inadequate nature of these indicators. In England and Wales only 1.9 per cent of households lacked a fixed bath, 2.7 per cent lacked an inside toilet and 4.5 per cent were forced to share either or both the above amenities. The figures for the black population in a relatively depressed metropolitan area, Bradford, are higher, but not exceptionally so (2.0 per cent, 5.8 per cent and 9.6 per cent respectively; City of Bradford Metropolitan Council 1984).

Other measures provide a clearer picture. The recent and substantial Policy Studies Institute survey of *Black and white Britain* (Brown 1984), noted that 26 per cent of black households (and 56 per cent of Bangladeshi households) had no garden, compared with 11 per cent of white households. Similarly while 43 per cent of white households did not have central heating, this figure rose to 66 per cent for Pakistani households. Most significantly the 1981 census noted that only 4.3 per cent of all households were living at densities in excess of one person per room, compared with 35 per cent of Asians in the PSI survey (Brown 1984) and 57.8 per cent of black households in our survey city, Bradford (City of Bradford Metropolitan Council 1984).

For all the above reasons we dispute Saunders' (1984) diagnosis of the key housing cleavage being simply between a 'privatized majority' and a 'marginalized minority'. In practice the two supposedly distinct categories overlap. Despite the heavily subsidized privatization of a significant minority of the public-sector housing stock, not all rented property can be dismissed as marginal. More importantly for this analysis, the 'privatized majority' actually contains a section of the 'marginalized majority', some of whom are living in overcrowded or near-slum conditions in sub-standard environments. For us, the division within home-owners seems no less politically or socially significant than that between owners and renters.

In order to explain this internal cleavage – and in particular the location of the typical Asian owner below it – we now adopt a neo-Marxist approach. This must of necessity be exploratory rather than definitive. While there are well-developed perspectives on housing as social reproduction and on migrant labour, there is as yet little literature integrating these two elements into an analysis of the reproduction of migrant labour in the housing market (Cater & Jones 1987).

Housing and social reproduction

In Marxist class-conflict theory, housing is essentially regarded as one of the necessary means by which labour power is reproduced and hence it is central to the basic capitalist contradiction of accumulation versus reproduction. If Capital expropriates surplus value at an excessive rate (in the form of low wages/high rents), then Labour is denied its means of subsistence, and the reproduction of labour power is threatened by poor housing and the consequent high rates of morbidity and mortality. If on the other hand Labour is permitted to consume in excess of the subsistence level (high wages/low rents), then the rate of capital accumulation is reduced. Therefore workers' real wages will always tend towards the subsistence level.

An essential feature of the capitalist production system is that housing (along with certain other consumer items such as education and health care) cannot profitably be provided for the whole population at acceptable standards of quality (on the mechanics of this, see Harvey 1973, Saunders 1981). Consequently the period in which Marx and Engels were writing was characterized by abysmal housing standards, barely consistent with minimal biological survival. Quite logically, the founding fathers of Marxism saw this as a potentially revolutionary contradiction, with Capitalism nurturing the germ of its own downfall in the form of a dissatisfied working class. To date, of course, the predicted overthrow of Capital has failed to materialize, a non-event generally attributed to the increasing intervention of the Capitalist state, which has taken on widening responsibilities for the provision of improved housing and other consumption items in line with workers' rising expectations (Castells 1977, Taylor & Hadfield 1982, Preteceille & Terail 1985). Orthodox treatments of state involvement in the housing market tend to assume that its purpose is genuinely reformist (see Dunleavy (1980) on 'pluralist' and 'elite' versions of the theory of the state). In direct contrast to this, Marxist theory stresses that state intervention is ultimately designed to benefit Capital not Labour. It begins with 'the fundamental observation that the State in capitalist society serves the interests of the capitalist class' (Taylor & Hadfield 1982, p. 241) or, in the original words of the Communist Manifesto, 'the State is but a committee for managing the interests of the whole bourgeoisie'.

In Britain, state involvement in shelter provision has become extremely wide-ranging and includes a considerable emphasis on direct action such as slum clearance and the construction of council housing (Taylor & Hadfield 1982, p. 241). Yet, despite the significance of the public rented sector (27 per cent of all households in 1984), we should not be blinded as to the rôle of the state in the private sector. Arguably the commitment by postwar governments of both political parties to the stimulation of private house-building and the subsidizing of its occupants represents an equally decisive form of intervention (McKay & Cox 1979). Notwithstanding the rhetoric of 'private property', 'sturdy independence' and 'standing on one's own feet' that pervades it, owner-occupation is a vital element in the process of collective consumption: more so than council housing in the narrow sense that it confers greater rewards on the consumers themselves.

Where does all this leave the black minority owners who languish in the lower levels of the private sector? Typically their housing situation represents little more than a token note in the direction of material appeasement; not sufficient, one would imagine, to engender a sense of satisfaction or of loyalty to a system that in effect punishes them for 'good' behaviour. Collective consumption is a selective process which benefits some sections of Labour more than others and excludes some workers altogether. At root this is dictated by the financial limitations of the state: under capitalism the universal provision of decent consumption standards is seen to mean rampant inflation and unacceptably high levels of public borrowing. Consequently appeasement must be doled out selectively.

On the surface, this differential allocation of rewards would seem to pose a serious threat to social stability by creating a sense of relative deprivation and a mood of disaffection among the losers. But to argue this would be to reckon without the ideological attributes of owner-occupation. It seems that, whatever the level of material benefits, the status of property owner is sufficient reward in itself to engender a sense of identity with the capitalist order. Numerous writers have commented on the ideological effectiveness of home-ownership as a counter-revolutionary institution, a stabilizing mechanism for a potentially self-destructive social system (Boddy 1976, 1980; Castells 1977; Kemeny 1980; Merrett 1982; Forrest *et al.* 1986). Kemeny (1980, p. 372) writes of the 'conservatising effects of owner-occupation' and its value to Capital as a 'deterrent to social and industrial unrest'. For Forrest *et al.* (1986) the widespread diffusion of this form of tenure is no less than a 'Bulwark Against Bolshevism', a title derived from 1920s Tory propaganda. All these writers are pursuing the notion that property ownership has a crucial ideological dimension as well as political and material dimensions: beyond merely appeasing its consumers by material rewards, it actually incorporates them as loyal and acquiescent supporters of the capitalist order. It gives them a 'stake in the system', converting them from potentially dissident renters into debt-encumbered petty-property owners who perceive tangible losses from any disturbance in the established status quo.

Evidently home-ownership is part of the structure of capitalist 'ideological hegemony'. This concept refers to the suffusion of the entire culture with capitalist values and capitalist versions of history, so that ordinary workers cease to perceive themselves as exploited and come to accept capitalist domination as normal, inevitable, and even possibly benevolent. If legitimacy is the key to the survival of the capitalist order, then the significance of subnormal owner-occupation is that it represents legitimation on the cheap. As property owners, Asians and the other consumers of submerged housing are co-opted into a commitment to the prevalent social order (Agnew 1981, Miles 1984), even in the total or relative absence of state subsidization. Here we should note that Asians 'miss out' on their share of collective consumption in two ways: directly, because of their low take-up state-subsidized mortgage finance; and indirectly, because of their substantial under-representation in public-sector housing. In an attempt to quantify the latter, Jones (1981/2) estimates that there are 20 000 fewer Pakistani households in council housing than their social class profile would suggest. This benefits the state by lessening the demand for additional local-authority housing, reducing the level of public-sector housing subsidy and limiting a thorny managerial problem for local-authority housing departments.

Asian residential behaviour is thus a useful component in the current drive to reduce public expenditure which has bitten so painfully into housing and other social provisions. In Saunders' (1984) view, this 'rolling back' of the state is not a short-term policy specific to the post-1979 Conservative government: this administration has simply rendered explicit a trend that has been brewing for some years in response to an endemic fiscal crisis of the state. It appears that we have now moved into a phase in which consumption will become increasingly privatized, and socialized provision will be reserved only for the marginalized minority.

This, however, omits those who are both privatized and marginalized. If, as a result of diminished state provision, Labour is increasingly required to assume the costs of its own reproduction, then social stability may be threatened – witness the 1980s urban riots and the increase in petty crime and other submarginal adjustments to poverty. In these circumstances we should not underestimate the usefulness of a section of Labour sufficiently ideologically incorporated to undertake its own reproduction willingly.

But why do certain minorities, most notably Asians, occupy such a prominent position in this residential stratum? Our short answer to this is *racism*. Asian residential status is a product of their membership of a racialized fraction of the working class (Miles 1982); a group whose relationship to the means of production and whose objective interests place them firmly in the category of Labour but who are isolated from the mainstream of the working class by racist ideology (Phizacklea & Miles 1980). Modern structuralist theories of racism in advanced capitalist societies hinge centrally on the advantages gained by Capital and the state from the use of migrant labour 'imported' from the dependent neocolonies of the world periphery (Castles & Kosack 1973, Castells 1975, Nicholls 1980, Miles 1982). Such theories frequently emphasize that, historically, colonial labour has been subject to super-exploitation; the purpose of doctrines of racial inferiority was to legitimize the coercive and brutal practices necessary to achieve super-profits (Tinker 1977). Inhumanity was condoned by reclassifying the victims as some lesser kind of being. In this context, the postwar migration of black labour represents a further extension of the colonial labour principle. Though physically located within the capitalist metropolis, black migrants essentially perform those low-level under-rewarded tasks that have traditionally fallen to the lot of the colonial worker: and their presence in these reserved occupations continues to be legitimized by racism. Furthermore, and this is critical for our argument, they have a similar specific rôle in the sphere of reproduction.

Migrant labour and reproduction

In the sphere of reproduction the direct advantages of a migrant labour system are principally reaped by the state, which is absolved from part of its responsibility for reproducing labour in the sense that migrants either do not demand or do not qualify for their full share of collective consumption. This has been well summarized by Castells (1975) who points out that:

a As active workers unaccompanied by dependents, migrant populations require less educational expenditure for the young, pensions for the old, health care for the sick. In Britain studies of the early New Commonwealth population demonstrated its distorted demographic structure, heavily biased towards males in the economically active age groups, and the consequent low level of demands on the public purse (Eversley & Sukdeo 1969).

b As 'forced bachelors' and transients their claims on public housing are nil or negligible. The early literature on the 'lodging house' phase of black immigration to Britain (Butterworth 1967, Rex & Moore

1967) highlighted this, and its relevance to Asians today has been noted by Jones (1983/4).

c As 'Third Worlders', their standards and expectations of housing and other consumption items are lower than those of the native-born population (Dahya 1974); their conditions of reproduction are below the average standards of indigenous workers (Nicholls 1980).

Castells also argues that the politico-legal status of immigrants in most metropolitan countries creates a perpetual mood of insecurity which inhibits black political organization and even the search for individual rights, a point forcibly underlined by Sivanandan (1978) on the subject of British immigration restriction. The tightening of immigration law, together with the climate of Parliamentary debate surrounding it, has effectively transmitted the symbolic message that black workers are only here on suffrance as units of production and are not fully entitled to press their needs and rights as humans. Moreover, official policies of this ilk lend moral endorsements to the kind of popular racism which, in Newman's words, 'seeks to restrict and limit the social roles of minority group members and restrict their collective impact upon the larger society' (1973, p. 152). Seen in this light, minority residential behaviour appears not so much as self-willed but as a strategy of non-confrontation, a retreat from hostility. This essentially *reactive* nature of Asian segregation has been noted by Rex (1973) and more recently demonstrated by Phillips (1986), who found that the concentration of Bengalis in the worst council estates in Tower Hamlets was at least partly dictated by fear of the consequences if they attempted to move into more desirable areas. Hence, segregation in the unwanted residual areas of the housing market is actually a matter of physical and emotional security from the minority's own viewpoint: from the state's viewpoint it is a highly functional conflict-minimization device, one of the means by which racism can be managed. Without it, racism would become dysfunctional, leading to unacceptable levels of intergroup conflict, violence and social disorder. As we have remarked elsewhere:

> Until now the relative absence of tension . . . has been due not so much to good race relations as to the virtual absence of *any kind* of relations. Because Asian demands on social resources . . . have been minimal, white society has to a great extent ignored their existence (Cater & Jones 1978, p. 82).

At this point we should remind ourselves that the situation in postwar Britain has until recently differed markedly from the pure migrant labour system operated in states such as South Africa and a number of West European countries. Legal regulations which permit no permanent residence for migrants and require them to be repatriated at the end of their work contracts ensure that migrant labour is self-reproducing to the maximum degree possible. The British case differs in that until the 1962 Immigration Act, the status of 'Commonwealth immigrant' allowed unrestricted settlement both to migrants and their dependents. Subsequent legislation has reduced the black worker to a quasi-*gastarbeiter* status (Sivanandan 1978), but not before the establishment and growth of a permanent black population. On the surface, it may appear that the British state is now burdened with the full costs of black labour reproduction – and this at a time of high and rising unemployment, when the usefulness of such reserve labour is fast diminishing.

These, however, are precisely the conditions in which ethnicity is at its most functional for the Capitalist state. In the case of groups such as Asians, for whom institutions of mutual aid and collective self-help continue to play a real part in community life, it is both economically feasible and politically acceptable to hive off some of the responsibility for reproduction on to the community itself. Thus, while it may still be impossible legally to repatriate unwanted Asian labour to its country of origin (actual or ancestral), it is certainly possible to repatriate it to the 'Third World Within' – ethnic concentrations within the British city. In so far as the ethnic community actually does function as a self-help mechanism, it partially relieves the state of the burden of reproducing Asian labour, much of which is now surplus to requirements. Though permanent settlers in the literal sense, Asian populations continue to maintain patterns of consumption more typical of migrant workers than of a settled working class willing and able to make legitimate claims on consumption resources. The forms of housing behaviour reviewed in this chapter – subnormal home-ownership, avoidance of public housing, segregation in run-down under-resourced neighbourhoods – are all instances of the migrant-like contribution made by Asians to their own reproduction. Further instances might also be cited, notably the contribution of the Asian business sector in mopping up unemployed labour which would, in other working-class communities, be dependent on the state. Many of these proliferating small businesses are commercially so submarginal as to represent a species of disguised unemployment/underemployment (Aldrich *et al.* 1981). Officially this is recorded as self-employment, even though many (or even most) of its participants earn less than they would in semi-skilled factory work.

In the light of all this, the reader may be struck by the thought that the Asian population are the trendsetters in a Thatcherite Britain committed to dismantling the 'nanny State' and devolving its burdens on to the family and the community. This is not our view, which regards self-reproduction as a form of enforced self-exploitation. More accurately,

Asians might be seen as the contemporary successors to the traditional white working-class community in the style of Young and Willmott (1957), whose kin- and neighbourhood-based capacity for looking after its own members relieved Capital of the need to pay a decent wage and the state of responsibility for social welfare.

Conclusion

The central thesis of this chapter has been that ethnicity, to the extent that it flourishes as a real and dynamic force, must be analysed in terms of the interests of the dominant class. In arguing thus we have laid considerable emphasis on the material function of the ethnic community, that is continually to reproduce 'migrant' labour, to the comparative neglect of its political and ideological functions. Though space does not permit a full analysis, we would conclude with the brief observation that ethnicity is in many ways an apolitical force, at least in so far as class-based action is concerned. As we have implied in this chapter, ethnic-based organization offers an accessible alternative to class-based resistance in the matter of material survival. Beyond this, however, there are other ethnic mechanisms that militate against class action, two of the most important being:

a ethnic cultural exclusiveness, which re-inforces the mutual alienation of ethnic and non-ethnic fractions of the working class
b social control – the subjection of ethnic members to the discipline of family, religion and other traditional authority structures. (Rex 1982, pp. 59–62.)

We are well aware that these conditions are far from absolute or permanent; and that Asian resistance to racism and class exploitation is a growing force (Race Today Collective 1979, Sivanandan 1981/2). Nevertheless we remain convinced by Rex's assertion that it will 'be a long time . . . before the Indian culture is so weakened both externally and in the conscience of the young that changed patterns . . . will be adopted' (1982, pp. 61–2). In the meantime the ethnic community is best seen as an unwitting accomplice of the status quo rather than as any kind of real alternative to it.

Notes

1 The term 'Asians' is used here to refer to people of Indian, Pakistani or Bangladeshi origin or ancestry.
2 According to the latest published figures (for 1981), 75 per cent of Asian

householders were owner-occupiers compared with 55 per cent of the white population (Central Statistical Office 1983). The latter figure has increased by approximately five percentage points since this date, but we have no comparable current figure for Asian owner-occupation.

3 For example in the ten-year period from the end of 1974 to the end of 1984 average house prices increased by 320 per cent while the Retail Price Index increased by 260 per cent (Nationwide Building Society 1986).

References

Agnew, J. 1981. Home ownership and the capitalist social order. In *Urbanisation and urban planning in capitalist society*, M. Dear & A. J. Scott (eds). London, Methuen.

Aldrich, H. E., J. C. Cater, T. P. Jones & D. McEvoy 1981. Business development and self-segregation: Asian enterprise in three British cities. In *Ethnic segregation in cities*, C. Peach, V. Robinson & S. J. Smith (eds). London: Croom Helm.

Aldrich, H. E., J. C. Cater, T. P. Jones & D. McEvoy 1983. From periphery to peripheral: South Asian petite bourgeoisie in Britain. In *Research in the sociology of work II*, I. H. Simpson & R. Simpson (eds). Connecticut: JAI Press.

Aldrich, H. E., T. P. Jones & D. McEvoy 1984. Ethnic advantage and minority business development. In *Ethnic communities in business: strategies for economic survival*, R. Ward & R. Jenkins (eds). Cambridge: Cambridge University Press.

Anwar, M. 1979. *The myth of return: Pakistanis in Britain*. London: Heinemann.

Ballard, R. & C. Ballard 1977. The Sikhs: the development of South Asian settlement in Britain. In *Between two cultures*, J. L. Watson (ed.). Oxford: Blackwell.

Billig, M. 1978. *Fascists: a social psychological view of the National Front*. London: Academic Press.

Boddy, M. 1976. The structure of mortgage finance: building societies and the British social formation. *Transactions, Institute of British Geographers*, New Series **1**(1), 58–71.

Boddy, M. 1980. *The building societies*. London: Macmillan.

Bose, M. 1979. The Asians of Leicester: a story of worldly success. *New Society*, 16 August, 339–41.

Brown, C. 1984. *Black and white Britain*. London: Policy Studies Institute.

Brown, K. 1981. Race, class and culture: towards a theorisation of the 'choice/constraint' concept. In *Social interaction and ethnic segregation*, P. Jackson & S. J. Smith (eds). London: Academic Press.

Butterworth, E. 1967. *Immigrants in West Yorkshire*. London: Institute of Race Relations Special Publication.

Castells, M. 1975. Immigrant workers and the class struggle in advanced capitalism: the West European experience. *Politics and Society* **5**.

Castells, M. 1977. *The urban question: a Marxist approach*. London: Arnold.

Castles, S. & G. Kosack 1973. The function of labour immigration in West European capitalism. *New Left Review* **73**, 3–18.

Cater, J. C. & T. P. Jones 1978. Asians in Bradford. *New Society*, 13 April, 81–2.

Cater, J. C. & T. P. Jones 1979. Ethnic residential space: the case of Asians in Bradford. *Tijdschrift voor Economische en Sociale Geografie* **70**, 86–97.

Cater, J. C. & T. P. Jones 1987. *Social geography: an introduction to contemporary issues*. London: Arnold.

Central Statistical Office 1983. *Social trends*. London: HMSO.

Central Statistical Office 1985. *Social trends*. London: HMSO.

City of Bradford Metropolitan District Council 1984. *District trends*. Bradford: Policy Unit, City of Bradford MDC.

Crace, J. 1978. The bazaar on the corner. *Sunday Telegraph Magazine* **114**, 82–99.

Dahya, B. 1973. Pakistanis in Britain: transients or settlers? *Race* **14**, 241–77.

Dahya, B. 1974. The nature of Pakistani ethnicity in industrial cities in Britain. In *Urban ethnicity*, A. Cohen (ed.). London: Tavistock.

Darke, R. & J. Darke 1979. *Who needs housing?* London: Macmillan.

Duncan, S. S. 1977. *Housing disadvantage and residential mobility: immigrants and institutions in a northern town*. Working paper No. 5, Faculty of Urban and Regional Studies, University of Sussex.

Dunleavy, P. 1979. The urban bases of political alignment. *British Journal of Political Science* **9**, 409–43.

Eversley, D. & F. Sukdeo 1969. *The dependents of the coloured Commonwealth population of England and Wales*. London: Institute of Race Relations Research Publication.

Forester, T. 1978. Asians in business. *New Society* **23**, February, 420–21.

Forrest, R., A. Murie & P. Williams 1986. *Bulwark against Bolshevism*. London: Hutchinson.

Harvey, D. 1973. *Social justice and the city*. London: Arnold.

Hussain, M. S. 1975. The increase and distribution of New Commonwealth immigrants in Greater Nottingham. *East Midlands Geographer* **6**, 105–29.

Jones, T. P. 1981/2. Small business development and the Asian community in Britain. *New Community* **9**, 467–77.

Jones, T. P. 1983/4. Residential segregation and ethnic autonomy. *New Community* **11**, 10–22.

Jones, T. P. & McEvoy D. 1974. *Residential segregation of Asians in Huddersfield*. Paper presented to the Institute of British Geographers' Annual Conference, Norwich.

Jones, T. P. & D. McEvoy 1986. Asian enterprise, the popular image. In *Survival of the small firm*, J. Curran, J. Stanworth & D. Watkins (eds). Farnborough: Gower.

Karn, V. 1977/8. The financing of owner-occupation and its impact on ethnic minorities. *New Community* **6**, 49–64.

Kearsley, G. W. & S. R. Srivastava 1974. The spatial evolution of Glasgow's Asian community. *Scottish Geographical Magazine* **90**, 110–24.

Kemeny, J. 1980. Home ownership and privatisation. *International Journal of Urban and Regional Research* **4**, 372–88.

Khan, V. S. 1976. Pakistanis in Britain: perceptions of a population. *New Community* **5**, 222–9.

Lyon, M. 1972/3. Ethnicity in Britain: the Gujerati tradition. *New Community* **2**, 1–11.

McKay, D. & A. M. Cox 1979. *The politics of urban change*. London: Croom Helm.

Merrett, S., 1982. *Owner-occupation in Britain*. London: Routledge & Kegan Paul (with F. Gray).

Miles, R. 1982. *Racism and migrant labour*. London: Routledge & Kegan Paul.

Miles, R. 1984. Marxism versus the sociology of 'race relations'. *Ethnic and Racial Studies* **7**, 217–37.

Nationwide Building Society 1986. *House prices in 1985*. London: Nationwide Building Society.

Newman, W. M. 1973. *American pluralism: a study of minority groups and social theory*. New York: Harper & Row.

Nicholls, T. 1980. *Capital and labour: a Marxist primer*. London: Fontana.

Pahl, R. 1975. *Whose city?* Harmondsworth: Penguin.

Peach, C. 1983. Ethnicity. In *Progress in urban geography*, M. Pacione (ed.). London: Croom Helm.

Phillips, D. 1986. *What price equality? A report on the alllcation of GLC housing in Tower Hamlets*. London: GLC Housing and Research Policy Report No. 9.

Phizacklea, A. & R. Miles 1980. *Labour and racism*. London: Routledge & Kegan Paul.

Preteceille, E. & J. P. Terail 1985. *Capitalism, consumption and needs*. Oxford: Blackwell.

Race Today Collective 1979. New perspectives on the Asian struggle. *Race Today* **11**, 103–9.

Rakoff, R. M. 1977. Ideology in everyday life: the meaning of the house. *Politics and Society* **7**, 85–104.

Rex, J. 1973. *Race, colonialism and the city*. London: Routledge & Kegan Paul.

Rex, J. 1982. West Indian and Asian youth. In *Black youth in crisis*, E. Cashmore & B. Troyna (eds). London: Allen & Unwin.

Rex, J. & R. Moore 1967. *Race, community and conflict*. London: Oxford University Press.

Rex, J. & S. Tomlinson 1979. *Race, colonialism and the city*. London: Routledge & Kegan Paul.

Rex, J., S. Tomlinson, D. Heanden & P. Ratcliffe 1977/8. Housing employment, education and race relations. *New Community* **6**, 123–6.

Robinson, V. 1979a. Contrasts between Asian and white housing choice. *New Community* **7**, 195–201.

Robinson, V. 1979b. Choice and constraint in Asian housing in Blackburn. *New Community* **7**, 390–7.

Robinson, V. 1981. The development of South Asian settlement in Britain and the myth of return. In *Ethnic segregation in cities*, C. Peach, V. Robinson & S. J. Smith (eds). London: Croom Helm.

Saunders, P. 1981. *Social theory and the urban question*. London: Hutchinson.

Saunders, P. 1984. Beyond housing classes: the sociological significance of private property rights in the means of consumption. *International Journal of Urban and Regional Research* **8**, 202–27.

Sivanandan, A. 1978. From immigration control to 'induced repatriation'. In *A different hunger*, A. Sivanandan (ed.), London: Pluto Press.

Sivanandan, A. 1981/2. From resistance to rebellion: Asian and Afro-Caribbean struggles in Britain. *Race and Class* **23**, 111–52.

Smith, D. J. 1976. *The facts of racial disadvantage: a national survey*. London: Political and Economic Planning.

Taylor, P. J. & H. Hadfield 1982. Housing and the state: a case study and structuralist interpretation. In *Conflict, politics and the urban scene*, K. R. Cox & R. J. Johnston (eds). London: Longman.

Tinker, H. 1977. *The banyan tree: overseas emigrants from India, Pakistan and Bangladesh*. London: Oxford University Press.

van den Berghe, P. L. 1978. *Race and racism: a comparative perspective*. New York: Wiley.

Ward, R. 1982. Race, housing and wealth. *New Community* **10**, 3–15.

Werbner, P. 1980. From rags to riches: Manchester Pakistanis in the garment trade. *New Community* **8**, 84–96.

Werbner, P. 1985. How immigrants can make it in Britain. *New Society*, 20 September, 411–14.

Westergaard, J. & H. Resler 1975. *Class in a capitalist society: A study of contemporary Britain*. London: Heinemann.

Woods, R. I. 1979. Ethnic segregation in Birmingham in the 1960s and 1970s. *Ethnic and Racial Studies* **2**, 455–77.

Young, M. & P. Willmott 1957. *Family and kinship in East London*. London: Routledge & Kegan Paul.

9 *The rhetoric of anti-racism in public housing allocation*

DEBORAH PHILLIPS

After three decades of New Commonwealth and Pakistani settlements in Britain, striking differences in the housing conditions of minorities and whites prevail. It is true that the appalling standards faced by the earliest newcomers have vastly improved as minority households have moved from overcrowded often sub-standard private rental accommodation into better quality council and owner-occupied housing. Indeed, some minorities are now to be found in the more prestigious residential areas of the suburbs (Phillips 1981). However, in some respects the early picture of ethnic residential concentration, segregation and deprivation is little changed. The declining inner-city reception areas still provide a strong focus for minority clustering, black and white residential space often remain separate, and, in general, the same story of minority housing deprivation pertains. Blacks and Asians still occupy poorer quality property than whites, live in less desirable types of accommodation, and reside in the least popular locations, a pattern that is replicated throughout the different tenure categories (Brown 1984). Early writers portrayed such racial inequalities as a temporary phenomenon associated with newcomer status (Patterson 1965, Banton 1955). The social and spatial divisions have, however, so far survived the passage of time. More recent explanations have therefore sought to expose the deep structural divisions between minority and indigenous groups and the rôle of institutional racism in perpetuating inequalities between them (Miles & Phizacklea 1984).

This chapter argues that different forms of institutional racism have combined to sustain ethnic residential concentration and housing deprivation in Britain. Specific reference is made to racial inequalities within the public housing sector and to recent anti-racist strategies to combat them. The entrenched pattern of minority disadvantage in the allocation of council housing resources has been well documented. Early indications of discrimination emerged with the work of Burney (1967) and

Rex & Moore (1967), although Parker & Dugmore (1976) were the first to provide statistical evidence of the racial bias in council allocation procedures. Since then, detailed studies of authorities from Nottingham (Simpson 1981) to Bedford (Skellington 1981), Liverpool (CRE 1984a) to Birmingham (Flett 1979, Henderson & Karn 1984) and London (most notably, Hackney Borough Council: CRE 1984b) have uncovered evidence of systematic racial discrimination. All present a similar picture; black people tend to wait longer for housing, are more likely to be allocated poorer property and more frequently end up on less desirable estates. The process of institutional discrimination underlying this pattern of racial inequality is complex. Cumulative disadvantage arises, for example, from systematic discrimination in the competition for jobs, the provision of education, and inadequate protection by the police (especially over racial harassment), and this compounds discrimination inherent in the operation of the housing market itself.

The complexity of institutional discrimination within the housing market has been thrown into sharp relief by recent anti-racist initiatives within the public sector. Anti-racist programmes, devised and implemented at the local authority level, have in general aimed to promote greater housing equality by eliminating the racism embedded within public housing allocation rules, procedures and practices. Anti-racist initiatives have, however, encountered serious difficulties. First, the implementation of anti-racist strategies has often met with considerable resistance as established 'normal' institutional procedures have been challenged; even in the more enlightened housing departments, anti-racism is very much in its infancy. Secondly, and perhaps more importantly, the potential for local authorities to achieve greater housing equality through a deracialized allocation system must itself be questioned. Equality of opportunity within the public housing sector cannot eradicate the racial consequences of other sources of disadvantage, such as might arise from the disproportionately high minority need for council housing, the inability to pay for more expensive types of council dwellings, or the minorities' unwillingness to accept properties in certain areas because of journey-to-work difficulties or fear of racial harassment. In short, anti-racist initiatives within one sphere of institutional discrimination may be unable to transcend the wider, cumulative disadvantage arising from other sources of racism.

This chapter seeks to examine the forces underlying racial inequality within public housing and to evaluate the impact of current anti-racist initiatives in this sector. It begins with a brief look at the rise of the anti-racist movement in Britain and a detailed consideration of the institutional racism that it seeks to challenge. It then goes on to examine the specific consequences of institutional racism and anti-racism for those offered council housing in the London Borough of Tower Hamlets by the Greater London Council (GLC). The data presented span two years of housing allocation in 1983 and 1984, a period when the GLC was firmly committed to the promotion of equal opportunities and anti-racism. The data include an analysis of allocation records and a five-month period of participant observation in the local Tower Hamlets housing department. The discussion aims to highlight the complexity of the institutional discrimination process, the problems of anti-racist policy implementation and the limitations of local authority initiatives designed to redress minority housing deprivation.

Racism and anti-racism

Recent debate has focused particular attention on the material and ideological components of British racism in the 1980s (Miles 1982, Castles 1984). The prime concerns of this inquiry have been to locate the roots of racism within an historical context, to understand the mechanisms by which it is perpetuated and to identify the points at which racism may be tackled (Lawrence 1982, Morgan 1985). The rôle of the state and its institutions in producing and maintaining racist ideologies and racial divisions has been of central importance to this debate. Some, such as Sivanandan (1982), Miles & Phizacklea (1984) and such groups as the Workers Against Racism (1985), have gone so far as to argue that the state has been directly repressive of black minority rights, citing discriminatory immigration controls in support of their argument. Others (most notably Scarman 1981) have adopted the view that racial inequalites largely derive from the unintentionally discriminatory effects of institutional rules and practices and stressed the government's rôle in countering discrimination through race-relations legislation.

The legal standing of minorities in Britain over the past two decades has certainly improved, and blatant direct discrimination is now less common. Deeply entrenched mechanisms institutionalizing and sustaining racism nevertheless persist. Six years after the introduction of the 1976 Race Relations Act, which specifically outlaws both direct and indirect discrimination, the third Policy Studies Institute Survey (Brown 1984) found widespread evidence of continuing institutional racism. The report concluded (p. 318) that 'it was not enough to establish a formal legal framework of equality by outlawing discrimination'; 'vigorous positive action' would now be necessary to erode entrenched racial inequalities in Britain.

The 1970s and 1980s have seen a multitude of local action programmes. Some, presuming racial inequalities to derive from cultural differences, have mistakenly identified the 'problem' as a black one, and formulated black-orientated solutions in the belief that differences would disappear with time. They have thus often left untouched the forces perpetuating white institutional racism (Sivanandan 1983). The past decade has, however, witnessed the development of a different type of strategy with the emergence of anti-racism in Britain. Unlike previous multicultural approaches, the anti-racist movement has tried to attack racism at its roots by directly opposing racial categorization and the beliefs that sustain it. Anti-racism locates racial problems with white society and its institutions. Its main aims are to isolate the mechanisms sustaining inequalities and to call established ideologies into question. The formal state institutions and extreme right-wing groups have thus emerged as the focus of attention.

The anti-racist struggle has assumed two main forms. First, local anti-racist organizations have sought to expose racism within state institutions (Ben Tovim 1982). Although partly constrained by their own position in relation to the formal structures of the state (for example, by the need to liaise with the police over racial harassment), evidence presented later suggests that such groups can help to trigger institutional reorganization when conditions are conducive to change. Secondly, the formal state institutions have themselves initiated anti-racist strategies, particularly at the local authority level. As major state institutions, responsible for employment as well as service provision, local authorities are in a key position to influence the structure of minority opportunities (Young & Connolly 1981, CRE 1985). The strategic rôle of local authorities was explicity acknowledged in the 1976 Race Relations Act (Section 71), which specifically enjoins councils to make 'appropriate arrangements' to:

(i) eliminate unlawful discrimination;
(ii) promote equality of opportunity, and good relations between persons of different racial groups.

This rather ambiguous directive, however, was supported by neither sanctions nor incentives, and the local authority response was predictably varied. For example, by 1984 only 14 out of 32 London boroughs had an equal-opportunities policy supported by a 'race' committee; the remainder had no formal policy on 'race'. Local authorities as a whole have been extremely negligent of their responsibility in this area, often treating racial equality as a low-priority issue (Young & Connolly 1981). Anti-racist initiatives have thus been sporadic and local, and highly variable in terms of both financial and political commitment towards them. Many local authorities have adopted a complacent 'colour blind' approach and are still resistant to the idea of ethnic monitoring. For example, only 12 out of 32 London boroughs were keeping the ethnic records essential to anti-racist initiatives in 1984. Others have been guilty of implementing 'token' anti-racist strategies (AMA 1985), abusing Section 11 funds[1] and appointing race-relations officers with little power to effect change.[2]

There are, however, authorities who have taken the 1976 injunction seriously, as clearly exemplified by the Labour-controlled Greater London Council. Building on its commitment to promote equal opportunities, the Council first put forward its proposals for 'a major programme of anti-racist activity' in December 1982 (GLC 1982). Initiatives were developed and implemented over the following year, which culminated in the launch of one of the best publicized and most comprehensive anti-racist programmes to date. Strategies included the designation of 1984 as 'Anti-Racist Year', the declaration of London as an 'anti-racist zone', organizational changes within the Council, and public rallies and events. Throughout this time, the GLC maintained a high anti-racist profile, which successfully brought racial issues to the fore. The impact of the Council's anti-racist initiatives on the institutional racism it sought to challenge was, however, less clear. Progress was certainly undermined by internal resistance to anti-racist strategies. However, difficulties also arose in tackling the complex and often poorly understood process through which racism becomes institutionalized. It is to a discussion of this process that the chapter now turns.

Institutional racism: an appraisal of the concept

Racism manifests itself both through individual actions and through the unequal and cumulative effects of institutional rules and procedures. 'Institutional racism', a term first coined in the American literature (Carmichael & Hamilton 1967, Knowles & Prewitt 1969, Blauner 1972), has been commonly used to refer to the discriminatory effects of institutional operations which 'systematically reflect and produce racial inequalities' (Jones 1972) irrespective of the intentions of the individuals involved. A racist society, writes Ann Dummett (1973), 'has institutions which effectively maintain inequalities between members of different groups in such a way that open expression of racist doctrine is unnecessary' (p. 131).

There is now widespread acceptance of the crucial rôle of institutional mechanisms in reflecting, producing and sustaining racial inequalities in Britain. The term 'institutional racism' itself has become an integral part of the vernacular, widely quoted in academic and political discourse alike. Indeed, the anti-racist movement has exhorted the public to 'challenge institutional racism'. However, in spite of its common usage, confusion over the exact meaning of the term abounds. The concept is often used in a loose, descriptive manner and has come to embrace a range of meanings, which are often imprecise, sometimes contradictory and frequently lacking in theoretical rigour. Discussions of individual attitudes, stereotyping, implicit guidelines, explicit rules and procedures, organizational arrangements, power sharing and structural determinants of minority status have all been subsumed within analyses of institutional racism. It has been argued (most notably by Mason 1982a and Williams 1985) that the failure to pull these different dimensions together into a more coherent framework not only undermines the value of the concept as an analytical tool, but also hampers the formulation of effective anti-racist strategies.

A review of the literature suggests that confusion and disagreement arise in several important areas. There is strong dissension, for example, over the level at which institutional racism operates, that is, whether racial inequalities are produced and sustained by structural forces or reducible to individual actions (see Mason 1982a,b). Structuralist interpretations, as most explicitly formulated in the work of Sivanandan (1982), attribute little importance to the rôle of individual actors, locating institutional racism in a broader historical understanding of social, economic and political change instead. Other writers, particularly in the social policy tradition, have stressed the central importance of individual involvement in the institutionalization of racism through policy formulation and implementation (CRE 1984a,b). These divergent views have rather different implications for the formulation of anti-racist strategies. Some programmes, such as that of the GLC, have attempted to incorporate some understanding of both dimensions, although strategies often become weighted towards an emphasis on individual responsibility and blame.

An interrelated area of uncertainty surrounds 'intentionality' of outcome. Definitions of institutional racism commonly incorporate the notion of unintentional discrimination, placing emphasis on the discriminatory *effects* of institutional procedures. Some, such as Scarman (1981), have gone so far as to argue that racial inequalities are simply the 'unintended consequences' of normal institutional practices, such as resource rationing. This particular line of argument, however, fails to consider why institutional operations have come to reflect racist ideologies and why one particular group becomes the victim of discrimination rather than another.

The distinction between and the implications of intentional and unintentional discrimination are important for anti-racist strategy. The relationship between racist and non-racist intentions and racist outcome is not straightforward. Racist intentions do not inevitably lead to racial inequalities. Conversely, institutional discrimination may arise in the absence of racist beliefs.[3] The potential for discrimination arises when normal institutional practices are consistent with, and thus draw on and are in turn reinforced by, prevailing racist ideologies. Institutional structure thus facilitates outcome, a point that can get lost in anti-racist training programmes, which tend to overemphasize the link between individual reform and institutional change (Gurnah 1984). Individuals do nevertheless play an important rôle in the production and reproduction of institutional racism. Even non-racist individuals help to perpetuate racist practices by their uncritical participation in racist structures; although here, as Jenny Williams (1985) points out, there is an important distinction to be made between individual action and intent.

Our understanding of the mechanisms through which inequalities are sustained has been further hampered by the use of 'institutional racism' as a descriptive term. As the early American writers (for example, Carmichael & Hamilton 1969, Blauner 1972) were eager to point out, institutional racism is a process. This process involves a range of institutions, whose procedures combine to produce a mutually reinforcing pattern of racial inequality. The term has, however, largely become synonymous with discrimination within a single institution. This rather narrow use of the concept is prevalent within social-policy-orientated research, keen to isolate racist procedures and practices as targets of change. This is not to undervalue this research; it is essential to the formulation of anti-racist strategies at the local level. However, the potential for racial inequality in housing, for instance, cannot be evaluated in terms of the presence or absence of discrimination within the housing market alone. Unequal opportunities in other areas can also produce residential segregation and deprivation along racial lines. Anti-racist strategies must therefore take account of the interrelationship between institutions and of the cumulative disadvantage arising from them.

The persistence of racial inequalities leaves few in doubt that institutional racism exists. Even Lord Scarman's rejection of an 'institutionally racist' Britain was founded on a very narrow definition of a 'society which knowingly discriminates' (Scarman 1981, p. 11), and was qualified by the assertion that 'unwitting discrimination' deserves careful investigation. It is, however, important to clarify the institutional mechanisms through which inequalities are sustained, especially if anti-racist challenges are to prove effective. Current debate points to several dimensions for consideration, namely:

(a) the normal process of institutional operation, most especially the procedures and practices involved and the ideologies that underpin them;

(b) the rôle of the individual in perpetuating institutional racism;
(c) the interrelationship between different institutions in producing and sustaining racial inequalities.

It is within this context that the discussion now turns to a more specific evaluation of institutional racism and anti-racism within the GLC. The analysis looks at anti-racist policy and institutional practice within an authority expressly committed to the promotion of equal opportunities and at the implications of this for the allocation of public housing resources. The data presented draw on an investigation of GLC housing policy and practice in Tower Hamlets, in the East End of London (Phillips 1986).

Anti-racism: GLC policy and practice

Anti-racist initiatives

The GLC's anti-racist policies appear to have been founded on a clear appreciation of the processes underlying the entrenched pattern of racism. Discussion documents acknowledged the historical roots of racism, the structural dimensions to racial inequalities, the rôle of individual prejudice and institutional arrangements, and referred to the interrelationship between different institutions in the production of inequalities (for example, GLC 1984). Anti-racist initiatives thus sought to challenge racist ideologies and their institutional forms as well as the way individuals thought and behaved.

The GLC not only aimed to tackle its own areas of administration, particularly resource allocation and employment, but also sought to engage in broader struggles, involving the police, the media and civil rights. The Anti-Racist Year programme, launched in 1984, vigorously promoted images of a battle to be fought. The programme itself went under the campaigning slogan of *London Against Racism*, and details were given of 'practical ways of opposing racism'. The public were informed of the GLC's 'all-out effort to challenge racism' and assured of the Council's resolve to 'fight racism'.[4] Londoners themselves were admonished to 'challenge racism' in all spheres and to take up the 'fight for equality'.

The GLC's comprehensive (if somewhat ambitious) anti-racist programme was clearly sensitive to the complexity of the task in hand. However, if the 'battle' against racism is to be won, programmes must progress well beyond the realms of rhetoric and rallies. What of the impact on specific racial inequalities in such areas as public housing? How successful was the GLC in moving from its broad, well-conceived policy statements to effective anti-racist strategies? These questions are posed in the light of the GLC's housing record in Tower Hamlets.

Anti-racism: the record

Past GLC policies and practices have had a significant impact on the structure of housing opportunities in Tower Hamlets. Until July 1985, when the GLC's housing management rôle was transferred to the borough council, the GLC was responsible for 32 000 properties in Tower Hamlets; over half (53 per cent) of the total accommodation in the borough. There is no precise record of the ethnic composition of GLC housing applicants, but the pattern of offers during the survey period of 1983–4 broadly reflected the ethnic mix of the borough. Local residents included Asians, West Indians, Vietnamese and Somalis as well as working-class whites. At the time of the 1981 Census, 13 per cent of the total population were of New Commonwealth and Pakistani origin, although recent estimates put the proportion much higher. For example, Bengalis, who constitute by far the largest group, are now estimated to number between 30 000 and 50 000 and to constitute approximately a quarter of the total population.

Asian demand for council housing in Tower Hamlets is unusually high. The structure of the local housing market, where over 80 per cent of the dwellings are council-owned, has prevented the Bengalis from fulfilling the well-known Asian predeliction for owner-occupation (Brown 1984). The Bengalis have, however, fared poorly in the competition for public housing resources and have tended to end up on the poorer quality, less desirable estates in the borough. Ten years ago, Parker & Dugmore (1976) suggested that GLC housing-allocation rules and procedures had contributed to this pattern of racial inequality. The ensuing years, however, brought minimal change and local community groups continued to criticize the Council for its racism. The most vociferous of these groups was the Spitalfields Housing and Planning Rights Service (SHPRS), who in 1982 provided yet more evidence to suggest that Bengalis were being allocated to the worst GLC estates in the borough (SHPRS 1982).

Housing inequalities in Tower Hamlets were therefore an obvious target for the GLC's anti-racist challenge and a number of policy and procedural changes were recommended. These included the introduction of a Race and Housing Action Team to monitor allocations and to deal with racial harassment, community consultative procedures, and anti-racist training for officers. Steps were taken to implement these initiatives within a broad framework of institutional reorganization, designed to enhance the priority given to racial issues. All provided an apparent infrastructure for change.

By the beginning of 1984, however, there were clear indications that all was not well. Following a local survey, SHPRS reiterated their earlier allegations of racial discrimination by the GLC, stating that 'two years ago ... immediate action was called for to end the racism of this allocations policy ... fundamentally, the situation on estates has not changed' (SHPRS 1984). The local GLC housing department in Tower Hamlets rejected the allegations, but the central Ethnic Minorities Committee, responsible for the implementation of the Council's anti-racist initiatives, called for further investigation.

The resulting evaluation of GLC housing policy and practice in Tower Hamlets revealed a considerable gap between anti-racist policy formulation and implementation (Phillips 1986). Analysis of an 18-month period of computerized housing records from the beginning of 1983 revealed significant racial inequality of outcome. The findings, which were based on an analysis of housing needs, preferences, offers and refusals, clearly vindicated the claims of SHPRS: Asians (most of whom were Bengali) had been offered a different and more limited range of housing than non-Asians,[5] often on older, less popular estates. Table 9.1 indicates the proportion of offers made to Asians and non-Asians on estates with most vacancies in the 18-month period. While approximately one-fifth (21.5 per cent) of the total vacancies were allocated to Asians, the proportion of vacancies offered to Asians on each estate varied greatly. For example, most of the vacancies on Berner (90 per cent) and Solander Gardens (86 per cent) were allocated to Asians; but in the case of Crossways and Exmouth, Asian offers comprised less than 5 per cent of the total. The spatial distribution of offers largely reflected, reproduced and sustained the ethnic character of the estates. It also served to maintain inequalities in housing condition. For example, Asians were less likely to be offered newer (post-1969) properties (23 per cent of the Asians receiving any kind of offer; as opposed to 34 per cent of all the non-Asians who received offers), or to get access to a garden (9 per cent of Asians, compared with 13 per cent for non-Asians). They also received fewer offers with central heating, for example: 40 per cent of Asians were offered such properties compared with 56 per cent of non-Asians.

Housing allocation involves the matching of applicants to vacancies on the basis of housing needs and preferences. Differences in Asian and non-Asian demand that might have accounted for the offer outcome were therefore examined. The uneven distribution could not, however, be explained by the different property requirements of the groups. As indicated in Table 9.2, even when variations in size and floor-level requirements were taken into account, significant differences in the distribution of offers by estate remained. Differences in stated area preferences for accommodation did, however, partly underlie the pattern. For example, Asians were more likely to opt for locations in the

Table 9.1 Distribution of offers to Asians and non-Asians on estates with most vacancies.

	Offers (total number)	Offers made to Asians (%)	Offers made to non-Asians (%)
Avebury	141	8.5	91.5
Berner	71	90.1	9.9
Boundary	99	36.4	63.6
Brownfield	80	3.8	96.2
Burdett	339	19.8	80.2
Chicksand	93	58.1	41.9
Collingwood	189	37.6	62.4
Coventry Cross	100	34.0	66.0
Crossways	71	4.2	95.8
Exmouth	69	4.3	95.7
Holland	95	56.8	43.2
Lansbury	397	7.6	92.4
Leopold	96	6.3	93.7
Lincoln	181	18.8	81.2
Ocean	361	34.6	65.4
Ranwell	152	9.9	90.1
Royal Mint Square	191	20.4	79.6
Samuda	141	5.0	95.0
Solander Gardens	64	85.9	14.1
St George's	93	24.7	75.3
Stifford	79	15.2	84.8
Wapping	121	35.5	64.4
Will Crooks	77	42.9	57.1
Wellington	77	3.9	96.1
total: all estates	5294	21.5	78.5

west of the borough than whites, particularly in Spitalfields. This reflected both positive forces for ethnic association, and constraints on minority housing choice, such as fear of racial harassment and, given the findings of other similar studies, possible racial bias in the presentation of housing alternatives at the time of application (Henderson & Karn 1984; CRE 1984a,b). Locational preferences, however, by no means explained the whole picture. Asians and non-Asians requesting the same sub-areas of the borough tended to be offered accommodation on different estates, as did those with no recorded locational preferences. In an attempt to clarify the effect of racial status in the allocation process, offers to Asians

Table 9.2 Distribution of offers below the fifth floor to Asians and non-Asians on estates with most 2- and 3-bedroomed vacancies.

	Offers (total number)	Offers made to Asians (%)	Offers made to non-Asians (%)
2-bedroom offers under the fifth floor	1460	24.3	75.7
Lansbury	81	2.5	97.5
Burdett	86	12.8	87.2
Ocean	84	44.0	56.0
Lincoln	31	3.2	96.8
Samuda	26	—	100
Avebury	48	10.4	89.6
Collingwood	56	51.8	48.2
Wapping	52	30.8	69.2
Holland	51	84.3	15.7
Ranwell	39	10.3	89.7
St George's	6	16.7	83.3
3-bedroom offers under the fifth floor	976	43.2	56.8
Burdett	56	41.1	58.9
Lansbury	64	20.3	79.7
Lincoln	53	58.5	41.5
Ranwell	47	10.6	89.4
Ocean	50	82.0	18.0
Royal Mint Square	38	52.6	47.4

and non-Asians with similar property and locational requirements were compared. Table 9.3 ranks the main estates on which these comparable group of Asians and non-Asians received offers; no overlap at the top of the range occurs. The potentially discriminatory effect of racial status in the GLC allocations process at this time is clear.

The overall effect of the GLC's housing allocation process during this period was thus one of racial concentration, segregation and deprivation. In general, Asians tended to be directed towards the west of the borough, especially towards the traditional reception areas of Spitalfields, Whitechapel and Shadwell (see Figs 9.1 & 9.2). Within these areas, particular estates such as Berner were more likely to be offered to Asians, while others remained predominantly white (for example, Exmouth). A similar pattern of segregation emerged within some estates, where

Table 9.3 Locations offered to comparable Asians and non-Asians, by property size.

The five locations most often offered to Asians (in rank order)	The five locations most often offered to non-Asians (in rank order)
For 2-bedroom vacancies	
Ocean	Burdett
Collingwood	Lansbury
Berner	Samuda
Boundary	Lincoln
Solander Gardens	Ranwell
For 3-bedroom vacancies	
Ocean	Ranwell
Lincoln	Crossways
Coventry Cross	Burdett
Collingwood	Samuda
Will Crooks	Wellington

particular blocks became dominated by Bengalis or whites (for example, on the Ocean estate). Divisions between Asian and non-Asian residential space frequently followed a quality divide, with the Bengalis concentrated in the poorer, less popular housing.

The investigation revealed that, despite the GLC's anti-racist commitment, racial discrimination remained formally and informally institutionalized within the housing allocation process. There were several points at which discrimination occurred. First, there were still rules which, although non-racialist in intent, remained racially discriminatory in effect. For example, the homeless were entitled to only one offer of 'hard-to-let' accommodation for most of this period.[6] Since Bengalis were significantly over-represented within this group (for example, 40 per cent of the Asians housed were homeless compared with 4 per cent of the non-Asians), they were disproportionately disadvantaged by this policy. Secondly, procedures for sorting and matching applicants and properties incorporated elements of racial bias. For example, during this period, applicants were graded for particular qualities of property early in the matching process. Although gradings largely paralleled priority categories, this subjective assessment produced discrepancies which tended to work to the Asians' disadvantage. Thirdly, discretionary areas of decision-making, in which racial stereotypes were used to judge the suitability of applicants for vacancies, emerged as an important source of bias. This clearly underlay the variation in offer pattern in Table 9.3. Locational preference stereotypes, based on the assumption that Bengalis and whites preferred to live in their 'own areas', broadly structured the distribution of offers. Explicitly negative racial stereotypes, categorizing the Bengalis as less respectable than the whites, frequently deprived the black minority of access to better quality housing, especially in predominantly white areas.

External pressure for change in the GLC's discriminatory housing allocation policies and practices found internal support from the Race and Housing Action Team and the Ethnic Minorities Committee. In the

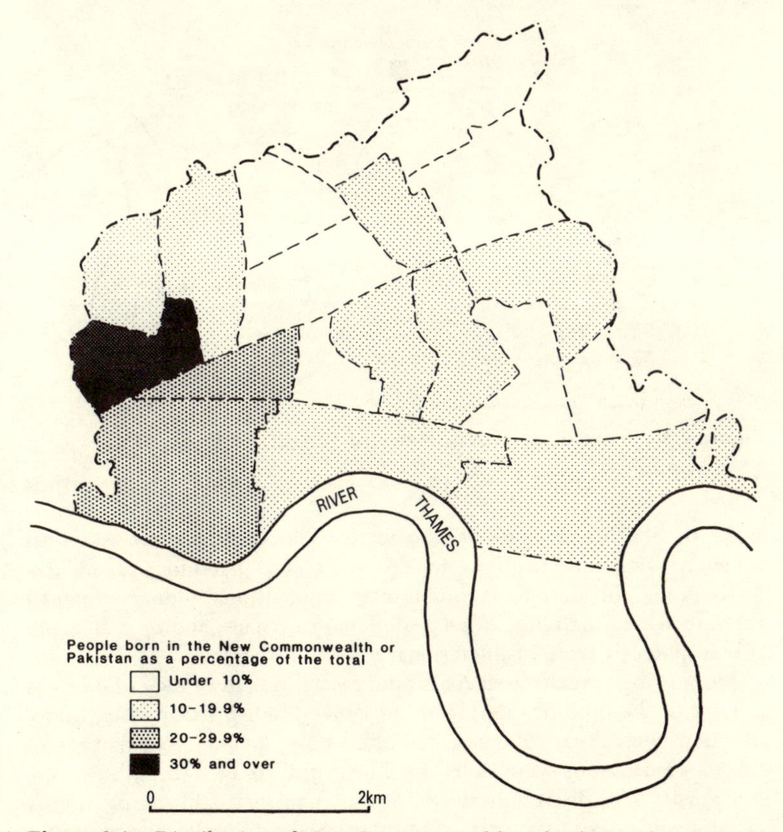

Figure 9.1 Distribution of New Commonwealth and Pakistani-born population in Tower Hamlets, by Ward, 1981.

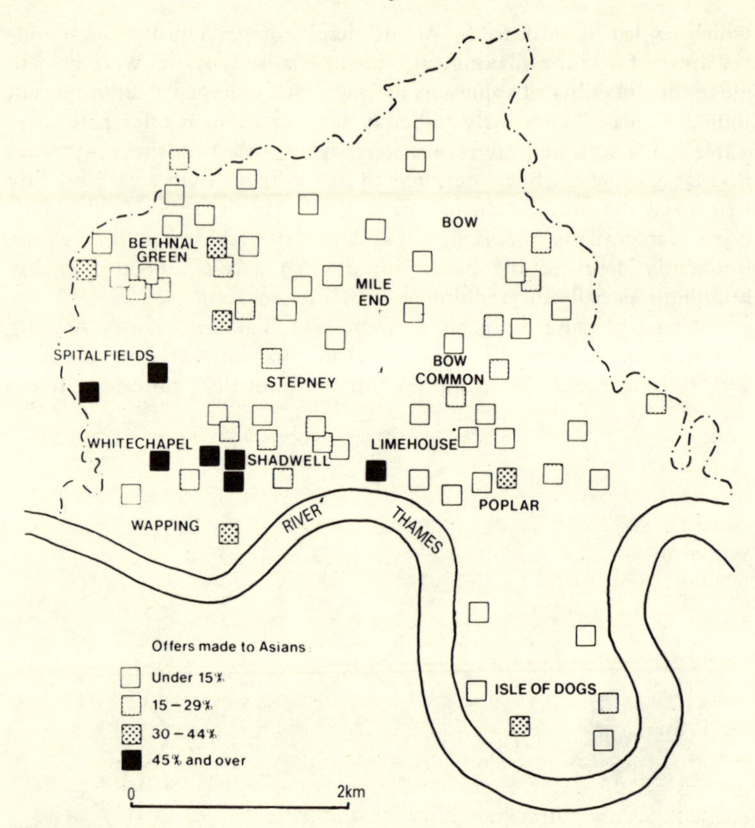

Figure 9.2 Proportion of vacancies offered to Asians on GLC estates, January 1983–May 1984.

wake of SHPRS' allegations, the formal allocation process was thus reviewed and modifications to the rules and matching procedures introduced. The aims were to eliminate unintentionally discriminatory rules, to widen the range of estates offered to Asians, and to reduce the rôle of potentially racist discretionary decision-making.

Monitoring revealed that Asians did receive a better range of offers as a result of the modifications (Phillips 1984). High-priority cases, many of whom were Asian, also received offers more quickly now that fewer subjective decisions were called for. Two disturbing factors, however, remained. First, discretionary decision-making was still giving rise to racial inequalities in the pattern of offers. Second, applicant acceptance and rejection of offers was tending to reinforce the established pattern of racial segregation and inequality in the borough. A small number of Asians did accept offers on traditionally white estates in the east of the area, but most were loath to settle outside the confines of their ethnic territory. Bengali-dominated estates, which were generally of poorer quality, were also unpopular with the whites. Asian unwillingness to break with their established pattern of residence was partly related to cultural factors and the pull of ethnic facilities. It was also, however, clearly related to the incidence of racial hostility in the borough. Offers on estates with a reputation for racial violence were often refused without even viewing the property. Intimidation by white tenants at the time of viewing also prompted refusals. Most Asians, however, simply feared the consequences of living outside the relative safety of Spitalfields.

The GLC's record of housing allocation in Tower Hamlets has not been good. Despite anti-racist initiatives, the Council did not manage to eradicate institutional racism from within its own structure. It was, however, also undermined in its attempts to promote greater racial equality by a range of external factors largely beyond its control. It is within this context that the potential success of the anti-racist challenge to housing inequalities in Tower Hamlets is evaluated.

Anti-racism: the challenge

Anti-racism set out to challenge institutional racism and to eradicate racial inequalities. Anti-racism itself, however, faces two serious challenges. First, there is a need to move from an understanding of the process of institutional racism to a strategy through which it can be tackled. Second, strategies must be implemented effectively if racial inequalities are to be eliminated. The GLC's anti-racist housing programme fell short on both counts.

This section turns first to the implementation of anti-racist strategies in the local Tower Hamlet's housing department. By 1984 a number of anti-racist initiatives, consistent with CRE recommendations, had been taken. Resistance to change was, however, apparent. The housing department's response to the anti-racist programme was slow, it was often reactive (for example, to outside pressure) rather than innovative, and it was also characterized by tokenism. Agreed strategies were not always properly implemented, while some areas remained untouched by the Council's anti-racism. For example, inadequate record-keeping rendered ethnic-origin information unusable, thus hampering the monitoring of housing allocation. The Race and Housing Action Team of ethnic minority officers proved effective in dealing with racial harassment, but felt isolated and uncertain of support from white colleagues on other issues. There were also few bilingual officers amongst those needing direct contact with the public. External criticism and internal pressure did, rather belatedly, bring improvements. The overall picture, however, was one of good intentions frustrated by bureaucratic inefficiency, local conservatism, lack of interest, and absence of political will. The result was that, at the outset at least, racism in housing was managed rather than challenged.

The second area of concern surrounds the gap between the Council's theoretical appreciation of institutional racism, as expounded in its political commitment to anti-racism, and the formulation of anti-racist practices. Difficulties appeared to arise in translating broadly conceived policies into effective local strategies. The problems, as apparent in the housing context, derived largely from two sources. First, the anti-racist housing programme failed to recognize fully the key discriminatory mechanisms within its own organizational structure. Secondly, the rôle of external factors in maintaining the entrenched pattern of housing inequality in the borough was perhaps underestimated. In order to elucidate these points further, the discussion now returns to the three dimensions of institutional racism introduced earlier, namely, the rôle of normal institutional processes, the rôle of the individual, and the interrelationship between different institutions in producing and reproducing racial inequalities.

Normal processes of institutional operation Housing management generally involves the allocation of a limited resource of variable quality. Allocation necessitates a process of rationing, sorting and matching as a means by which demand can be tailored to supply. A major objective in the allocation process is to let properties quickly so as to minimize vacancies. Such procedures and objectives are a well-established part of housing department routine and, as such, their potentially discriminatory effects present anti-racism with a difficult challenge. The consequences of these normal institutional processes for specific groups achieve particular salience under conditions of scarcity and decline, as in Tower Hamlets.

Tower Hamlets is one of the poorest boroughs in London, ranked second only to Hackney in terms of indices of deprivation and decline (London Borough of Tower Hamlets 1984, GLC Intelligence Unit 1985). Public housing here is both limited and poor, and spending cuts have exacerbated the mismatch between supply and demand. For example, in 1984, there were six households for every GLC vacancy, long waiting lists for homeless families, and a shortage of large accommodation. The GLC housing stock itself embodied huge inequalities in terms of design, standards of maintenance, heating provision (only about half of the properties had central heating) and estate reputation. Although most properties had all the basic amenities, general housing conditions were poor. Most were flats, few had garden access, and there was much vandalism and grafitti. Despite a homelessness problem, a quarter of the GLC's stock was classified as 'hard-to-let'. The relatively small proportion of better quality properties in desirable locations was therefore highly prized. Under such conditions, the normal process of sorting, rationing and matching becomes a powerful force in the structuring of housing opportunities. Any systematic negative bias in the routine application of allocation rules or in the use of discretionary decisions is likely to have grave consequences for the groups concerned.

The GLC did tackle the bias inherent in its formal rules and procedures. It was, however, less successful in eradicating the directly and indirectly discriminatory effects of racial stereotyping in discretionary decision-making. The act of stereotyping itself was rightly challenged through racism-awareness training, but the *need* to stereotype in the housing allocation process was not questioned. Individual prejudice is not sufficient an explanation. The incentive in part lay in the response to management pressure to fill vacancies quickly by keeping the offer acceptance rate high, that is, in a 'normal' institutional objective. Decisions had to be made about the most efficient way to allocate vacancies and how best to fill the less desirable housing. A range of preference and social stereotypes underlay these decisions. For example, where no clear preferences were given, Bengalis and whites tended to be offered properties in their 'own' areas, on the premise that these would be least likely to be rejected. It was also considered more efficient to offer poorer quality accommodation to Bengalis on the grounds that whites would probably refuse it. Officers were able to justify this unfair pattern in terms of their belief that whites were more deserving of the better quality resources than the lower-status, Bengali 'outsiders'.

Racial equality in GLC housing allocation was contingent upon greater anti-racist challenges to the normal processes of institutional operation. The sorting and matching of applicants competing for vacancies is an essential feature of the allocation process, but the possible consequences needed to be better understood. Explicit racial stereotyping was but one source of potential racial bias. A range of social attributes, such as class, family status and life-style play a rôle in the routine sorting process (Henderson & Karn 1984). Thus, even if race were to be eliminated as an explicit basis for differentiation, racial minorities could still be disadvantaged by their low class position, their family status (for example, Asian extended families and West Indian single parents are often held in low esteem) and their different life-style.

Furthermore, it was clear that attempts to minimize the vacancy rate through prejudgements based on stereotypes were entirely inconsistent with the anti-racist aim of widening minority housing opportunities. Such conflicts between housing management and anti-racist policies only brought confusion, with the result that the less clear-cut, 'lower priority' anti-racist goals were ignored.

The rôle of the individual The process of institutional racism is not reducible to the individuals who formulate and implement institutional policies. Organizational structures are shaped by wider social, economic and political processes (as reflected in national housing policies, for example) and institutional practices do not simply depend on the involvement of racist individuals for their discriminatory effect. The institutional structures facilitating racism are therefore of prior concern. Individuals do, however, act to sustain racism within institutions. Their rôle may be passive, as in a colour-blind approach; unintentional, as in locational stereotyping; or explicitly racist.

In the Tower Hamlets housing department, any officers involved in advisory rôles or discretionary decision-making had the power to discriminate. The propensity to discriminate along racial lines was tackled through a racism-awareness training programme. Housing officers, however, remained largely untouched by the anti-racist message of the trainers. There was still a tendency to stereotype Bengalis as dirty, irresponsible tenants, who would overcrowd and damage the Council's property. There was also resentment of their competition for the scarce public housing resource in Tower Hamlets. Interracial conflicts on the estates earned Bengalis the reputation of being 'troublemakers' (despite being the victims), while differences in life-style branded them as a nuisance to other residents. The consensus view represented the Bengalis as unfair competitors for resources and less deserving of the limited, better quality stock than the whites.

The GLC's anti-racist programme could not hope to eradicate the entrenched racism of officers overnight. It should, however, have sought to identify the way in which 'common-sense' racist attitudes (Lawrence 1982) gained expression through institutional structures. The anti-racist programme failed to identify the points at which individual officers had the power to discriminate or to consider the conditions under which discrimination became a necessary part of the job (as, for example, in the matching of applicants to vacancies). It also failed to recognize that some conditions even provided an incentive to discriminate racially. For example, officers responsible for the management of 'white' estates had a vested interest in discouraging Bengalis from settling there, since the inevitable interracial conflicts could only serve to increase their workload.

The failure to tackle the interface between institutional structure and individual racism in the allocation process not only hampered the progress of the anti-racist programme, but also directly undermined GLC anti-racist policies. For example, the Council declared itself opposed to the designation of 'no-go' areas for minorities. The cumulative effect of Bengali exclusion from better quality housing in Tower Hamlets, however, was to create virtual 'no-go' estates. While this may partly be explained in term of the unintentional racist consequences of normal institutional processes, the intention to exclude Bengalis from certain estates was explicit. According to an internal memo from the senior allocations officer, Bengalis were excluded from particular estates of better quality housing because of the

> difficulties which result from housing families who enjoy a different social and culinary style in blocks ... where the aroma of more savoury cooking tends to permeate the immediate area. Many (white) tenants find this situation unacceptable and their objections are certainly not racist but merely traditional. (*Bengalis in better quality housing*)

The intention to preserve white privilege at the expense of Bengali housing choice was clear.

The principle of differentiating on the basis of racial status in the allocation of resources was widely supported by GLC officers and brought internal resistance to anti-racist initiatives. Their racism was tacitly condoned by the lack of political will to enforce change and it also attracted explicit public sympathy. As Wellman (1977) and Karn (1983) have argued, racial discrimination constitutes a culturally sanctioned basis for resource allocation, especially in times of shortage. Anti-racist attempts to move towards greater equality have thus often met with strong public opposition (Platt 1985). This was evident in Tower Hamlets through the daubing of better quality properties allocated to Asians with racist slogans, racist correspondence to the Council from white tenants, and an arson attack on the Ethnic Minority Unit at the GLC.

The interrelationship between institutions It was argued earlier that racial inequalities reflect the cumulative disadvantage of institutional racism in various spheres. The implications of this for the improvement of housing opportunities within the public sector are enormous. It may be contended that, even if housing departments were to deracialize their policies and practices, spatial inequalities would persist along racial lines. The evidence for continuing racial segregation and deprivation in Tower Hamlets is strong.

At a time when the GLC were beginning to offer minorities a wider choice of properties, the Bengalis were not in a position to accept. There are major constraints on Bengali housing choice, which serve to maintain the status quo. Obvious disadvantage stems from the Bengali's weak position in the local labour market, where they are predominantly concentrated in the garment trade (Shah 1975, SHPRS 1980). Low wages, high unemployment[7] and disadvantages in the allocation of social security benefits (Gordon & Newman 1985) all constrain the Bengalis' ability to pay rent or to meet heating costs. Both restrict housing choice, while rent arrears can block opportunities for transfers within the council stock. Employment patterns, financial constraints and non-ownership of cars also combine to restrict area choice, such that Bengalis are more likely to opt for locations in the west of the borough, close to their main area of work. Furthermore, a heavy dependence upon shift-work increases the need to travel at night, a time when Bengalis feel most vulnerable to attack. This reinforces the desire to live in Spitalfields.

The East End has a long history of interracial hostility and conflict, as clearly exemplified by the presence of extreme right-wing organizations such as the National Front (Husbands 1983). Deteriorating local conditions have fuelled racial tensions in recent years as residents have competed for increasingly scarce resources. In the late 1970s, a Commission for Racial Equality report (1979) referred to the Bengalis in Spitalfields as a 'community under seige'. Since then the situation has deteriorated further, such that in a 15-month period, the GLC recorded 361 racial attacks (mostly on Bengalis) on its own estates alone. The fear of attack has given spatial expression to the structure of Bengali housing options in Tower Hamlets. Bethnal Green, National Front territory adjacent to the Spitalfields community, is largely avoided, its boundaries defended by hostility and violence. Offers on 'unsafe' estates with a history of violence are also often refused. The Race and Housing Action Team has attempted to combat racial harassment on GLC estates by protecting and supporting Bengali tenants, removing racist grafitti and taking steps to prosecute and evict perpetrators of the violence. However, increasing numbers of Asians have opted over the years to transfer from isolated locations to the safety of Spitalfields.

The local authority's ability to combat the effects of racial harassment on minority housing opportunities is limited. Progress rests in the hands of the police, who have hitherto failed to assure the Bengali community of protection or support in incidents of racially motivated attacks

(Stevens & Willis 1982). Checks on immigration papers and complaints that Bengalis are made to feel more like criminals than victims have undermined the community's confidence in justice. There has been no public outrage over racist attacks and no effective central government strategy to tackle them. Media images and government statements on immigration have instead continued to reinforce the image of black minorities as aliens, illegal immigrants and unworthy competitors for scarce resources (Hartmann & Husband 1974, Barker 1981). Popular racist activity and sentiment thus continue to sanction and reinforce the discriminatory institutional practices of those such as the police, housing managers and employers. All have served to produce and reproduce an entrenched and self-sustaining pattern of racial inequality.

Conclusion

The commonplace and rather imprecise use of the concept of 'institutional racism' belies the complexity of the processes encompassed within this single term. Institutional racism is both expressive and constitutive of the racism endemic in British society. As such it may be seen to reflect a wider historical process of racial categorization, domination and exploitation, to manifest structural features of society, and to be indicative of a wide range of ideologies, some of which are explicitly racist. Its tangible expression is to be found in the pattern of racial inequality, which despite its specificity in time and place, depicts a recurrent theme of social exclusion based on a negative evaluation of minority social standing and worth (Parkin 1979).

Any progress towards greater equality of opportunity within specific institutions is to be welcomed, especially as part of a wider anti-racist campaign. The benefits for minority housing opportunities within the public housing sector are certainly measurable. However, it must be acknowledged that specific anti-racist strategies, whether in housing, education or employment, will be undermined by other forms of institutional discrimination, especially in the present social, economic and political climate. Progress will be slow while anti-racist initiatives lack central-government co-ordination and backing, and while racial exclusion continues to be sanctioned through the racist ideologies supported by both the government and the media.

Notes

1 A recent survey indicated that Section 11 funds, intended for ethnic minority projects, were often being used for routine funding purposes (*Guardian*, 13 March 1985).
2 Evidence for this stems from first-hand information from talking to ethnic monitoring officers and from personal communications with the CRE. Young & Connolly's (1981) report details a similar picture for 1981.
3 Ibrahim (1984) has pointed out that the same sets of institutional structures and practices could be operated by blacks without any necessary improvement in minority opportunities.
4 All these quotations are drawn from publicity material included in the GLC's *London Against Racism* information packs.
5 Lack of information about ethnic origin prevented a more detailed breakdown by subgroup. Asians, who were by far the largest group of minority housing cases, were identified by name.
6 The GLC were following instructions from Tower Hamlets' borough council, who were responsible for the borough's homeless but could nominate them for GLC housing. The GLC were, however, still guilty of complying with racially discriminatory instructions.
7 For example, in 1980, 25 per cent of the Asians in Spitalfields were registered as unemployed compared with 11.4 per cent of the non-Asians here.

References

AMA 1985. *Housing and race: policy and practice in local authorities*. London: Association of Metropolitan Authorities.

Banton, M. 1955. *The coloured quarter*. London: Jonathan Cape.
Barker M. 1981. *The new racism: Conservatives and the ideology of the tribe*. London: Junction Books.
Ben-Tovim, G. *et al*. 1982. A political analysis of race in the 1980s. In *Race in Britain; continuity and change*, C. Husband (ed.), 303–16. London: Hutchinson.
Blauner, R. 1972. *Racial oppression in America*. New York: Harper & Row.
Brown, C. 1984. *Black and white Britain: the third PSI survey*. London: Heinemann.
Burney, E. 1967. *Housing on trial: a study of immigrants and local government*. Oxford University Press for the Institute of Race Relations.

Carmichael, S. & C. V. Hamilton 1967. *Black power*. London: Jonathan Cape.
Castles, S. 1984. *Here for good: Western Europe's new ethnic minorities*. London: Pluto Press.
CRE 1979. *Brick Lane and beyond: an inquiry into racial strife and violence in Tower Hamlets*. CRE Publications.
CRE 1984a. *Race and housing in Liverpool: a research report*. CRE Publications.

CRE 1984b. *Race and council housing in Hackney: report of a formal investigation.* CRE Publications.

CRE 1985. *Annual report of the Commission for Racial Equality, 1984.* CRE Publications.

Dummett, A. 1973. *A portrait of English racism.* London: Penguin.

Flett, H. 1979. *Black council tenants in Birmingham.* Research Unit on Ethnic Relations, University of Birmingham, Working Paper No. 12.

GLC 1982. *Race equality and ethnic minorities in London: future strategy report* (EM 164).

GLC 1984. *Race relations in London.* Unpublished report to the GLC Ethnic Minorities Committee (EM 754).

GLC Intelligence Unit 1985. *London data.*

Gordon, P. & A. Newman 1985. *Passport to benefits? Racism in social security.* London: Child Poverty Action Group & Runnymede Trust.

Gurnah, A. 1984. Racism awareness training. *Critical Social Policy* **11**, 6–15.

Hartmann, P. & C. Husband 1974. *Racism and the mass media.* London: Davis Poynter.

Henderson, J. & V. Karn 1984. Race, class and the allocation of public housing in Britain. *Urban Studies* **21**, 115–28.

Husbands, C. 1983. *Racial exclusionism and the city: the urban support of the National Front.* London: Allen & Unwin.

Ibrahim, F. 1984. Racism: class or non-class. *NATFHE Journal*, October.

Jones, M. M. 1972. *Prejudice and racism.* New York: Addison Wesley.

Karn, V. 1983. Race and housing in Britain; the role of major institutions. In *Ethnic pluralism and public policy*, N. Glazer & K. Young (eds). 162–83. London: Heinemann.

Knowles, L. K. & K. Prewitt 1969. *Institutional racism in America.* New York: Prentice-Hall.

Lawrence, E. 1982. Just plain common-sense: the 'roots' of racism. In *The Empire strikes back: race and racism in 70s Britain.* Centre for Contemporary Cultural Studies (ed.), 47–94. London: Hutchinson.

London Borough of Tower Hamlets 1984. *Planning research and information: a profile of Tower Hamlets.* Planning Department, Directorate of Development.

Mason, D. 1982a. Race relations, group formation and power: a framework for analysis. *Ethnic and Racial Studies* **5**, 423–39.

Mason, D. 1982b. After Scarman: a note on the concept of institutional racism. *New Community* **10**, 38–44.

Miles, R. 1982. *Racism and migrant labour.* London: Routledge & Kegan Paul.

Miles, R. & A. Phizacklea 1984. *White Man's country: racism in British politics.* London: Pluto Press.

Morgan, G. 1985. The analysis of race: conceptual problems and policy implications. *New Community* **12**, 285–94.

Parker, J. & K. Dugmore 1976. *Colour and the allocation of GLC housing*. GLC Research Report, 21.

Parkin, F. 1979. *Marxism and class theory: a bourgeois critique*. London: Tavistock.

Patterson, S. 1965. *Dark strangers: a study of West Indians in London*. London: Tavistock.

Phillips, D. A. 1981. The social and spatial segregation of Asians in Leicester. In *Social interaction and ethnic segregation*, P. Jackson & S. J. Smith (eds), 101–21. London: Academic Press.

Phillips, D. A. 1984. *Monitoring of the experimental allocations scheme for GLC properties in Tower Hamlets*. Unpublished GLC report (TH173A).

Phillips, D. A. 1986. *What price equality? A report on the allocation of GLC housing in Tower Hamlets*. GLC Housing Research and Policy Report, 9.

Platt, S. 1985. I'm not racialist, but. . . . *New Society* **21**, 289–91.

Rex, J. & R. Moore 1967. *Race, community and conflict: a study of Sparkbrook.* Oxford: Oxford University Press.

Scarman, Lord 1981. *The Brixton disorders, 10–12 April, 1981*. Cmnd 8427. London: HMSO.

Shah, S. 1975. *Immigrants and employment in the clothing industry: the rag trade in London's East End*. London: Runnymede Trust.

SHPRS 1980. *The Spitalfields survey: housing and social conditions in 1980.* Spitalfields Housing and Planning Rights Service and the Catholic Aid Society.

SHPRS 1982. *Bengalis and GLC housing allocations in E1*. Spitalfields Housing and Planning Rights Service.

SHPRS 1984. *Bengalis and GLC housing allocations in E1: an update*. Spitalfields Housing and Planning Rights Service.

Simpson, A. 1981. *Stacking the decks: a study of race, inequality and council housing in Nottingham*. Nottingham Community Relations Council.

Sivanandan, A. 1982. *A different hunger: writings on black resistance*. London: Pluto Press.

Sivanandan, A. 1983. Challenging racism: strategies for the 1980s. *Race and Class* **25**, 1–11.

Skellington, R. S. 1981. How blacks lose out in council housing. *New Society*, 29 January.

Stevens, P. & C. Willis 1982. *Ethnic minorities and complaints against the police.* Home Office Research and Planning Unit.

Wellman, D. 1977. *Portraits of white racism*. Cambridge: Cambridge University Press.

Williams, J. 1985. Redefining institutional racism. *Ethnic and Racial Studies* **8**, 323–48.

Workers Against Racism 1985. *The roots of racism*. London: Janius.

Young, K. & N. Connolly 1981. *Policy and practice in the multi-racial city*. London: Policy Studies Institute.

10 *Ethnic minorities and racism in welfare provision*

MARK R. D. JOHNSON

There is a long tradition in social geography of concern and interest in the field of migration, 'race relations' and segregation (Jackson & Smith 1981, Clarke *et al.* 1984). Much of this has been descriptive, attempting to establish, describe and explain patterns of settlement (e.g. Davies & Newton 1972, Newman 1985). Increasingly, however, there has been a trend towards consideration of social problems, including those arising from these differential patterns. Social geographers have also become increasingly interested in the topic of 'service accessibility', particularly with reference to the needs of disadvantaged groups. Recent examples include Whitelegg (1982) on health care; Guy (1985) on the shopping behaviour of disadvantaged consumers; and various conference papers on access to justice (in rural Britain) and 'public service provision' which included race and health care, and spatial inequalities in education (Mohan 1985). Thus we are, perhaps, moving away from an interest in patterns of social segregation towards a concern for the immediate effects and indirect consequences of segregation. This chapter attempts to extend that movement into the little-charted waters of social-welfare provision.

Unlike most other services, such as housing, welfare is not one that can be termed universal or for which there is a clearly defined demand – use of welfare services is almost entirely voluntaristic; normative need is more than usually liable to subjective interpretation, and some clients frequently reject official overtures while others reputedly use services to which their entitlement may be dubious. On the side of provision, there is a great variety of services and benefits available which may be delivered in the community, in institutions, to individuals, families or groups, and for various age groups. The providers of many of these services regard themselves as professional carers rather than administrators, and consequently record-keeping and analysis take a low priority – and as stigma may attach to receipt we also find that survey data can be misleading or at least sparse. Given the size of the 'business', with social-security payments representing nearly a third of government expenditure, it is perhaps surprising and regrettable that we know so little about its operation in the population.

Race, welfare, economy and demography

The welfare state, as a functioning system for social maintenance and equitable redistribution, is perhaps rightly regarded as a 'jewel in the crown' of British society. Certainly it has been given as a reason why people from other countries should wish to come and live in Britain (as for example in a *Sunday Express* front-page article, 'Irish young flood in to get dole', 8 September 1985). There is no evidence that it has been significant in attracting black migrants – economic and domestic reasons being perfectly adequate to explain their migration (Peach 1968, Jones & Smith 1971). However, the present government has consistently acted in a fashion that suggests that at some levels credence is given to allegations of misuse and 'living off the welfare', seeking to restrict the access of black people (and others who may be less easily identified as migrants, and hence may escape challenge) to the services of the welfare state.

At the least, this may be construed as acknowledging the racism of the electorate. For a long time now there has existed a condition of entry in the immigration officer's handbook that those admitted should not require or intend to make 'recourse to public funds' (see para. 42 HC 394, 1980. Indeed, it dates back to 1656 in some form (see Waterman & Kosmin, Chapter 11 below). This general phrase was not until recently clearly defined although it has now been specified as including supplementary benefit, housing benefit and family income supplement following the European Court of Justice ruling on UK immigration laws (HC 503, July 1985). Evidently it could be used as a way of inhibiting those legitimately settled here from gaining their just desserts and paid-for benefit.

Another example of the same form of effective racist exclusion was the attempt in 1981 to impose charges for use of NHS facilities on all those (in categories that largely excluded European and most white immigrants) who had less than three years' residence in the UK. This at once had the effect of causing black British-born people, of Asian or Afro-Caribbean descent as much as of Arabian origin, to be asked for passports or proof of eligibility on needing health-service treatment. The fact that this misguided regulation was amended to a 'six month' eligibility after a couple of years of campaigning protest (and evident failure to raise substantial sums: GLC, 1985) is little consolation.

The health service, and questions relating to needs met (or unmet) through it, is a separate issue which cannot adequately be discussed in this volume. There is an extensive literature on health needs and service provision (Johnson 1983, 1984a, b; Donovan 1984) with respect to ethnic minority issues, and from the perspective of other geographical authors (Rathwell & Phillips 1986). Even more recently there has been a growth of material to encourage anti-racist action or appropriate service provision (Training in Health and Race 1984, Mares *et al.* 1985). All of these contain references to a large and expanding professional literature, from a clinical or service-providing perspective, which suggests that the issues have at last received attention, perhaps because of the growth of community-based and black-led organizations fighting for change. It is not therefore within the scope of this chapter to discuss these aspects of welfare, important though they may be in affecting the totality of the 'quality of life' experienced by Britain's black population.

Social work and race

The dominant impression one obtains in reviewing the literature on race and welfare is how little and how late has been the development of writings in the academic or professional press (Johnson 1986). Prior to the 1971 report by Jones & Smith on the *Economic impact of Commonwealth immigration* there are only a handful of articles and books in the social-work literature. It is perhaps not surprising that, with a low level of knowledge and ability to assist minorities among social-work practitioners, they found a low level of usage among minorities – and that subsequent researchers expressed concern about minority failure to utilize services that were provided (Foren & Batta 1970, Derbyshire 1981).

As it is, apart from a seminal examination by Bessie Kent (1964) of the social worker's cultural pattern and 'how it affects casework with immigrants', the majority of reports implied that the roots of the problem for the service lay not with its own characteristics but with the 'immigrant'. This approach, in effect blaming the client, led to a number of descriptions of 'immigrant culture' (such as Morrish 1971 or Croucher 1972) which, while they may have had some value in training and understanding, were liable to create new stereotypes and reinforce disadvantage arising from migrant labour status and societal racial prejudice. Particularly, one may note that such practice-orientated writings as there were focused upon those specialized services such as fostering and adoption, where there was for whatever reason considerable use of social-work services or where it was considered that black clients might present a problem to the services (Fitzherbert 1967).

In all this there was very little concern for the client's view, or for the use of the general, generic services and income-support facilities. Indeed, from time to time one comes across the view expressed in this field, as in others, that to treat black clients as in any way different, to pay attention to their needs in a cultural context, would in itself be discriminatory or wrong in some way.

This view, of course, stems from the European Social Work belief that the individual is unique and in some way that this individuality has to be upheld – which conflicts with the understanding in most Asian cultures that the community is super-ordinate – and perhaps this may be related to some workers' desire to promote 'integration' by stressing this 'European' value. And of course it totally ignores the particular, collectively experienced hazard of racism. Further, it must be realized that, often with some justification, at least some sections of the community regard the social-welfare apparatus as being part of the state's system for social control (Ouseley *et al.* 1981).

Amazingly little attention has been paid to this 'mismatch' between, as it were, supply and demand. However, that does not mean that it has gone totally unnoticed among the minority black communities. They, as the Jewish groups before them (see Waterman & Kosmin, Ch. 11 below) have taken action through the churches, mosques or temples and local advice centres, creating 'a self-help movement that fills the gap left by statutory services unable or unwilling to care for the needs of black people' (*Faith in the city* 1985). As the black communities have become organized and made their presence felt, at least the practitioners have begun to prepare themselves better to provide some sort of service. The growth of black community organizations and professional associations having a campaigning rôle is relatively new, and has yet to be properly documented. Examples of such action may be instanced (BBC 1985, *Black children in care* 1983, Trought 1986), and the 'Fowler Review' (Cmnd 9517–9) has stimulated further activity such as the bulletin of the Committee for Non-Racist Benefits, based on the Community Information Project in Bethnal Green. However, many barriers still remain before Britain's black communities can fully enjoy the benefits of the welfare state to which, after a substantial period of settlement (if migrants) or, in increasing numbers of cases a lifetime, they have more than adequately contributed through wealth creation, labour power and taxation. It is to this more structural aspect (sometimes referred to as institutional racism) rather than to caseworking and welfare-provision reforms, that I wish to devote the remainder of this essay. Some of these structuring factors are geographical, others economic, and others more demographic; but the conclusion would appear to be that despite the rhetoric that has developed over time the position of black people in the welfare state is deteriorating rather than improving.

Demography

It is well established that most migrant labour comes from a particular group or generation – usually in the mid-20s to 30s – and that with the exception of more recent refugee movements most black migration to this country has been of this type, supplemented by the completion of families, which usually means spouses and children rather than dependent elders. Secondly, we are aware that the majority of migrants of Caribbean origin came to Britain in the 1950s and early 1960s, while the migrants of the Asian communities followed a decade later. Consequently, we are now faced with a cohort of Afro-Caribbeans approaching retirement age (Barker 1984) as their descendents enter economically active life to face unemployment, as demonstrated in survey after survey. This must mean that we are about to see a crushing change in the ratio of economically dependent to economically active – and even more so if that denominator were for those 'gainfully active', for which the data are not available (see Table 10.1).

This approach is that taken by the DHSS in their recent Green Papers on the reform of the social-security system (Cmnd 9517–9) in order to establish that a 'crisis of financing' will shortly face the national insurance system, and has been used to establish the need for Inner City Partnership funding. From the table it is clear that while the numbers of pensionable age in the white population have grown somewhat, and the dependency ratio (i.e. the number of economically active who are at least in theory available to support dependents outside that age range) has worsened slightly, the situation with regard to the minority or black community has changed even more markedly. Initially, up to 1971, this was largely due to changes in the proportion of children, and the apparent fall in pensioners may be in part related to white people of colonial birth, but now it is apparent that pensioners are becoming a significant part of the dependent sector.

Table 10.2 unpacks these figures for three of the major categories of Britain's black population and compares them with the best estimate of the white population. From these data it would initially appear that *at present* the black communities are in a better position theoretically than the white population to care for their dependents – notably those of pensionable age – with the possible exception of those of Pakistani origin, where a sort of 'baby bulge' can be observed. However, these figures are of course misleading because of the differentials in unemployment observed in the 1981 Labour Force Survey and even more dramatically in the 1984 data (Table 10.3), and because in some of the Asian communities (notably the Muslims, which includes virtually all those of Pakistani origin) it is incorrect to assume that any major

Table 10.1 Changes in dependency ratios, 1961–81.

	1961[a] (%)	1966[b] (%)	1971[c] (%)	1981[d] (%)
NCP population from 'New Commonwealth' or Pakistan				
under five	2.05	3.66	13.96	10.73
from five to school leaving age	8.30	14.35	21.94	22.70
pensionable	3.04	4.27	1.71	3.93
dependency ratio	6.5	3.5	1.66	1.68
'white'[e] population				
under five	7.80	8.52	7.83	5.89
from five to school leaving age[f]	15.16	14.48	15.57	16.15
pensionable	14.88	16.13	16.75	18.15
dependency ratio[g]	1.64	1.56	1.49	1.48

[a] 'Commonwealth immigrants in the conurbation' (England & Wales): N.B. Pensioners only, all over 65.

[b] 'Commonwealth immigrants tables' (England & Wales). Pensioners includes women over 60.

[c] 'Country of birth tables' (Great Britain). Parents born in 'New Commonwealth' or Pakistan.

[d] 'Country of birth tables' (Great Britain). Head of household born in 'New Commonwealth' or Pakistan.

[e] 'Total white population (1961, 1966). Both parents UK born (1971); head of household UK born (1981).

[f] School leaving age: 14 in 1961–71 and 15 in 1981.

[g] Number of those in economically active age group per 'Dependent'.

Source: Census volumes indicated above.

proportion of females aged 16–60 are gainfully economically active. Given that fact, it is certain that each black person gainfully employed is supporting a much higher number of dependents than the average white person – and therefore surprising that such studies of benefit uptake as do exist have not found a much higher level of reliance. Yet this is not the case – our own survey in the Midlands found that only 37 per cent of low-income Asian families had applied for supplementary benefit compared to 54 per cent of low-income whites and 57 per cent (virtually the same) of low-income Afro-Caribbean households.

Looking to the future, it is interesting to note on the bottom line of Table 10.2 that while the numbers of 'potential pensioners' (i.e. those aged between 45 and pensionable age) in the white population are roughly equal to the number of actual pensioners, for the black communities between three and six times as many will become pensioners by 1995. This must have consequences for the future, in terms of

Table 10.2 Dependency ratios for selected groups, 1981.

	Birth place of head of household			
	UK (%)	Caribbean (%)	India (%)	Pakistan (%)
Population				
under five	6	7	10	19
5–15	16	25	23	26
of pensionable age and over	18	3	6	2
dependency ratio[a]	1.48	1.9	1.56	1.12
pensioner support ratio[b]	3.3	22.0	10.4	30.3
potential pensioner ratio[c]	1.1	6.3	2.6	6.6
proportion of households 'lone pensioner'	14.8	2.4	4.9	1.3

[a] Calculated by reference to numbers not of working age relative to those aged between 16 and pensionable age.

[b] Ratio of pensioners to those of working age (cf. Cmnd 9519).

[c] Ratio of those aged between 45 and pensionable age to those of pensionable age, as a measure of future potential pensioners.

Source: 1981 Census, calculated from data in 'Country of birth' volume.

the need for public provision. Social services departments *are* becoming aware of the need for provision of places in old people's homes and day-care services, including meals-on-wheels (Liddiard 1980), but have traditionally appeared to rely on the expectation that the 'extended family' and 'Third World cultural values' will provide for this group's needs (Fenton 1985). Bhalla & Blakemore's (1981) survey of the Elders in Ethnic Minorities has already shown that this was optimistic, and placing considerable strain on black families. Much of the under-utilization was due to a lack of knowledge about services, or to the fact that the services provided were inappropriate (e.g. in dietary terms – both for Asian *and* Afro-Caribbean clients), and to a failure of the service providers to investigate in depth the needs of black clients. This 'selective blindness' (or deafness) on the part of service providers is commented on by Ouseley *et al.* who say: 'there appears to be a myth among social services that low takeup . . . by the Asian community is because of extended family [networks]. This is not so because pressures of immigration laws and new environment [by which I understand him to mean problems in housing] have prevented the *existence* of extended families' (Ouseley 1981, p. 70). The lack of adequate housing is not a problem that is going away – indeed, the evidence is that since large

Table 10.3 Unemployment rates in Great Britain, by ethnic origin.

1981	White (%)	West Indian or Guyanese (%)	Indian, Pakistani or Bangladeshi (%)	Other (%)	All (%)
male	9.7	20.6	16.9	13.9	9.9
female	8.7	14.5	17.9	14.7	8.9

1984	White (%)	West Indian (%)	Indian (%)	Pakistani or Bangladeshi (%)	Other (%)	All (%)
male	11.0	28.5	13.1	33.6	17.4	11.5
female	10.1	16.6	18.3	39.6	20.2	10.5

Source: Labour Force Survey, 1981 and 1984.

units of accommodation are so lacking it will become increasingly difficult for families to house their elderly relatives, and this situation will be aggravated precisely by the move into local authority housing that has been observed among the Asian community (Cross & Johnson 1982, Robinson 1980).

Economy, society and geography

From the demographic, we can now turn to economic and geographical aspects to examine their impact, modulated by the effect of race, on the welfare of the black communities. I have already alluded to the significance of selective unemployment, but perhaps Table 10.4 will make the point with greater force, using two West Midlands boroughs which rarely appear in studies of this kind but which have significant minority communities and have been generally badly affected by unemployment. It will be evident that the black groups are much more heavily concentrated in areas worst affected – and in which it may be presumed that other services and opportunities are equally depressed. An analysis of the ward-level distribution of ethnic minorities and unemployment rate data for Birmingham, Coventry and Wolverhampton undertaken by Malcolm Cross produces Spearmans Rank Correlation Coefficients in excess of 0.7, which is highly significant (Cross 1985). For Birmingham, the so-called 'Core Area' of the Inner City Partnership scheme, which has been designated on the grounds of its environmental

Table 10.4 Proportions of population subgroups living in high-unemployment wards, February 1985.

	Walsall[a] (%)	Sandwell[b] (%)
'white' population	47.9	41.6
'white' under fives	40.8	44.2
'white' pensioners	56.2	42.6
'NCP'[c] population	86	75
'NCP' under-fives	86	77
'NCP' pensioners	87	76

[a] Walsall – worst 10 wards out of 20, all above borough average of 16.5%.
[b] Sandwell – worst 11 wards out of 24, all above borough average of 19.7%.
[c] 'NCP' – population in households where head of household was born in the 'New Commonwealth' or Pakistan.
Source: West Midlands County Council reports.

and social disadvantage, contains 76.6 per cent of the black population but only 18.9 per cent of the white population. The same area has *twice* the city unemployment rate; 308 of its 600-odd enumeration districts fall into the most deprived 2½ per cent of the national total; and the rates of overcrowding (at over 12 per cent) and youth unemployment (over 40 per cent in 1981) are the worst even among the eight Partnership areas of urban Britain.

None of these data suggest that the black community of the future will find it easy to take on the burden of managing and financing its own welfare. Yet, as I have already cited Ouseley as observing, there is a politically convenient myth that 'self-reliance' is actively desired by Asian communities and this generates the 'triple jeopardy' of the ethnic minority elderly (Norman 1985). Allied to this is the insistence that service providers are faced with 'special needs' requiring special provision which is apparently most effectively dealt with by reliance on voluntary agencies and community groups – 'the traditional way of pioneering and developing services' (Geoff Ward, in Norman 1985). As Ward goes on to say, 'we must beware that this is not just a cop-out!' But there is undoubtedly (and for good financial reason) an increasing reliance by policy makers upon the use of central and local government funding of voluntary agency activity as a means of stretching budgets (Elliott *et al.* 1984, DoE 1985). Having rejected the notion of special provision (except though the 'facilitating' and transitory rôle of Section 11 money, which rarely finances service delivery posts carrying any supporting resource) we discover time and again that the agencies wish in a sense to 'privatize' provision, that services for black-community specific needs are 'best provided through black-community voluntary agencies' (e.g. Liddiard 1980[1]). This does not seem to have been an option considered for the white suburbs or 'hard-to-service' rural communities. Once again we find that the state, by changes in policy or by the coincidence of factors that are not explicitly racist in themselves, is systematically producing detrimental effects for black communities – which I and others term 'structural racism', a more insidious but nonetheless real and experienced form of deprivation.

Legislation, citizenship and benefit

But the state has more direct ways of affecting black people's access to welfare services. I have already alluded to the provisions in the Immigration Rules, which restrict those given entry clearance subject to the condition that 'the parties will thereafter be able to maintain themselves and their dependents adequately without recourse to public funds' (HC 503). The most recent revision to those rules has laid down that this includes not only housing under the Homeless Persons Act 1977, but also supplementary benefit – which gives access to many other benefits such as free school meals. One wonders what proportion of black people who in our survey said they did not apply for benefits on the grounds of ineligibility did so because of their consciousness of this restriction. And it must be remembered that because many immigrants came here having become skilled workers in their country of birth, an estimated '350 000 male ethnic elderly will fail to make full National Insurance Contributions (i.e. at least 40 years) before they retire and therefore be dependent on Supplementary Benefit' (Glendenning 1980). This would appear to be a *prima facie* case of the state enabling Capital to avoid the full costs of the 'social reproduction of labour', by calling on a reserve army.

More recently, the whole benefits system is undergoing a process of review, for which a series of tribunals took evidence, leading to the publication of the Green Papers referenced earlier (Cmnd 9517–9). Surprisingly perhaps, only two organizations out of over 60 giving evidence were identifiably concerned with ethnic minority needs – these did not include the Commission for Racial Equality – and the only reference in these documents to the black community was the statement that the 'Which Benefit' leaflet is now available in six Asian languages. But a number of changes are proposed which will affect black people either directly or indirectly (WIW 1985). There is to be greater liaison with other (i.e. local) social-welfare personnel – which may, if reforms in social-worker training enable them to overcome their problems in communication with black clients, assist black people. Some 'social fund' payments will be available on a wider basis – albeit sometimes 'recoverable' – instead of only being available to those drawing supplementary benefit; but others, including the presently automatic maternity grant, will disappear or only be available to those 'eligible for income support'. That same income support will be set at a lower level for the under-25s – the precise section of the black community currently most affected by unemployment. There was also to be instituted a 'presence test' although it was unclear how many years' residence would have been required to qualify – or the effect of extended visits abroad to Asia or the Caribbean. The White Paper, following resistance by various campaigning groups (many under the umbrella of the Committee for Non-Racist Benefits), withdrew the presence test but implies new checks in its place (para. 3.30, Cmnd 9691).

The revised programme for action incorporates a number of other changes, reflecting no doubt the impact of some 7000 submissions of written evidence, and these have improved the outcome of the review for the black population. For example, family income supplement (another benefit included under the restrictions of HC 503) was to have required a longer period of employment in order to qualify for a shorter duration of payment, disadvantaging those in the most marginal or unstable jobs. As 'family credit', with a *wider* population eligible, it may prove of greater benefit although some reservation needs to be entered concerning those employed in such marginal occupations as the garment industry ('rag trade'), in which many Asian women find work, but where health and safety legislation (and probably other employment laws) are reportedly more honoured in the breach than the observance (WMLPU 1984 gives some evidence on this). Cuts in Housing Benefit (again, see HC 503), originally intended to save some £500 million, will undoubtedly affect most severely those in the worst accommodation, which includes many black citizens. Finally, and potentially most serious, special-needs payments will now come out of a 'social fund' which will be discretionary and cost-limited – i.e. subject to a fixed budget ceiling.

What evidence we have, particularly from other areas such as housing or promotion and recruitment, suggests that increasing discretion enables more direct discrimination on the basis of racial prejudices, and so works against the interests of black people. This may be contentious, and benefit officers may well exercise their discretion in favour of black clients; but the fixing of a budget suggests that even if they do so, if budgets are allocated on a local (or even regional) basis, then areas with a high black population may more rapidly exhaust their budget. The revisions in the White Paper are intended to reassure the public over this, stating that local area budgets will be assessed on local need partly based on current demand, and will be guaranteed some flexibility. However, fixing of budgets implies some ceiling, and the need for economies is likely to prevent this being set at a generous level. Certainly if the economy continues to decline, the principal areas of black settlement will continue to be those most affected; and it is uncertain how far local budgets will be assessed in line with some measure of local unmet need. Given that there are at present relatively low levels of take-up of benefits, it would seem that strategies to encourage the needy to obtain their entitlement would merely benefit the earliest claimants. Accessibility, or proximity to a benefit office, may become a critical geographical factor.

Conclusion

In conclusion, it may be seen that there are many aspects of the welfare state in which racism – personal or political/ideological responses to individual prejudice, institutional, or structural – affects the quality of life for Britain's black citizens. Unfortunately, while the victims are only too aware of this (even if sometimes denied the opportunity to express their grievances), the system and its professionals – and the academics who study both – are or have been relatively slow to explore these issues and to seek remedies. The situation is changing, in response to considered political pressure and organization among the black communities and perhaps more quickly in response to crises of social order, exemplified by rioting. Equal opportunity statements are being written, and ethnic monitoring to collect evidence to promote policy changes is being undertaken. Social workers *do* now have the opportunity to learn about aspects of culture, so they can no longer claim not to have the information to distinguish between West Indian and West African cultural beliefs or between Asian religions. Some even undergo anti-racism training, although few enough carry it into practice (Richards *et al.* 1985). Service delivery and communications or consultation procedures are improving slowly, but there is a lot more room for improvement. And continually, while reform is going on at a surface level, the structural pressures are worsening, threatening to undo all the achievements of 'multiculturalism'. Over ten years ago Boss & Homeshaw (1975) wrote that

> The services are there but only help the people who are lucky enough to know of their existence or who are referred to them by other formal agencies. Unfortunately these agencies . . . seem mostly to be concerned with blackness only if it appears to be a threat to the existing social fabric. The rest of the black population, providing it remains quiet, will continue to subsist in conditions of . . . overall environmental poverty (p. 337).

It is ironic that the conference to which *this* essay was originally addressed took place less than a week before the 1985 disturbances in Handsworth (and subsequent events in Brixton, Tottenham and Toxteth) demonstrated that at least some sections of the 'black community' were no longer prepared to tolerate these conditions. The future will demonstrate to what extent this 'threat to the social fabric' generates a more lasting set of responses in the employment and welfare fields than did the 1981 disturbances.

Note

1 Evidence to support this trend in policy orientation is largely verbal or 'unpublished' – sources include GLC briefing documents, interviews with key personnel, and the reports of voluntary agencies themselves.

References

Barber, A. 1985. Ethnic origin and economic status. *Employment Gazette* **93**(12), 467–77.

Barker, J. 1984. *Black and Asian old people in Britain*. London: Age Concern England.

BBC 1985. *Black going grey* (Television programme: video and leaflet). Open Space, BBC TV.

Bhalla, A. & K. Blakemore 1981. *Elders of ethnic minority groups*. Birmingham: Commission for Racial Equality/All Faiths for One Race.

Black children in care 1983. Evidence to the House of Commons Social Services Committee. Association of Black Social Workers and Allied Professions, London.

Boss, P. & J. Homeshaw 1975. Britain's black citizens. *Social Work Today*, 18 September, 334–7.

Clarke, C., D. Ley & C. Peach 1984. *Geography and ethnic pluralism*. London: Allen & Unwin.

Cmnd 9517–9 1985. *Reform of social security*. HMSO.

Cmnd 9691 1985. *Reform of social security: programme for action*. HMSO.

Cross, M. 1985. Black workers, recession and economic restructuring in the West Midlands. *CRER Conference*, University of Warwick.

Cross, M. & M. Johnson 1982. Migration settlement and inner-city policy. In *Migrant workers and the metropolis*, J. Solomos (ed.), 117–33. Strasbourg: European Science Foundation.

Croucher, D. 1972. A British social worker's impression of Jamaican life. Reprinted in BASW 1978, *Studies in intercultural social work*. Birmingham: British Association of Social Workers.

Davies, P. & K. Newton 1972. The social patterns of immigrant areas. *Race* **14**(1), 43–57.

Derbyshire 1981. *A study of social provision to ethnic minority groups*. Derbyshire County Council.

DoE 1985. *Five-year review of the Birmingham Inner City Partnership*. Inner Cities Research Programme 12, Department of the Environment.

Donovan, J. L. 1984. Ethnicity and health: a research review. *Social Science & Medicine* **19**(7), 663–70.

Elliott, S., G. Lomas, & A. Riddell 1984. *Community projects review: a review of voluntary projects receiving urban programme funding*. Department of the Environment.

Faith in the city 1985. Report of the Archbishop's Commission on Urban Priority Areas. London: Church House.

Fenton, C. S. 1985. *Race, health and welfare*. Bristol: Department of Sociology, Bristol University.

Fitzherbert, K. 1967. *West Indian children in London*. London: Bell.

Foren, R. & I. Batta 1970. Colour as a variable in the use made of a LA Child Care Department. *Social Work* **27**(3).

GLC 1985. *Ethnic minorities and the NHS in London*. London: Greater London Council.

Glendenning, F. 1980. Another turn of the screw. *New Age*, Spring, 27–9.

Guy, C. 1985. The food and grocery shopping behaviour of disadvantaged consumers. *Transactions, Institute of British Geographers* **10**(2), 181–90.

HC 394 1980. *Statement of changes in immigration rules (under Section 3(2) of Immigration Act 1971)*. London: HMSO.

HC 503 1985. *Statement of changes in immigration rules*. London: HMSO.

Jackson, P. & S. J. Smith (eds) 1981. *Social interaction and ethnic segregation*. London: Academic Press.

Johnson, M. R. D. 1983. *Race and health: a select bibliography*. CRER, University of Warwick.

Johnson, M. R. D. 1984a. Ethnic minorities and health. *Journal of the Royal College of Physicians* **18**(4), 228–30.

Johnson, M. R. D. 1984b. Setting the scene. In *Unequal and under 5 in the West Midlands*. East Birmingham CHC, 5–10.

Johnson, M. R. D. 1986. *Race and care: a select bibliography*. CRER, University of Warwick.

Jones, K. & A. D. Smith 1971. *The economic impact of Commonwealth immigration*. Cambridge University Press/National Institute for Economic and Social Research.

Kent, B. 1964. The social worker's cultural pattern and how it affects casework with immigrants. Reprinted in Triseliotis, J. 1972 *Social work with coloured immigrants*, 38–54. London: Oxford University Press for the Institute of Race Relations.

Liddiard, R. 1980. Background to the world of immigrants, *Health and Social Service Journal*, 812.

Mares, P., A. Henley & C. Baxter 1985. *Health care in multi-racial Britain*. Cambridge: National Extension College.

Mohan, J. 1985. Urban politics and public service provision. *Area* **17**(2), 180.

Morrish, I. 1971. *The background of immigrant children*. London: Allen & Unwin.

National Association of Citizens Advice Bureaux 1983. *Immigrants and the Welfare State*. Chapeltown Citizens Advice Bureau/Harehills and Chapeltown Law Centre.

Newman, D. 1985. Integration and ethnic spatial concentration. *Transactions, Institute of British Geographers* **10**(3), 360–76.

Norman, A. 1985. *Meeting the needs of older people from the ethnic minorities*. Conference Report, Centre for Policy on Ageing.

Ouseley, H., D. Silverstone & U. Prashar 1981. *The system*, Runnymede Trust/ South London Equal Rights Consultancy.

Peach, G. C. K. 1968. *West Indian migration to Britain*. London: Oxford University Press for the Institute of Race Relations.

Rathwell, T. & D. Phillips 1986. *Health, race and ethnicity*. London: Croom Helm.

Richards, J. K., S. Griffiths & M. Nicholas 1985. *Survey of racism awareness*. Trent Polytechnic Papers in Education.

Robinson, V. 1980. Asians and council housing. *Urban Studies* **17**(3), 323–31.

Training in Health and Race 1984. *Providing effective health care in a multi-racial society*. Training in Health and Race/National Extension College.

Trought, A. 1986. Self-help and the Chinese community. *Self-Help News* **5**, 1.

Whitelegg, J. 1982. *Inequalities in health care*. Retford: Straw Barnes.

WIW 1985. Report in *West Indian World*, 7 August.

West Midlands Low Pay Unit 1984. *Below the minimum – low wages in the clothing trade*. West Midlands Low Pay Unit.

11 *Ethnic identity, residential concentration and social welfare: the Jews in London*

STANLEY WATERMAN AND BARRY KOSMIN

Whereas it was once an ideal that immigrant groups entering a host society would disappear into it within a generation or two, reality has shown that many remain distinct and distinctive. Certain ethnic groups maintain high levels of concentration long after they have ceased to be immigrants, a result of spatial segregation or social prejudices which militate against their full integration or assimilation (Kantrowitz 1981, Jackson 1981). Some others appear to occupy an intermediate position, creating ethnic clusters but avoiding segregation, while at the same time integrating functionally into the host society.

The Jews are one of the most long-established ethnic groups within the United Kingdom. Several generations have now passed since the mass immigration of East European Jews first made its mark on cities in Western Europe and North America after 1880. Though rarely the first Jews to arrive, the immigrants in that wave are largely responsible for providing the present-day composition and tone of these communities, especially in the United Kingdom.

At first involved in petty trade and manufacture, Jews in general have been upwardly mobile, moving into more sophisticated manufacturing, trading and services, the professions and the academic world (Pollins 1982). This has been accompanied by demographic and residential changes. Birth rates have dropped, resulting in small families and an ageing population; there is greater secularization and a greater opportunity for assimilation. Spatially, there has been a marked suburbanization. Many of these trends and the issues they bring in their wake have been recognized (Gould & Esh 1964, Lipman & Lipman 1981).

Although Jews have become better off there has been a conscious effort to retain a cultural distinctiveness. Synagogues (often doubling as community centres) and educational institutions have retained an overall importance, while foodstores and recreational facilities reinforce the distinctiveness. Thus, although the Jews are visually little different from their host community three or four generations after immigration, attachment to certain aspects of culture remains strong.

As most Jewish communities exhibit high levels of spatial concentration, it appears that cultural distinctiveness might act as a restraint on the dispersion of the majority of Jews through the total metropolitan area. Why do Jews continue to cluster as a group and what implications does this phenomenon carry for understanding similar processes amongst other middle-class ethnic groups in Britain? It can be hypothesized that upwardly mobile, middle-class ethnic groups that elect not, or are unable, to assimilate fully into the host society need spatial clustering to allow viable ethnic institutions to flourish. At the same time, they avoid segregation which might isolate them from the benefits to be accrued from full functional integration into the society.

The Jewish population

Identity and identification

In some contemporary societies the problem of accurately locating the Jewish population is aided by the national census bureau which asks a question on religious affiliation and ethnic affinity or identity. This is the case in such countries as Canada (where conflicting interpretations can be gleaned as both questions are asked separately), Ireland (in both the Republic and the North), Australia, Israel and the USSR. Neither the United States nor Great Britain ask a religious question in their national census, so their Jewish voluntary social service, educational and community agencies receive no direct informational aid from the census. As a consequence, they must generate their own data by surveys or use surrogate and indirect indicators for estimating the size and locations of the populations they serve (Kosmin 1978; Kosmin & Levy 1983b, 1984).

How do we define a Jewish population? Is there a difference between Jewish self-identification and identification by the majority society or official state system? Is there a reciprocity between Jewish self-identification and the official system? If there is, then the way in which Jewish populations are defined will vary from country to country.

The problem of Jewish definition is longstanding, for it has implications for the Jewish community itself and for its relationships with the Gentiles who form the majority population in all states throughout the world with the exception of Israel.

In the 20th century alone, Jews have been defined by several different criteria. There exists a strictly Orthodox definition, derived from Jewish religious law (*halacha*). In the *halachic* definition, any person born of a Jewish mother or converted to Judaism to the satisfaction of the established Orthodox authorities is Jewish. All others are not.

This, of course, can lead to confusion. For instance, a woman who converts from Judaism to Christianity remains Jewish *halachically*, as do her children. This can reach a state of the absurd in which, after two or three generations, individuals on the maternal line who had had no contact with Judaism might nevertheless be considered by certain Jewish groups to be Jews. Fortunately, this situation seldom arises as, by this time, such individuals are often ignorant or complacent about their Jewish backgrounds and are thus lapsed, from the viewpoint of the Jewish community. Children in households in which the father is *halachically* Jewish but where the mother is not are not considered by Orthodox Jews to be Jewish. Where the elements of Jewish religious practice are followed within the household, they are accepted by Progressive Judaism.

A variant of this example, and one that can lead to great friction within the Jewish community, concerns conversions conducted outside Orthodox authority. Yet for practical purposes, a more pragmatic and pluralistic approach is adopted by agencies at the operational level.

Nazi Germany defined the Jews on the basis of their own conceptions of 'race' and descent. In pre-World War II Poland, Jews were regarded as a language community. In this regard, speakers of Yiddish were regarded as Jews whereas those Jews who spoke other languages at home were regarded as members of that specific national language group (Ruppin 1934). In the Soviet Union, the Jews emerged as a non-territorial nationality and today carry their stamp of identification '*Evrei*' in their internal passport. These three ways of defining Jews are not related directly to religion.

In contrast, where there is an established state religion, the definition is based upon religious factors. In the Islamic world, Jews were traditionally considered a separate nation on the very basis of their religion, as they were in medieval Christendom. In many parts of continental Europe, the Jewish community was an autonomous corporate body with legal and fiscal authority of its own. Even today in Denmark and the Federal Republic of Germany, Jews are taxed separately from other citizens and part of the taxes is returned to communal institutions to maintain their own welfare and educational facilities. In Alsace-Lorraine and Belgium the Napoleonic system of state-employed Jewish religious functionaries remains. The *consistoire* system which separates Church and state is in force today in the rest of France (Kosmin 1982). In the best tradition of the French Revolution, the French state does not recognize differences amongst individual citizens, but accommodates Jews as a distinct cultural group (Benguigui 1969). In the words of Count Stanislas de Clermont-Tonnerre, speaking in the debate on the Eligibility of Jews for Citizenship in the French National Assembly on 23 December 1789, 'The Jews were to be denied everything as a nation but granted everything as individuals' (quoted in Mendes-Flohr & Reinharz 1980).

In the United States and other former British colonies (termed euphemistically the Old Commonwealth), the separation of Church and state follows the French model but without the Concordat with the Papacy. These countries are secular states. Religious or ethnic institutions are of a decidedly voluntary nature, membership constituting a voluntary act. Nevertheless, Australia, New Zealand, and South Africa have a religious census and Canada has both a religious and ethnic-origin question. Interestingly, in the United States census there are language and minority ethnic questions, as well as questions about 'race' and ancestry, none of which are directly relevant to the Jews today. Therefore, the Jews do not fall into an officially recognized category and, moreover, the Jewish organizations vigorously oppose any official categorization which might affect equality of individual citizenship or impair the constitutional separation of Church and state. In keeping with the federal and pluralistic nature of American society, they oppose formation of a defined national Jewish collectivity. As a result, the mechanisms for identification as 'Jewish' in the United States are pluralistic, flexible and open-ended.

The situation in the United Kingdom is different yet again. Like France, and in contrast to the United States, society and history have recognized the Jews as a religious group and encouraged the formation of a religious establishment. The modern Jewish community in Britain dates from 1656 when Cromwell's government granted a petition by Jews from Holland to be allowed to enter England and practise Judaism. The return of the Jews did not involve any corporate status or special privileges and thus there were no restrictions regarding residence or occupation. The only stipulation was that the Jewish immigrants should not be a charge on the state or parish, an obligation and convention which was upheld into the 1930s when, in the context of absorbing 50 000 refugees from the Nazi Reich, the representative leaders of Anglo-Jewry promised the government that 'all expenses, whether in respect of temporary or permanent accommodation or maintenance, will be borne by the Jewish community without ultimate charge to the State' (Wasserstein 1979).

The Cromwellian stipulation, the existence in England of an Established Church, and elements of a negative perception of Jews in

Christian society in the early 19th century, led to the establishment of separate welfare organizations such as the Jewish Blind Society (1814) and the Board of Guardians for the Relief of the Jewish Poor (1859). This situation which compelled the Jews to look after the needs of their own coreligionists independently of the religiously-based state system has forced them to think of themselves principally as a minority based on religious identification. This is reinforced by the status granted to the Jewish religious establishment and to the Chief Rabbi in emulation of the hierarchy of the Established Church of England. However, Jewish religious identification in Britain differs from that in, say, the Islamic states in that association of the individual with the religious group and synagogue is purely voluntary.

Complicating all these various methods for identifying Jews is a more recent political phenomenon resulting from World War II and the Holocaust and the creation of the state of Israel. The Zionists, as Jewish nationalists, sought to solve the 'Jewish problem' by normalizing the Jews and defining them in purely nationalistic terms, as a 'people' or 'nation'. Interestingly, Israeli Jews are defined in their identity papers as having both Jewish religion (*dath*) and Jewish nationhood (*leom*). As most Diaspora Jews have developed some identity with Israel in the wake of the European Holocaust, a secular, national definition of 'Jewish' has become more commonplace (see Waterman 1983). It is therefore now more acceptable among many British Jews to choose to identify on a criterion that is not strictly religious.

Jewish social services

Identification of the population at hand is not simply an academic problem, for it has direct practical implications on agencies that serve the Jews. Welfare and service agencies are less concerned with legalistic or cultural definitions of a Jewish population but instead are more concerned with a broadly-based functional identity. In the cases of the United States and the United Kingdom, such a definition would be understood to encompass people who either define themselves or are defined by others as Jews (de Lange & Kosmin 1979, Goldstein 1982).

There are 73 Jewish agencies concerned with social service delivery to the Jewish population of Greater London in the mid-1980s, although four major agencies provide the bulk of the welfare services (Wolkind 1985). Most of these agencies have staffs of under 20 people; nevertheless, as a whole, they employ over 2000 people and have a combined budget in excess of £20m per annum. Considerable effort in research and planning is required to avoid inefficiency and duplication in service provision (Community Planning and Research Committee 1985).

The variety of Jewish social and associated services arose not only as a result of specific historic circumstances but through other tangible concerns. In the Jewish case, religious concerns are not simply matters of language of prayer or form of liturgy: they impinge directly upon life-styles, and include food taboos, a separate calendar, and parallel legal systems affecting family and personal status. The outcome of these differences is a tangible separate group identity which is commonly accepted in terms of cultural mores, social outlook, humour, food consumption, and even attitudes to alcohol. As a result, there is a comprehensive system of Jewish social services which includes social work, meals-on-wheels, sheltered housing, day centres, old people's homes, marriage-guidance and divorce counselling, youth clubs, orphanages, and mental health provision – in fact the whole gamut of services which complement or supplement the welfare state. The rise of the welfare state after 1945 did not do away with separate provision. It only had a direct effect on charity, in the form of direct financial donations to the needy and on the Jewish hospitals which were incorporated into the National Health Service.

The definitional problem is only the first stage in locating the population to be served by Jewish agencies, for data on Jews, however they are defined, are scarce. The locational problem will undoubtedly become an increasingly significant factor in the work of these agencies as they enter a period of declining budgets and an increasing dependent population in the second half of the 1980s and the following decades.

In 1985, the effects of reductions in the budgets of London Jewish social service agencies had begun to be felt as local and national government reduced their financial grants and contributions. There have been closures of some institutions, with the threat of further closures and reductions in services in the future.

In addition to the need for centralization and co-ordination of operations in this field, there is another consequence of this situation which has a direct spatial implication. With the closure of institutions and the increasing likelihood that reduced budgets will mean less capital available for opening new ones, service provision will become more home-based than is presently the case. The implication is that a further, spatial dimension will be incorporated into the planning needs of these agencies in the coming years. For not only will it be necessary to locate all those individuals in need of aid; it will also be necessary to locate their next of kin. But as the residential situation is not static, it will also become an increasing necessity to understand Jewish residential mobility patterns and processes, and to establish an effective monitoring system to this end.

Jewish social trends

Although the Jews are frequently cited in the social-science literature as an example of a population maintaining a segregated pattern as the urban matrix expands, and although the relationship between segregation and group conservation is recognized as important, little work on Jewish residential patterns has been carried out and even less on the processes of residential change.

Much of the work on Jews in Britain has been historical, often ending at World War I (Lipman 1954, Gartner 1973). However, some historical and sociological work has taken a broad look at the modern community (Gould & Esh 1964, Freedman 1955, Lipman & Lipman 1981, Pollins 1982).

Sociological research on the present-day Jewish community has succeeded in bringing to light several of the problems of a Jewish community in a modern secular environment. Krausz' study of Jews in Edgware (1968; 1969a,b) concentrated on demographic trends, Jewish identification and social mobility of the interwar generations of Jews. Cromer's (1974) study in Wembley looked at the problems of intermarriage and survival of the community.

Considerable work has been carried out over the past 20 years on demographic trends and population estimates, using a variety of sources such as synagogue memberships, synagogue marriages, Jewish burials and circumcisions (Prais 1972, 1974; Prais & Schmool 1968, 1970, 1973; Kosmin 1982; Kosmin & Levy 1983b; Haberman *et al.* 1983; Kosmin & Waterman 1986; Waterman & Kosmin 1986c). Although some doubt may be cast on the accuracy of the estimates, the trends are more reliable.

Jewish residential patterns

Studying changing patterns

The main difference between studying social trends and the changing residential patterns of the Jews is the need in the latter case to be able to locate the population with reasonable accuracy. Several different methods have been tried with varying degrees of success. Most frequently, use has been made of synagogue memberships, although this data source only accounts for an estimated 75–80 per cent of the population. The resulting map (Figure 11.1) indicates only an institutionally-affiliated population and not a residential population. Moreover, there are considerable doubts about the accuracy of such figures and their comparative value (Waterman & Kosmin 1986a).

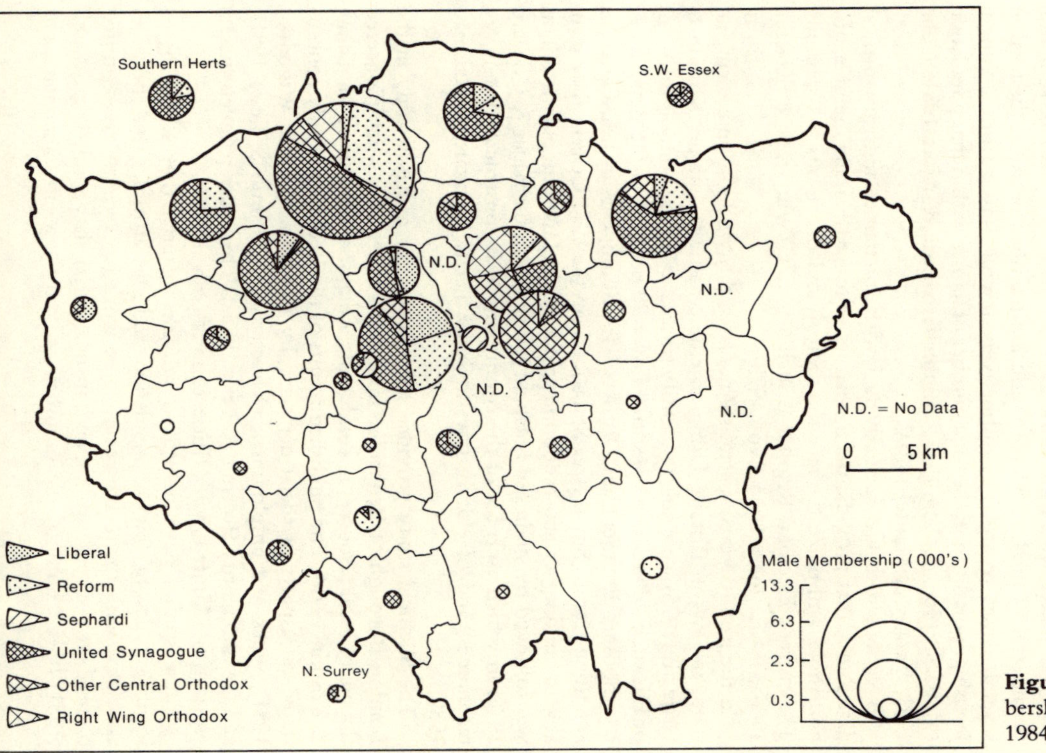

Figure 11.1 Synagogue membership, by London boroughs, 1984.

Names that have been recognized as indicating a distinctive ethnic affinity have frequently been used in social studies involving distinctive ethnic groups (Boal 1969, Smith 1982). Although this method is not without its pitfalls (Kosmin & Waterman 1987) – as people adopt new names to suit their surroundings or to hide their ethnic origins, especially in upwardly mobile social settings – variants of it have been used in Jewish social research for over 40 years. This has been particularly true in North America where much use has been made of a Distinctive Jewish Names (DJN) method. Here, a ratio is established between the total of several names regarded as distinctively Jewish drawn from a known Jewish population and the total number of names within that population (Massarik 1966, Cohen 1981, Himmelfarb Loar & Mott 1981, Cohn 1981, Varady & Mantel 1981).

A variant is to use distinctively Jewish forenames and surnames. A data source that provides the forenames of more than one member of a household will make the task less difficult and the results less spurious. The current Electoral Register is such a source (Waterman & Kosmin 1986a, 1987a; Jones & McEvoy 1978).

In the past 10 years, several studies have thrown light on the current status of Jews in Britain. Using an ethnic-name technique, Kosmin & Grizzard (1975) were able to identify small areas in the London Borough of Hackney where Jews constituted over 70 per cent of the population. By choosing Enumeration Districts with homogeneous housing types and using Census Small Area Statistics from the 1971 Census, they were able to build up a socio-economic profile of the Jewish population. This was followed by studies of Sheffield Jewry (Kosmin, Bauer & Grizzard 1976) in which a profile of the community was constructed by survey techniques, and of elderly Jews in Inner Leeds (Grizzard & Raisman 1980).

The Sheffield study provided the impetus for a major study in the London Borough of Redbridge some years later. This work pioneered the use of local resources in conducting a major social survey (de Lange & Kosmin 1979). The published results of the Redbridge study concentrated on aspects of social demography (Kosmin, Levy & Wigodsky 1981), work and employment characteristics (Kosmin & Levy 1981), and Jewish identity (Kosmin & Levy 1983a). This work parallels similar work in the United States (Ritterband & Cohen 1984).

These studies have helped to provide material that allows us to identify types of Jewish population and the attitudes and practices associated with them. However, some work has recently concentrated on residential change among Jews from a spatial viewpoint. Varady *et al.* (1981) have examined the suburbanization of Jews in Cincinnati, and

Waterman (1981) has shown the sectoral nature of Jewish urban migration in Dublin. In a later study, Waterman (1983) has shown the rôle of social contacts and community institutions in influencing the decision of several young families in Dublin to relocate in suburbs closer to the city centre. At the same time, this work indicated the importance in micro-scale locational decisions of knowing of Jewish neighbours living in close proximity. Studies in the United States also indicate the continuance of Jewish suburban clustering (Jaret 1979, Klaff 1983, Goldscheider 1986).

Jews in London

The concentration of Jews in north-west London is a phenomenon which has been documented over the years (Lipman 1954, Prais & Schmool 1968, Waterman & Kosmin 1986d, Newman 1985) with varying degrees of accuracy. The major concentration centres on the London Borough of Barnet, including the Golders Green–Hendon–Finchley complex, and Edgware, with extensions into the neighbouring boroughs of Camden, Brent, Harrow, the City of Westminster, and southern Hertfordshire. Further major Jewish concentrations exist in Hackney, north London, and in Redbridge, in the north eastern part of the metropolis (Waterman & Kosmin 1986b). The population estimate, based on a DJN methodology, yielded a total of 215 000 in the London area for 1984, a figure that corresponds with a total of 224 000 for 1977 based on mortality methods in the same area (Haberman, Kosmin & Levy 1983).

A more detailed estimate using the 1984 Electoral Register was carried out for the London Borough of Barnet. This permitted the construction of a more detailed map than did the DJN count for Greater London, which used the London Telephone Directory as its data base (Waterman & Kosmin 1986a).

At ward level, it was found that 60 per cent of all Jews in Barnet are located in six of the 20 wards – Edgware, Garden Suburb, Golders Green, Hendon, Finchley and Childs Hill. Yet, in no ward are they in the majority and only in Edgware do they constitute over 40 per cent of the total population. It is worth reiterating that the Jews fail to constitute a majority in even a single ward, although they are about 17 per cent of the population of the Borough. What is more, at the Polling District level, it is estimated that they form a majority in only four of the 101 subdivisions; and in only a single instance do they achieve a proportion as high as 60 per cent of the population. A further check of estimated Jewish populations in the Polling Districts shows that at the level of six-digit postal districts (or street sections) there is also a noticeable clustering of Jews. In fact, it is at the street level that clear ethnic majorities emerge in the most concentrated wards (Waterman & Kosmin 1987b, Waterman 1987).

These results represent the most detailed practicable estimate obtainable by desk research. It is the questions raised by the mapping exercise, rather than the results themselves, that seem to be significant, especially when considering the possible implications for social processes active amongst the Jews and other ethnic groups in London.

The literature in urban ecology and urban social geography suggests that congregation offers an excellent medium for conservative grouping (Boal 1978, Timms 1971, Lee 1977, Driedger & Church 1974, Knox 1982, Mesinger & Lamme 1985). The apparent anomaly whereby the 48 000 Jews in Barnet appear to be highly clustered yet fail to form a majority in the wards (with an average population of 14 000) or polling districts (most with populations of between 2500 and 3000) can be termed concentration without segregation. Explanation must be sought in the desire of many Jews to live as a 'community' in areas which offer a wide variety of housing types to suit the needs of a variegated group. It is necessary for such housing types to be in close mutual proximity so that the Jewish population can operate also as discrete geographical communities. However, the homogeneity of suburban housing estates probably precludes even higher levels of clustering because of the varied housing needs of the individual households that constitute the 'community'.

Variations at ward level in the siting of synagogues and the settlement of the Jewish population, and thus in the membership–population ratios, suggest that the Jews of Barnet are not uniform, either in their cultural backgrounds or in their life-styles. Comparing the distribution of synagogues by membership with the population estimates, by ward, it appears that in Hendon, in parts of Golders Green and in Finchley the congregation of Jews is primarily associated with religion and religious institutions, whereas in other areas (Garden Suburb and Childs Hill), it is related more to proximity of social contacts and to social interactions. Edgware exhibits both of these trends and represents a largely independent, self-contained concentration of Jews. At the same time, some 40 per cent of the Jews in the borough live in wards in which they constitute a small minority, usually less and often much less than 10 per cent, although in many instances it appears that there is still clustering of Jews at the scale of individual streets. In this context, it would be interesting to understand the perceptions of Jews and of others of what constitutes a Jewish neighbourhood or area so as to acquire a better comprehension of the territorial meaning of such areas (Waterman 1983).

Research questions

What the mapping exercise has not succeeded in doing is to provide answers to several basic questions on the social processes involved in Jewish residential change. The answers to these questions, which have direct practical implications for the planning and implementation of future social services for the Jewish population, can only come through detailed field research. Some of the questions in need of further research are:

(1) To what extent is the spatial congregation of a group perceived by individuals within that group as a conscious conservative measure against assimilation and to maintain group identity? The relationship between spatial concentration and assimilation has been discussed widely in the literature over the past 30 years (Duncan & Lieberson 1959, Driedger & Church 1974, Boal 1978).

(2) What implications might congregation have on subsequent experiences of members of the group, such as in upward social mobility, the formation of social ties and out-marriage?

(3) How is the choice of residential location among the Jewish community affected by income, social class, social ties and other factors? Is it possible to measure the extent to which prestige within the community (a social factor) is traded off against more material things (personal factors) in restricting the search areas of Jewish movers? In short, to what extent can residentially segregated patterns be explained and predicted by structural, discriminatory and pluralist factors?

(4) Research has indicated a distinct Jewish preference for metropolitan living and a corresponding reluctance to leave London for other types of urban centres (Kosmin & de Lange 1980). This was true for all social classes. Is there a distinctive Jewish residential ideology which seeks to unite the anonymity of a metropolitan life with the intimacy of an ethnic suburb?

(5) Although it is customary to refer to the Jews as if they constitute a uniform population, this is not the case. The spectrum ranges from the ultra-Orthodox, through Liberal Jews, to those who are nominally Jewish but are unaffiliated. Is it possible to identify subpopulations and do these correlate with varying moving decisions and patterns? (See Elazar & Medding 1983.)

(6) Can a typology of ethnic movers and a model for ethnic moving be formulated, encompassing such groups as 'early movers', 'followers', 'fillers-in', 'successors', etc?

(7) How and when is a new independent Jewish community perceived and recognized? What is the minimum size and the appropriate combination of factors needed for the constitution of a separate Jewish area?

Although intrinsically interesting as a group, it is worthwhile asking whether a study of Jews could contribute to a better understanding of some of Britain's other ethnic minorities. Despite the existence of large ethnic-minority populations, census data reveal that over the whole of London only a handful of wards have a majority of any single-minority group. Thus it could be argued that the general Jewish settlement model is not unique and that the overall London pattern is congregation not segregation.

It is our contention that the Jews can provide a useful model for achieving a better understanding of Britain's upwardly mobile ethnic minorities. Contemporary British Jews are white, middle-class and relatively long-settled but the majority have chosen to maintain a separate identity while integrating into most aspects of the general society. On the other hand, most other white immigrant groups tend to lose this separate identity within a single generation of immigration. Some upwardly mobile New Commonwealth groups are less free than the Jews have been to dissolve into the general population (Krausz 1972). Recent research (Dhanjal 1977/78, Cohen 1984/85) appears to indicate that many of the problems faced by the Jews in the past are being tackled in similar ways by these groups in suburban environments. As the Jews are two generations longer established than other ethnic groups, they might be able to provide a model for some of these, especially those middle-class ethnic groups who but for skin colour would find it easier to assimilate into the general society.

These groups, in ideology, in outlook and in current practice, require a similar combination of residential concentration, but reject and regard as dangerous complete residential segregation. They, too, need a balance between segregation and assimilation: a combination of sufficient local ethnic concentration to allow the establishment of viable ethnic institutions while not simultaneously detracting from the benefits of economic and political integration in the wider society. This probably requires concentrated suburban residential settlements within the variegated metropolis and this, in turn, has policy implications for a pluralist society, with each group having its own specific agenda.

The problems facing Britain's ethnic groups in planning for their future are numerous. Are members of ethnic groups to be locked into those groupings forever or will they be eventually absorbed into society at large? Do they, in fact, *wish* to be assimilated or to remain different (see Yancey *et al.* 1985)? Whether these questions are answered in the affirmative or the negative has a direct bearing on planning to answer the future demands and needs of the groups. How does living as a minority group in a given society affect the self-identification of the group? Will we see the emergence of overarching Asian or Afro-Caribbean ethnic groups in British society as the real differences between the immigrant groups originating in different regions diminish to be replaced by a new identity provided by their position within British society? What people will be placed into these new categories? How will mixed families be classified?

It is questions like these that British Jews have attempted to face at the political and operational levels over the past century. Their experience in British society in making the transformation from an immigrant to an ethnic minority group is worthy of examination since there is often a clear case for similar analyses of other minorities as well. A data base appears to be needed in order to record the experiences of people who differ from the majority in values, life-styles, and consequently, social needs. If such distinctive life-styles are identified, there may be a need to recognize culturally specific needs for certain social services. Failure to do so might leave members of minority groups permanently disadvantaged (Ward 1984). However, this dilemma is highlighted by the fact that amongst a given minority there may be some elements pushing for recognition as a distinct subunit within society, appreciating the social benefits this might bring; yet others are fearful of the social costs and of the uses that might potentially be made by the majority to their disadvantage.

It is up to each minority group to negotiate amongst themselves and with the majority in order to work out a satisfactory *modus vivendi* and *modus operandi*. It is to be hoped that social science will have a practical rôle to play.

Note

This essay was prepared when Stanley Waterman was an Academic Visitor in the Department of Geography at the London School of Economics and Political Science and Barry Kosmin was the Director of the Research Unit of the Board of Deputies of British Jews.

References

Benguigui, G. 1969. *Aspects of French Jewry*. London: Vallentine Mitchell.

Boal, F. W. 1969. Territoriality on the Shankill–Falls Divide. *Irish Geography* **6**, 30–50.

Boal, F. W. 1978. Ethnic residential segregation. In *Social areas in cities*, D. T. Herbert & R. J. Johnston (eds), 57–95. London: Wiley.

Cohen, G. 1984/85. Ethnicity in a middle-class London suburb. *New Community* **XII**(1), 89–100.

Cohen, S. M. 1981. *UJA demographical/attitudinal survey kit*. New York: United Jewish Appeal.

Cohn, W. 1981. What's in a name. *Public Opinion Quarterly* **47**, 660–5.

Community Planning and Research Committee 1985. *Interim Report (1), Jewish Social Services: A review of current provision in Greater London and adjacent areas*. London: Central Council for Jewish Social Service.

Cromer, G.1974. Intermarriage and communal survival in a London suburb. *Jewish Journal of Sociology* **16**, 155–69.

Dhanjal, B. 1977/78. Asian housing in Southall: some impressions. *New Community* **6**(1/2), 88–94.

Driedger, L. & G. Church 1974. Residential segregation and institutional completeness. *Canadian Review of Sociology and Anthropology* **11**, 30–52.

Duncan, O. D. & S. Lieberson 1959. Ethnic segregation and assimilation. *American Journal of Sociology* **64**, 364–74.

Elazar, D. J. & P. Y. Medding 1983. *Jewish communities in frontier societies*. New York and London: Holmes & Meier.

Freedman, M. 1955. *A minority in Britain*. London: Vallentine Mitchell.

Gartner, L. P. 1973. *The Jewish immigrant in England 1870–1914*, 2nd edn. London: Simon Publications.

Goldscheider, C. 1986. *Jewish continuity and change: emerging patterns in America*. Bloomington: Indiana University Press.

Goldstein, S. 1982. Population movement and redistribution among American Jews. *Jewish Journal of Sociology* **24**, 5–24.

Gould, J. & S. Esh 1964. *Jewish life in modern Britain*. London: Routledge & Kegan Paul.

Grizzard, N. & P. Raisman 1980. Inner-city Jews in Leeds. *Jewish Journal of Sociology* **22**, 21–33.

Haberman, S., B. A. Kosmin & C. Levy 1983. Mortality patterns of British Jews 1975–79: insights and applications for the size and structure of British Jewry. *Journal of the Royal Statistical Society, Part 3* **146**, 294–310.

Himmelfarb, H. S., R. M. Loar & S. H. Mott 1981. Sampling by ethnic surnames: the case of American Jews. *Public Opinion Quarterly* **47**, 247–60.

Jackson, P. 1981. Paradoxes of Puerto Rican segregation in New York. In *Ethnic segregation in cities*, C. Peach, V. Robinson & S. Smith (eds), 109–26. London: Croom Helm.

Jaret, C. 1979. Recent patterns of Chicago Jewish residential mobility. *Ethnicity* **6**, 235–48.

Jones, T. P. & D. McEvoy 1978. Race and space in cloud-cuckoo land. *Area* **10**, 162–6.

Kantrowitz, N. 1981. Ethnic segregation: social reality and academic myth. In *Ethnic segregation in cities*. C. Peach, V. Robinson & S. Smith (eds), 43–57. London: Croom Helm.

Klaff, V. Z. 1983. The urban ecology of Jewish populations: a comparative analysis. *Papers in Jewish Demography 1981*, 343–61. Jerusalem: Institute of Contemporary Jewry.

Knox, P. L. 1982. *Urban social geography*. London: Longman.

Kosmin, B. A. 1978. Demography and sampling problems. In *Communautés Juives (1880–1978), Sources et Méthodes de Recherche*, D. Bensimon (ed.), 258–70. Paris: Le Centre Inter-Universitaire des Hautes Etudes du Judaïsme Contemporaine.

Kosmin, B. A. 1982. The Jewish experience, Part 4. E354, Block 3, Units 8–9 *Minority experience*, 79–96. Milton Keynes: Open University Press.

Kosmin, B. A., P. Bauer & N. Grizzard 1976. *Steel City Jews*. London: Board of Deputies of British Jews.

Kosmin, B. A. & N. Grizzard 1975. *Jews in an Inner London Borough*. London: Board of Deputies of British Jews.

Kosmin, B. A. & D. de Lange 1980. Conflicting urban ideologies: London's New Towns and the Metropolitan preference of London's Jews. *The London Journal* **6**, 162–75.

Kosmin, B. A. & C. Levy 1981. *The work and employment of suburban Jews*. London: Board of Deputies of British Jews.

Kosmin, B. A. & C. Levy 1983a. *Jewish identity in an Anglo-Jewish community*. London: Board of Deputies of British Jews.

Kosmin, B. A. & C. Levy 1983b. *Synagogue membership in the United Kingdom 1983*. London: Board of Deputies of British Jews.

Kosmin, B. A. & C. Levy 1984. *Jewish emigration from the United Kingdom – a working paper*. London: Board of Deputies of British Jews.

Kosmin, B. A. & C. Levy 1985. Jewish circumcisions and the demography of British Jewry. *Jewish Journal of Sociology* **27**(1), 5–11.

Kosmin, B. A. & S. Waterman 1986. Recent trends in Anglo-Jewish marriages. *Jewish Journal of Sociology* **28**(1) 49–57.

Kosmin, B. A. & S. Waterman 1987. The use and misuse of ethnic names in Jewish social research. In *Papers in Jewish demography 1985*. Jerusalem: Institute of Contemporary Jewry (forthcoming).

Kosmin, B. A., C. Levy & P. Wigodsky 1981. *The social demography of Redbridge Jewry*. London: Board of Deputies of British Jews.

Krausz, E. 1968. The Edgware survey: demographic results. *Jewish Journal of Sociology* **10**, 83–100.

Krausz, E. 1969a. The Edgware survey: occupation and social class. *Jewish Journal of Sociology* **11**, 75–95.

Krausz, E. 1969b. The Edgware survey: factors in Jewish identification. *Jewish Journal of Sociology* **11**, 151–64.

Krausz, E. 1972. Factors of social mobility in British minority groups. *British Journal of Sociology* **23**, 275–86.

de Lange, D. & B. A. Kosmin 1979. *Community resources for a community survey.* London: Board of Deputies of British Jews.

Lee, T. R. 1977. *Race and residence.* Oxford: Oxford University Press.

Lipman, S. L. & V. D. Lipman 1981. *Jewish life in Britain 1962–1977.* New York: K. G. Saur Publishing.

Lipman, V. D. 1954. *A social history of the Jews in England, 1850–1950.* London: C. A. Watts.

Massarik, F. 1966. New approaches to the study of the American Jew. *Jewish Journal of Sociology* **8**, 175–91.

Mendes-Flohr, P. R. & J. Reinharz 1980. *The Jew in the Modern World.* New York: Oxford University Press.

Mesinger, J. S. & A. J. Lamme III 1985. American Jewish ethnicity. In *Ethnicity in contemporary America*, J. O. McKee (ed.), 145–68. Dubuque, Iowa: Kendall Hunt.

Newman, D. 1985. Integration and ethnic spatial concentration: the changing distribution of the Anglo-Jewish community. *Transactions of the Institute of British Geographers* (N.S.) **10**, 360–76.

Pollins, H. 1982. *Economic history of the Jews in England.* London: Associated University Presses.

Prais, S. J. 1972. Synagogue statistics and the Jewish population of Great Britain 1900–1970. *Jewish Journal of Sociology* **14**, 215–28.

Prais, S. J. 1974. A sample survey of Jewish education in London 1972–1973. *Jewish Journal of Sociology* **16**, 133–54.

Prais, S. J. & M. Schmool 1968. The size and structure of the Anglo-Jewish population 1960–1965. *Jewish Journal of Sociology* **10**, 5–34.

Prais, S. J. & M. Schmool 1970. Statistics of Milah and the Jewish birth-rate in Britain. *Jewish Journal of Sociology* **12**, 187–93.

Prais, S. J. & M. Schmool 1973. The fertility of Jewish families in Britain, 1971. *Jewish Journal of Sociology* **15**, 189–204.

Ritterband, P. & S. M. Cohen 1984. The social characteristics of the New York area Jewish community, 1981. *American Jewish Yearbook 1984*, 128–61.

Ruppin, A. 1934. *The Jews in the Modern World.* London: Macmillan.

Smith, G. 1982. *The geography and demography of South Asian languages in England: some methodological problems.* Linguistic Minorities Project Working Paper No. 2, University of London, Institute of Education.

Timms, D. W. G. 1971. *The urban mosaic.* London: Cambridge University Press.

Varady, D. & S. J. Mantel 1981. Estimating the size of Jewish communities using random telephone surveys. *Journal of Jewish Communal Service* **57**, 225–34.

Varady, D. P., S. J. Mantel Jr, C. Hinitz-Washofsky & H. Halpern 1981. Suburbanization and dispersion: a study of Cincinnati's Jewish population. *Geographical Research Forum* **3**, 5–15.

Ward, R. 1984. Race and housing: issues and policies. *New Community* **11**, 201–5.

Wasserstein, B. 1979. *Britain and the Jews of Europe 1939–1945*. Oxford: Clarendon Press.

Waterman, S. 1981. Changing residential patterns of the Dublin Jewish community. *Irish Geography* **14**, 41–50.

Waterman, S. 1983. Neighbourhood, community and residential change decisions in the Dublin Jewish community. *Irish Geography* **16**, 55–68.

Waterman, S. 1987. *Jews in an outer London suburb: Barnet*. London: Board of Deputies of British Jews.

Waterman, S. & B. A. Kosmin 1986a. Mapping an unenumerated ethnic population: Jews in London. *Ethnic and Racial Studies* **9**, 484–501.

Waterman, S. & B. A. Kosmin 1986b. The distribution of Jews in the United Kingdom, 1984. *Geography* **71**, 60–5.

Waterman, S. & B. A. Kosmin 1986c. *British Jewry in the eighties: a geographical and statistical guide*. London: Board of Deputies of British Jews.

Waterman, S. & B. A. Kosmin 1986d. The Jews in London. *The Geographical Magazine* **58**, 21–7.

Waterman, S. & B. A. Kosmin 1987a. Residential change in a middle-class suburban ethnic population: a comment. *Transactions of the Institute of British Geographers* **12**(1), 107–12.

Waterman, S. & Kosmin, B. 1987b. Residential patterns and processes: a study of Jews in three London boroughs. *Transactions of the Institute of British Geographers* **12** (in press).

Wolkind, J. 1985. *London and its Jewish community*. London: West Central.

Yancey, W. L., E. P. Ericksen & G. H. Lean 1985. The structure of pluralism: 'We're all Italian around here, aren't we, Mrs. O'Brien?' *Ethnic and Racial Studies* **8**, 94–116.

PART IV

Ideology and resistance

The final three chapters are all concerned with the relationship between racist ideologies and the politics of resistance. For dominant ideologies can be challenged in a variety of ways, ranging from relatively autonomous cultural strategies to more instrumental forms of protest.

The urban riots of the early 1980s, for example, comprise a particular problem of interpretation. Admitting that all forms of explanation are ideological, **Michael Keith** argues that most accounts of the riots suffer from an 'empirical emptiness'. He argues that by failing to specify what actually took place during particular episodes of disorder, academics have simply misappropriated the symbolic power of the riots for their own ideological ends. Rather than attempting a causal explanation of the riots, he offers a preliminary description of particular events, revealing the inadequacies of other accounts that fail to take the problem of explanation seriously.

The rôle of academic social science is also questioned by **Peter Jackson** in his analysis of US attitudes to Puerto Rico. Given the existence of such wide discrepancies in power between the island and the mainland, it is not surprising that US political discourse about Puerto Rico is systematically distorted. The academic social science literature is, however, no less distorted, serving a crucial ideological rôle in perpetuating Puerto Rico's colonial subordination. The chapter concludes with a brief discussion of the emergence of a critical Puerto Rican scholarship which has begun to challenge the hegemony of North American social science.

Finally, **John Silk** explores the representation of blacks in the American cinema, focusing on films set in the American South. He shows how changes in the portrayal of blacks on the screen can be related to changes in black–white relations in general, to changes within the motion-picture industry, and to changes in the economic and political position of the South in particular. While racist stereotypes predominate, a minority of films have employed the medium for the expression of anti-racist sentiments, confirming the potentially positive rôle of culture in articulating resistance to racism and exploitation.

12 'Something happened': the problems of explaining the 1980 and 1981 riots in British cities

MICHAEL KEITH

This chapter aims to demonstrate the ease with which the symbolic power of rioting is misappropriated in academic discourse, the possibility of realistic description of that rioting, and the problems of explanation arising from the limits to such description.

The disorders of 1980 and 1981 have been used by very different groups in civil society, using very different arguments, as an ultimate cautionary symbol. This is partly a strategic ploy in the competition for scarce resources, partly an inevitable product of the process by which history is endowed with a meaning that does not disrupt a particular view of the world, a particular form of 'common sense'. Imposing order on chaos, individuals and institutions make sense out of confusion that lends coherence to life. In this process of conceptualization the riots are at one time both a social product that lives on as part of the cultural present and a manifestation of violent conflict that remains firmly part of the historical past. In trying to distinguish between these two facets of a phenomenon it is essential to recognize the difficulty of any attempt to recapture lost times.

All forms of explanation are ideological in that they have their own philosophical background and history external to any particular usage. This does not preclude the possibility of realistic explanation. Analysis may try to probe the historical reality as well as the cultural conceptualization of rioting. Indeed, if it cannot do so then the expositions of the systematic distortion manifest in the media presentation of riots would appear sanctimonious (CCCS 1982, Murdock 1984, Burgess 1985). Hence, although no academic discourse is value-free, it is not all equally value-loaded: 'objectivity is a direction not a terminus' (Harré 1979, p.

119). It is the contention of this chapter that a failure to take on board the problems of explanation have led to manifestly biased significations of rioting as scholars too compete to claim 'the riots' for their own discipline.

An explanation can be broken down into an explicans and an explicandum.[1] The explicandum should consist of descriptions and reportage of the phenomenon to be explained (Runciman 1983). It is suggested here that rioting is so easily misappropriated in academic discourse because the 'explanations' advanced tend to make no attempt to establish an explicandum, even in their own terms. In plain English, everybody can say 'why there were riots' but few ever mention 'what the riots were'. The chapter goes on to suggest that description alone can never touch on the causal processes of violent conflict. It is possible to produce a portrait of the sort of people involved in rioting, whatever that nebulous verb implies. It is possible to produce a semantic description, revealing the perceptions of those involved in the conflict. It is also possible to focus on the context of the riot, the unacknowledged conditions of action that precede conflict. However, descriptions can only be linked with causes, explicandum with explicans, in the matching of the interdependent domains of the theoretical and the empirical.

There is no attempt in this chapter to advance a causal explanation of rioting in British cities in 1980 and 1981, only to formulate a preliminary description of those events which must precede it in a realistic analysis. This process of description is alone difficult enough to devalue the truth claims implicit in much academic discourse. Unless some attempt is made to qualify 'what happened' the rioting will remain no more than a rhetorical symbol; an ideological construction, polemically deployed.

'Academic' explanations of rioting

Three illustrations are taken here as symptomatic of the way in which the riots have been used in this symbolic fashion. Representing three different explanatory paradigms, they display the same characteristics of misappropriation.

For Cashmore & Troyna (1982, pp. 17–33) the British rioting in 1980 and 1981 is the historical property of the social-problem group black youth; indeed it is the most lurid manifestation of crisis from which their study draws its legitimacy.[2] They are keen to stress that alongside the well-documented 'passive' influence of institutional racism young black people have consciously adopted 'postures in relation to the rest of society'. This 'active' influence, it is suggested, takes the form of manifold rejection, viz.: 'dissatisfaction with society generally . . . which translates into a desire to have nothing to do with it', 'the idea of Babylon and all its implications', 'a fissure which developed between first-generation West Indians and their sons and daughters' and 'a reluctance to take employment'. It is not intended to analyse this 'cultural' model on its own terms, rather to examine the postulated connection between black youth and collective disorder[3] (see Fig. 12.1).

Cashmore & Troyna suggest that because of the pervasive idea of Babylon which 'cannot be conquered through conventional political measures', black political action will not gather mass political support except on specific one-off issues and so it is improbable that there can be political mobilization of young blacks in the immediate future. Two other responses to this situation are considered. Young black people can try to carry on in their stigmatized position or they can adopt 'strategies for survival', one of which will coincide with Hall's conception of crime becoming politically viable. These are short-term solutions only and the reason for the use of the term 'crisis'.

In the production of collective disorder two further notions are advanced. The first is a conception of violence which has its antecedents in the psychological behaviourism of the 1950s and 1960s. Violence is

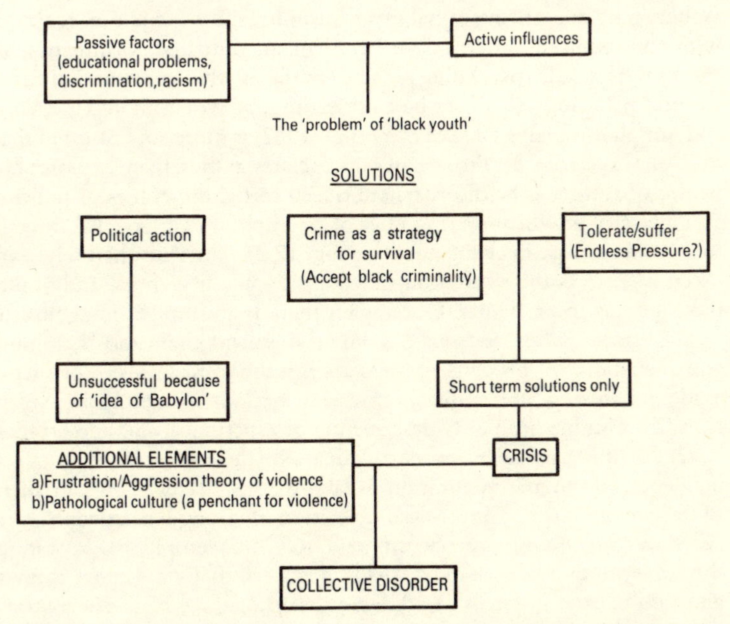

Figure 12.1 Cashmore's & Troyna's explanation of rioting (adapted from Cashmore & Troyna 1982).

seen as a 'strategy for venting frustrations', a form of pressure release; aggression as an automatic response to continual frustration. This is paired with a second notion, the cultural context of this process, an alleged 'penchant for violence within the West Indian culture possibly stemming from the days of slavery when the only method of retaliation was doing physical damage to the overseer'. When put together with the crisis of black youth one product is rioting, in which 'young blacks are responding not to specific targets but to the system generally, the system they call Babylon. In all the episodes of the early 1980s, black youths chose to attack the institutions that symbolized their entrapment within the system: houses, shops, cars, the police.' The police are characterized as the victims of an explicitly vicarious violence (along with the other Aunt Sallies); the conflict is not with the police force as such but with the society that force represents. Incidentally the rationality of the people on the streets is thus discredited: the rioters do not 'really' know what they are doing, even if implicitly they are all doing the same thing. Placing what interests them most into a causal relationship with what interests them least, Cashmore & Troyna attempt to explain away rioting in terms of 'black culture'. Even if their generalizations about the latter were correct this would be a tendentious argument as there is no convincing causal mechanism to link the explicans (culture) with the explicandum (riots) other than an outdated formulation of stimulus–response psychology. The result is the construction of a cultural pathology that is at best misleading, at worst invidious.

Though the analysis of Lea & Young (1982) is more sophisticated than that of Cashmore & Troyna, it too dictates rather than explores the nature of rioting. The disorders are traced to the difficulties of policing an inner city in which soaring rates of unemployment have led directly to a rapid increase in crime rates (see Fig. 12.2). The riots themselves are taken *a priori* as the expression of young people in general and young black people in particular. Once again there is an implicit image of the 'typical rioter' whose persona has only two salient characteristics: youth and alienation. In this context these are not innocent terms – they serve as dehumanizing preconditions for explanation. Rather than explain riots, the authors need only describe the production of the stereotypical rioter. In this particular format of such explanation 'race' is, in effect, incidental to the marginalization of labour: 'Race *becomes* an ingredient of this vicious circle. The *economic* alienation of young black people gives rise to a culture with a propensity to crime' (my emphasis). In spite of the contentious evidence in this field,[4] black criminality is taken as given and rationalized in terms of a liberal notion of the effects of economic disadvantage. Consequently, the conflict between black people and the police is described as the product of this criminal culture. 'The police

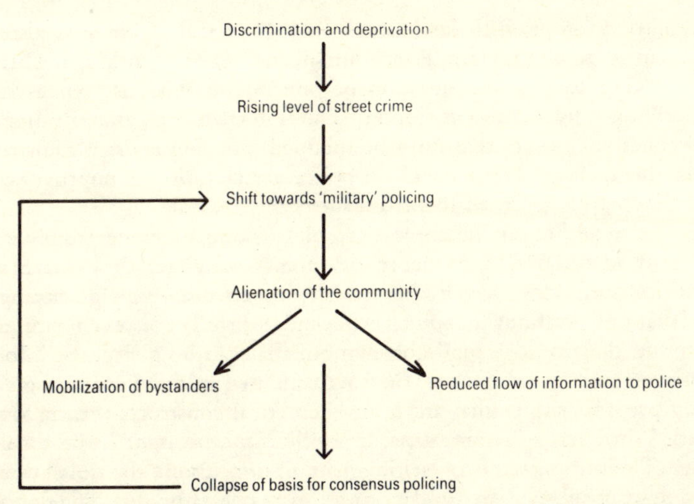

Figure 12.2 The collapse of consensus policing (adapted from Lea & Young 1982, and Cowell *et al.* 1982).

make the initial connection between race and crime.' There is an elementary historical flaw in such a deterministically phrased causal connection, in that this conflict quite clearly predated any suggestion of black over-representation among offenders. As early as 1972 the House of Commons Select Committee on Police/Immigrant Relations stated categorically (1972, p. 69) that 'It was made clear by all witnesses . . . that relations between police and younger West Indians are fragile, sometimes *explosive*' and that (p. 70) 'Of all the police forces from which we took evidence not one had found that crime committed by coloured people was proportionately greater than that by the rest of the population. *Indeed in many places it was somewhat less*' (my emphasis). Poor 'police/black relations' stretch back a long way; Lea & Young do not so much revise history as forget it.

An alternative can be seen in Thackrah's study of *Reactions to terrorism and riots* (1985, pp. 150–5). Here the tension between theory and practice is particularly marked and the relationship between academic discourse and public policy is of particular concern as the author is a lecturer at Bramshill Police College. Thackrah does not advance an explanation of the rioting as such, yet in the assumptions he makes about the disorders he throws light on a conception of the phenomenon that may structure policy-orientated reactions to it. The policing problem is twofold. There is a question of public order: 'In the 1981 riots unprovoked attacks with firebombs resulted in the police needing to take a fresh look at their capability to cope with lawlessness on such a scale.' There is also a problem of public image: 'Police intelligence has to work against the Left, trying to link the question of policing to what is seen as the underlying causes of recent rioting.' The rioters are seen as an irrational and cohesive 'crowd' that must be subdued; policing *per se* is removed from the agenda as the attacks on police are classified as unprovoked, and the focus is returned to British society as a whole.

In short, although the three examples examined come from very different ideological and theoretical points of view, they share an empirical emptiness which characterizes much discussion of rioting;[5] explanations without genuine explicanda, fitfully or even smugly adopting the riot as a malleable symbol that can be easily tied into a preferred schema. Moreover, the language used in the discussion gives away presuppositions that are more theoretical constructs than proven 'truths'. In effect, theoretical predispositions are brought to bear on a series of 'common-sense' or 'self-evident' notions about 'the riots' which tend to be unsubstantiated and, at times, even contradictory. The rioting is considered to be a single generic activity carried out by a holistic unit ('the crowd' or 'the rioters') and is consequently susceptible to a relatively straightforward single explanation.[6] This unit is made up of people each of whom can be characterized by the descriptive term 'average rioter', who is variously either black, young and 'alienated' or either black or white, young and 'alienated'. Reference to the events that constituted any single disorder in Britain in 1980 or 1981 belies such tidy classification. Even where evidence is relatively scarce it is possible to obtain a broad picture of the actions of 'the crowd'.

Evidence I: Faces in the crowd

A possible source for empirical analysis of 'the crowd' itself is the records of those arrested during disturbances.[7] The most obvious flaw in such a method is that the behaviour of many of this group will be misrepresented. Several will be found not guilty of any offence; several others will be convicted of offences they never committed, having been picked out almost at random from those present. Similarly, there will be allegations that the chance of arrest in incidents of public disorder is disproportionately high for some groups victimized by the police; men and black people in the case of the British riots. There is no way that the significance of either trend can be confirmed or refuted, although both intuitively and from personal experience I would suggest that in the chaos of collective violence the former problem, which is in many ways the antithesis of the latter, is more prevalent.

The value of arrest records in this context rests on two basic assumptions. The first is that those arrested provide a sample of the sort of people 'on the streets' during the riots. The second is that, however strong or weak the link, the actual offence charges against individuals will more often than not be indicative of an action or form of behaviour carried out by that individual during the disturbance. In a perfect legal system this connection would be explicit; in the circumstances this is manifestly not the case. Nevertheless, given that there is in all probability some link it should be possible to identify, with some degree of certainty, the sort of people that were carrying out certain activities during the rioting. There is no suggestion that it is possible to produce a rigorous statistical analysis of the behaviour or 'definitive pathology' of rioters, but rather the contention that such data can identify 'faces in the crowd', and can give a descriptive hint about the behaviour of various individuals.[8]

Brixton, April 1981

Although the Brixton rioting can be validly considered as a single event, there were important differences between the violence of Friday 10, Saturday 11 and Sunday 12 April. On Friday the rioting consisted entirely of a conflict between a group of black people (mostly young) and the police, and lasted for a few hours only. On the Saturday much larger numbers of both police and rioters were involved in disturbances that lasted from four o'clock in the afternoon until late at night; significantly, looting and arson were spread over a much wider area. On the Sunday well over a thousand police officers were deployed in a high-profile occupation of part of Brixton, and although there were both looting and attacks on the police, the disorders were not as serious as the night before.

In total 145 shop premises were damaged, 28 properties by fire. The targets within this group were not randomly chosen. Predictably, suppliers of consumer durables (particularly clothes, shoe and electrical-equipment shops) and off-licences proved 'favourites', whereas shops run by popular local figures in Railton Road, at the centre of the trouble, escaped unscathed. Such rational actions stand in contrast to the 'mindless hooligans' that often populated Fleet Street's rioting world. Similarly, the occasion was used by some to pay off old scores. The landlord of *The George*, a pub that was a target for arsonists, had been reported to the Race Relations Board in 1966, and throughout the 1960s and 1970s the treatment of black people at the pub had been a specific protest issue in several local marches. Even the *South London Press*, not noted for such local sensitivity, and which had taken an editorial line supporting the police at the time of the riots, remarked that this burning was 'undoubtedly an act of revenge for years of racial discrimination'. Similar tensions also quite possibly lay behind arson in a nearby newsagent. Apart from these two, and a few other exceptions, the most serious damage to property occurred some 200–300 m away from the conflict with the police on the Saturday night. Lord Scarman even went so far as to say that 'While the centre of the disorder was Leeson Road and the northern end of Railton Road, its effects were being felt over a wide area of central Brixton. In the commercial area of Brixton Road, the northern half of Atlantic Road, Electric Avenue and Coldharbour Lane, widespread looting had developed since about 6 pm. Both whites and blacks – some of them very young – were involved. To several witnesses, the whites appeared to be older and more systematic in their methods. It also appears that the looters were, in the main, quite different from the people who were attacking the police in Railton Road. Several witnesses had the impression that many of the looters came from outside Brixton and were simply taking advantage of the disorders for their own criminal purposes.' (Scarman 1981, 3:61.) Nevertheless, Scarman's suggestion is based on only a few eye-witness reports.

Although it was not possible to obtain comprehensive details of all those arrested during the disturbances, overlapping sets of data were gathered from the local press, court records and personal research. Age distribution, offence and address were known for a group of 193, ethnicity[9] and offence for a group of 101, and all four characteristics for a subset of 46: 253 arrests were made between the Friday and Sunday, with a further 29 on the Monday. In Table 12.1 the data have been categorized to facilitate comparison with the Home Office data later

Table 12.1 Age distribution for a sample of those arrested in the Brixton rioting, April 1981, by offence type.

Offence	Under 17	17–20	Over 21	Total
violence against person	0(0%)	17(30.9%)	38(69.1%)	55(100%)
burglary/theft	5(7.5%)	37(55.2%)	25(37.3%)	67(100%)
criminal damage	1(20%)	3(60%)	1(20%)	5(100%)
threatening behaviour	1(1.9%)	20(37.7%)	32(60.4%)	53(100%)
obstruction	0(0%)	3(23.1%)	10(76.9%)	13(100%)
'riot'	1(0.8%)	40(33%)	80(66.2%)	121(100%)
'loot'	6(8.3%)	40(55.6%)	26(36.1%)	72(100%)

Source: See text.

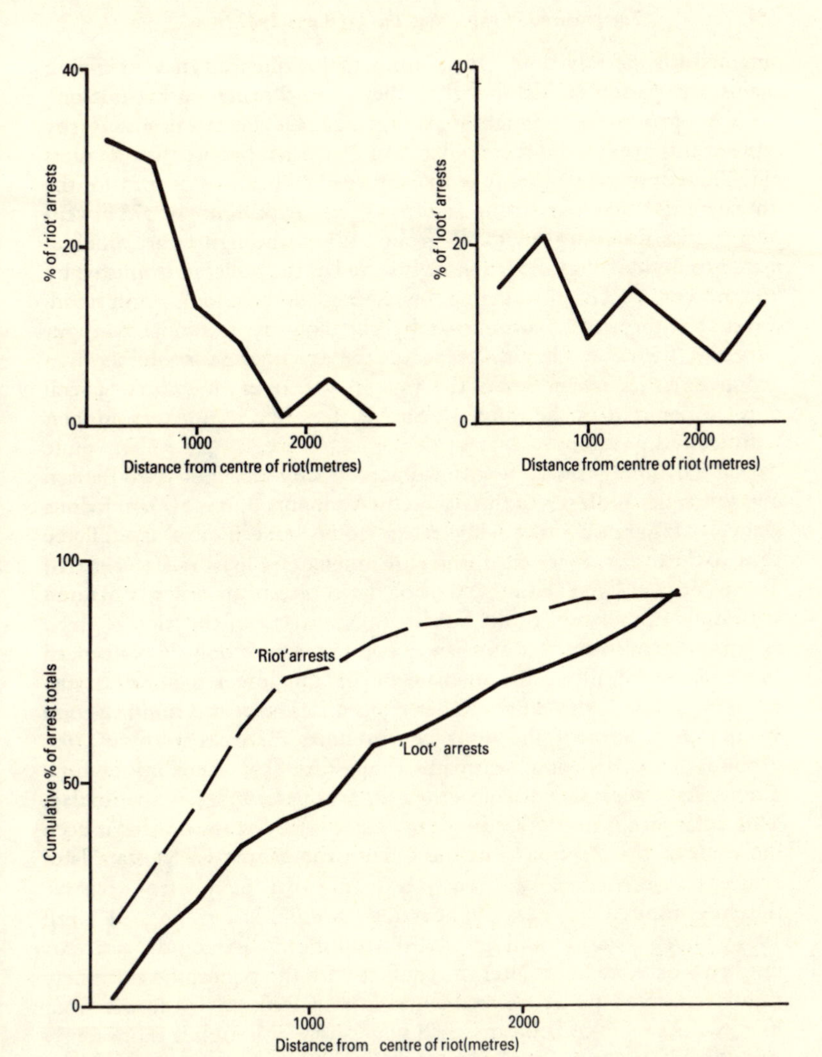

Figure 12.3 Residence pattern of arrest samples, Brixton, April 1981 (Home Office data).

obtained for the July riots. Of the three major offence types – violence against the person (VAP), burglary/theft, and threatening behaviour – the VAP group has the highest average age. Of the 38 (almost 70 per cent) of this group who were older than 21, 13 were more than 30 years old. This compares strikingly with a figure of 60 per cent over 21 for the 'threatening behaviour' group and the much lower figure of 37 per cent over 21 for the 'burglary/theft' group. When the figures are amalgamated to divide offences into those directed at the police, committed by 'the rioters' (VAP, threatening behaviour, obstruction), from those directed at property, committed by the 'looters' (criminal damage, burglary, theft),[10] a clear difference in the age distribution of the two groups emerges: 66 per cent of the 'rioters' were over 21, compared with only 36 per cent of the 'looters'. Such a disparity is, not surprisingly, statistically significant.

Marked differences were again observed between these two groupings when the addresses of this data set were mapped. It is evident from a glance at Figure 12.3 that while most 'rioters' and 'looters' lived fairly close to the disturbances, the former are much more localized: 18 per cent of the 'rioters' lived within 200 m of the centre of disorder compared with only 1.5 per cent of the 'looters'; 63 per cent of the 'rioters' lived less than 800 m from the disturbances compared with only 36 per cent of the 'looters'. Similarly the median, upper and lower quartiles of the dispersions reveal the much higher concentration of the rioting group around the centre of the disorders. In short there seem to be clear grounds for supporting Scarman's suggestion that there might be a difference between the distances the two groups travelled to the rioting. This difference is important more in relative than in absolute terms. At the time of the Brixton riots the Commissioner of the Metropolitan Police, David McNee, was quoted as saying that 'people from outside the area inspired Saturday night's riot' (*South London Press*, 14 April 1981). Given that, chronologically, the looting in Brixton did not start until two or three hours after the conflict with the police, the extremely tightly clustered pattern of residence around Railton Road, 'the front line', would seem to belie any such suggestion, although it is obviously not possible to evaluate the significance of single instances in statistical generalization (see Fig. 12.4).

Ethnicity also appears to be a crucial variable in differentiating between offence types (Table 12.2). Data sets that overlapped with those already cited facilitated both a general breakdown of the characteristics of a sample of 101 by police 'ethnic' classification, and of a 46-person subset of this group by both ethnicity and age. Again the most notable of the specific breakdowns occurs in the VAP category. Each of the 30 arrests in this group stemmed from a physical conflict between police

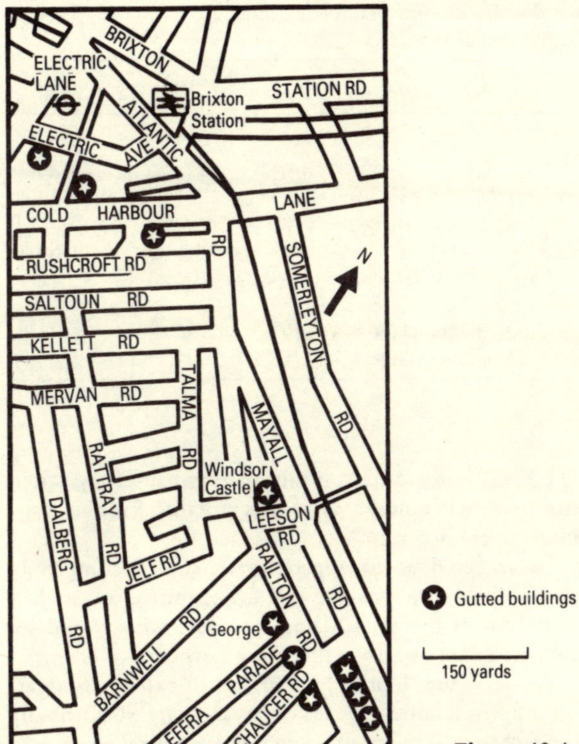

Figure 12.4 Central Brixton.

and the person charged: 80 per cent of the group were black. In marked contrast to this, those arrested for burglary or theft were still mostly black, but in this offence category 23 (41 per cent) were white. When the offence categories were amalgamated, the white group, accounting for 27 per cent of those arrested for 'rioting', make up a much larger proportion of the 'looters' (38.5 per cent). The multi-racial composition of the arrest figures has been represented as a cross-cultural conflict with the police,[11] yet on closer examination this seems to be almost exclusively a black preserve; only in the looting was there more general participation.

The breakdown of offence groups by both age and ethnicity is based on too small a sample size to yield any conclusive comment or statistical generalization. However, the trend that does emerge to some extent is that the majority of both blacks and whites arrested for 'looting'

Table 12.2 Ethnic distribution for a sample of those arrested in the Brixton rioting, April 1981, by offence type.

	Black	White	Asian	Other	Total
violence against person	24 (80%)	6 (20%)	0 (0%)	0 (0%)	30
burglary/theft	31 (55.4%)	23 (41%)	1 (1.8%)	1 (1.8%)	56
criminal damage	4 (100%)	0 (0%)	0 (0%)	0 (0%)	4
threatening behaviour	4 (44.4%)	5 (55.6%)	0 (0%)	0 (0%)	9
obstruction	2 (100%)	0 (0%)	0 (0%)	0 (0%)	2
'riot'	30 (73.2%)	11 (26.8%)	0 (0%)	0 (0%)	41
'loot'	35 (58.3%)	23 (38.3%)	1 (1.7%)	1 (1.7%)	60

Source: See text.

offences were in the 17–20 age group, contrary to Scarman's suggestion; again black arrests for VAP emerge as a class notable for the large numbers in the black over-21 group.

If offence type is accepted as an approximate, if not completely reliable, indicator of behaviour, then the call for empirical research is vindicated by an analysis of the arrest data. For those who would see riots as an expression of greed or criminality the emergence of *two* valid descriptive classes ('rioters' and 'looters') is difficult to explain. Suggestions of an influx of 'troublemakers' into the area are substantially undermined by the fact that so many of those involved lived so close to the disorders. Not all local black people were 'rioters', not all the 'looters' were juvenile whites; and undoubtedly there were many involved in both activities. The type of descriptions that can be deduced from this sort of data should not be used to replace one set of stereotypical images of 'the average rioter' with another. Nevertheless, the differential involvement of black and white, old and young, is sufficient to suggest that there were really almost two riots, involving two sorts of people: one was a highly localized, full-scale confrontation with the police, involving a broad cross-section of people from a very small area of Brixton; the other, which occurred some distance away from this, was an opportunistic reaction to the collapse of public order.

London, July 1981

Using material gathered with the co-operation of the Home Office Statistical Unit, it can be demonstrated briefly that in large part the patterns that emerge from the Brixton data are replicated in London as a

Table 12.3 Age distribution of those arrested for selected offences in 'serious incidents of public disorder' in London, July 1981.

	Under 17	17–20	Over 21	Total
violence against person	30 (18.7%)	73 (45.6%)	57 (35.6%)	160 (100%)
burglary/theft	74 (31.1%)	100 (42%)	64 (26.9%)	238 (100%)

Distribution differs significantly at 0.05 level of significance, $\chi^2 = 8.54$.
Source: Home Office Statistical Unit.

whole in the many outbreaks of disorder that occurred across the city in July 1981.[12] Focusing again on the comparison between the offences of theft and VAP, the age of those arrested for the former offence is significantly lower than that for the latter (Table 12.3), the conflict with the police tending to be the preserve of the older people present at the scene of the riot. Ethnicity is also an important variable in distinguishing between both different incidents of disorder and different activities at each location.[13] Although for London as a whole the white group accounts for more than any other (503 out of 1050, or 48 per cent),[14] this is largely a result of large numbers of white youths being arrested at relatively trivial incidents of public disorder that would most probably not normally have been classified as 'riots'. The most serious incidents of rioting in London in July 1984 in Southall (25 per cent of those arrested white), Brixton (36 per cent), Hackney (28 per cent) and Wood Green (32 per cent) involved only a minority of white arrests.[15] In individual locations, in crimes of violence (Table 12.4), which consisted almost entirely of clashes with the police, blacks are consistently over-represented relative to their proportion of arrests for the whole incident. Again, while all ethnic groups were involved in all facets of the disturbances, the arrest data imply that confrontation with the police is primarily dominated by black people, many of whom were over 21. It was not the young alone who rioted. The disorders cannot be reduced to a 'natural' element in the traditionally fractious process of juvenile socialization, the riot as part of the late-20th-century rites of passage.

Fallacies

'On the streets'

In the 1981 riots there were certainly occasions when 'black and white youth stood shoulder to shoulder against the police' (Gilroy 1982), but if patterns found in the arrest data are representative it would be misleading to suggest that such stances were typical of white involvement in

Table 12.4 Involvement in conflict with the police in 'serious incidents of public disorder' in London, July 1981.

	Arrests (total)	Black (number)	Black (%)	Violence against people (total)	Black (number)	Black (%)
Southall	61	4	7	21	2	10
Wood Green	71	42	59	5	4	80
Brixton	257	161	63	19	15	78
Hackney	107	75	70	23	18	78
Battersea	79	24	30	14	12	86
Tooting	41	10	24	6	2	33
West Ham	49	16	33	7	4	57
Croydon	45	4	9	2	0	0
Penge	42	22	52	1	0	0
Notting Hill	28	12	43	12	6	50
Walthamstow	29	1	3	9	0	0

Representation differs significantly at 0.10 level of signficance using binomial test, $z=1.43$.

Source: Home Office Statistical Unit.

collective disorder generally. Those incidents that constituted the most serious breakdowns in public order in London involved the smallest proportions of white arrests, while the white group was consistently over-represented in those offences connected with looting. Conflict with the police was not the central focus of white involvement. It would also be simplistic to characterize this conflict in terms of 'black youth' alone when such a significant proportion of those arrested were over 21. The picture of the young black man as the quintessential member of the rioting crowd of 1981 is in many ways profoundly misleading, a powerful cultural symbol, easily manipulated, especially when the dubious stereotypical classification of 'the average rioter' is placed alongside the common dread of the rampaging mob.

'The crowd' as an analytical concept

People are often understandably frightened of large numbers of other people. There is a long academic and literary tradition that has flourished on such fear and shaped common perception of this mythical crowd.[16] Individuals in this tradition lose control of themselves in a large group of people, who are normally portrayed as suffering from some form of mass hysteria. In such images the crowd assumes its own identity, its own driving force. Yet as Berk (1974) has pointed out, there is really very little evidence that such 'deindividuation' occurs. Normatively sanctioned behaviour may be very different for those in a large crowd than it would be for the same people in a very different situation, but this is not readily distinguishable from the contextual protocols that inform our everyday life. Similarly, in the rioting in London in 1981 a very wide range of behaviour characterized each incident of public disorder; there is no evidence of a psychologically disturbed 'mob', or even a single cohesive unit. Yet this unit is often implied. A cultural picture, the image of the crowd as animal, is presented as a natural phenomenon, part of *la condition humaine*, and is then used as the raw material for ostensibly realistic explanation.[17] Indeed, there is something slightly insidious in depriving a group of people of historical agency in this way, reducing the considered to the instinctive or automatic and the human to the bestial. It is a politically powerful transformation, utilized in various forms across the spectrum of opinion. Once this 'supra-individual' is seen to exist, its persona can be identified by 'the specialist': in 1981 both Darcus Howe (1982) and Chief Constable James Anderton (*The Times*, 12 December 1981) saw revolution, while Joshua & Wallace (1983) claimed that the crowds were an expression of the political struggles of Black Britons, a revolt against racial subordination. Similarly, Scarman (1981), in his analysis, never rejected this sort of populist image of the crowd as animal and at times came close to propagating the picture himself. Although this might be considered understandable in an 'old man' talking about the relatively young, it took an even older one to put his finger on one tacit assumption of the Scarman report. C. L. R. James (1981) pointed out that 'Lord Scarman is terrified by the power of young blacks. If he understood the reasons for this power he would not exaggerate and elevate their revolt into a force for the destruction of British society.'

In short, it would appear that although the phrase 'the crowd' is culturally meaningful, it is of dubious analytical utility. Not only is there evidence that variations in behaviour are not irrational (as in the Brixton looting), but also different sorts of people are involved in very different sorts of activity. In this context unity is an illusion and it is not possible to identify exactly who is a member of such a collectivity.

Evidence 2: The escalation of violence

A second source of empirical analysis can be derived from comparison of the different incidents of rioting. Specifically, the most serious disturbances of 1980 and 1981 shared many similarities, particularly in the escalation of violence. Most descriptions of the major 'riots' consider the events that immediately precede trouble, the triggers to violence, as either irrelevant or inconsequential. They are normally characterized in terms of metaphoric combustibility, merely the spark that causes the inevitable fire.[18] While it would be facile to overestimate the importance of the single instance, any explanation of rioting is incomplete unless it can account for the manner in which seemingly trivial incidents develop into major forms of collective destruction. Central to such a thesis is a rejection of the behaviourist conception of violence as some form of pressure release and an assumption that in crowd situations violence is both rational and meaningful.

With the notable exception of Moss Side, Manchester, the basic chronology of most of the major disorders of 1980 and 1981 has been relatively clearly established, even where there is controversy over the sequence of events that followed on from the initial trigger.

In April 1980 a protracted police raid on the *Black and White* café, at the centre of black settlement in Bristol, led to the St Paul's riot. In April and July 1981 three separate incidents, widely interpreted as examples of wrongful arrest in Railton Road, the 'front line', in Brixton resulted in three occasions of massive 'mobilization' against the police. In July 1981 a well-known local young black man, suspected of stealing a motorcycle, was stopped in Princess Boulevard, close to the Amberley Street/Upper Parliament Street heart of Liverpool 8, an area that was later erroneously described as Toxteth. He was subsequently charged with two counts of grievous bodily harm and one of assault against the three police officers involved in his arrest. The motorcycle was his own; the scenes that surrounded his apprehension turned into full-scale rioting. In London, on the same day, the arrival of large numbers of National Front supporters at a centre of Asian settlement in Southall for a pop concert, and their subsequent behaviour, induced a violent reaction against both the arrivals and the police force, which tried to intervene in this conflict. In Hackney, the attempt to disperse a crowd that had gathered outside Johnson's cafe in Sandringham Road, an area of black settlement, resulted in a major conflict with the police and widespread looting. Only in one of the most serious incidents of disorder in London, that of Wood Green, where the smashing of a shop window prompted extensive looting, did the trigger not involve the violent confrontation of two parties. Any understanding of the escalation of violence obviously rests on an ability to divine the feelings and perceptions of those who reacted to these 'trigger events'.

There is a fundamental problem relating to all historical reconstruction that tends to be lost in the vagaries of methodological discussion. Both in understanding somebody's actions and evaluating a symbolic event one is effectively 'reading' the social world. Problematically, the social equivalent of the basic 'speech act'[19] or 'parole'[20] is surrounded by exactly the same complexities as those faced in the philosophy of language. Actions have preferred descriptions. Even if a major element of the rioting is the conflict between the black community and the police, the action of throwing a petrol bomb may be considered by one person as avenging a specific insult, by another as a blow against 30 years of racial harassment, and by yet another as a blow against white society in general. More pertinently, it would be unusual if any one individual in such a situation did not consider that he or she was doing more than one thing at the same time. Motives may be rationalized into neat lists *post hoc*, but tend to be much more complicated in realization. This is the power of 'mental direction', in essence the problem that intentionality sets to any study of the social world, prohibiting glib statements about the relationship between mental states and physical behaviour.

The arrests that precipitated rioting in Bristol, Liverpool and London may have been ostensibly commonplace events but, most importantly, they took place in a specific context which endowed the straightforward action of arrest with a far greater symbolic power than usual, a context defined by time and place. The police force represent in part the state's claim to a monopoly of legitimate violence; the rioting is in part a rare, if resounding, rejection of this claim; yet it is not a rejection that is made equally by all sections of society, rather it is dependent on such factors as age, gender and the scene in which subsequent events are triggered. It was significant that the rioting in 1980 and 1981 was confined to very small areas of cities and was not obviously linked to any consciously articulated social movement; those involved tended overwhelmingly to live very close to the scene of the disturbance.

Arguably, the trigger is interpreted metonymically by those who either see it or hear about it. It is an act which is representative simultaneously of both a general situation and particular grievances, and, most self-evidently, it is an act to which violence is a response sanctioned by a large enough crowd of people to constitute collective disorder. This does not mean that violence is sanctioned by all present. As has already been suggested the propriety of violence does not vary arbitrarily between different age groups, genders, or even times of day. Nor does it mean that this action of violence is considered the same act by all involved: the preferred descriptions of the same actions may be traced to differing intentional states. However, the reading of the signification of the trigger event must be sufficiently clear to induce collective action. Within the 'social language' of a particular area at a particular time the trigger is read similarly by a large number of people.

It is because the trigger incident is taken as a single item, which as a part symbolizes a much larger whole, that the question 'Do people riot because of the police, or unemployment, or greed?' – or any other neat reason – is quite literally meaningless. Such precise categorizations and partitions find no equivalents in the structure of action. So, although it is vital to understand the perception of those involved in rioting, it is very difficult merely to cite a collective that we call 'the rioters' and to try to discern some straightforward average or communal perception for this group, in the tradition of perception geography.[21] Perception studies have tended to be based on an assumption that perception may be generalized for 'homogeneous' groups. This assumption is a dangerous half-truth, both because any one actor belongs to many such groups (e.g. profession, gender, family, age group) and so such classification is at best partial, at worst stereotypical, and also more importantly because it misrepresents the fundamental problem of any language, be it literal, social or perceptual. For it is in the very nature of language that it facilitates communication, but equally guarantees communication breakdown between any two individuals. The medium itself only exists by its system of differences and similarities that is contingent upon both context and subjectivity. The negotiation and evaluation of space that informs the social language constitutes part of this context. Upper Parliament Street, Railton Road, Sandringham Road and Grosvenor Road (site of the *Black and White* café) were all community foci. The superficially different rioting in Southall occurred as a direct result of an influx of National Front supporters into the centre of an Asian community, an influx so widely read as one more racist violation that it prompted a violent response which rapidly turned on the police, not as vicarious target, but because it was considered by a large enough number of those present that the police had singularly failed to protect the Asian community from attacks; not only in London in general but in their own homes in particular. A crucial factor that seems to recur in promoting a widespread adoption of the abnormal strategy of violence is this theme of 'home', most succinctly summarized by the suggestion to Lord Scarman that 'the police were no longer protecting the area in a responsible manner. They were in fact a force of occupation within the Brixton area' (Brixton Rastafarian Collective 1981). It is not necessary to make a value judgement about such a statement, other than to establish its sincerity. Nor is there a suggestion that all rioters felt in this particular way: rather a contention that such feelings were common and crucially relevant to the transformation of situations of tension and occasional resistance to arrest into scenes in which many hundreds of people became involved in full-scale confrontations with the police. Such confrontations were in this crucial sense often parochial and defensive in nature. These were not examples of some post-Ardrey ethological territoriality (Ardrey 1969), but a more simple proprietorial evaluation of space that had evolved over many years and was set in the context of a similarly 'mature' historical conflict between the police force and black communities in Britain.

When papers and politicians ask 'What *causes* rioting?' it appears that what they expect is an answer that will present a recipe for disorder; so many parts police behaviour, so many parts unemployment, so many parts criminal behaviour, so many parts racial disadvantage, and so on. In one respect this is nothing new; the Greek *aitia*, from which we derive the word 'aetiology', covered both the concept of blame or guilt and that of causality – there was no distinction between the two. At times the sordid search for scapegoats appears to be the main function of the typical riot post-mortem. Underlying this 'common sense' approach is the more significant feature of Humean causality which attempts to frame explanation in terms of the constant conjunction of one set of phenomena with another. For Hume 'there is nothing in a cause except invariable succession' (quoted in Russell 1946), an attitude that is pervasive in the quantitative social science project which tends to try to 'explain' variance in one variable by a string of others, seeking to attribute to each a quantified causal status. There are numerous philosophical problems with such a model. Most importantly it is only by recognizing the prominent rôle played by theory in shaping what we call knowledge, thus acknowledging the need for a mechanistic theory of causality, that study of social action can be validated.[22] The only alternative is that in an attempt to *understand* behaviour, the notion of *causes* of action is replaced by an emphasis on the *meaning* of action. In this sense understanding the relationship between trigger events and the escalation of violence is at best a semantic exercise, uncovering the meaning of behaviour at the level of the individual social language(s), whether it is the feelings of a policeman or somebody 'on the streets' that are thus revealed. This can only be the basis for partial explanation, for while it is only by tracing the link between intentional states and the real world that explanations will be plausible, it is only by linking the nature of intentionality with social form that explanations will be useful.[23]

The limits to description

This is perhaps the problem Habermas refers to when he suggests that 'a sociology that accepts meaning as a basic concept cannot abstract the social system from structures of personality' (Habermas 1971), and it is a problem that is as much ethical as it is academic. There is a very real sense in which studies of the social world may descend into 'the recesses of society', collect tracts on the 'language' and understanding of those who live in such esoteric locations (be it Sloane Square or Skye) and then return with plunder to academe. Such potentially parasitic and powerful information-gathering brings into play the age-old tension between theory and practice as soon as the newly gleaned information is placed within a particular explanatory framework. If it is to go beyond the insights of the actors that form the 'objects of study' such a framework must be able to account for unacknowledged conditions of action and unintended effects of behaviour. In this sense Marx's suggestion that everyday appearances and modes of expression are inadequate basic units for the analysis of the social world is not an exclusively Marxian notion, and merely serves to highlight the impossibility of a purely empirical social science. The theorist's categorizations and divisions of society can be seen as utilitarian pictures of that society that serve as alternatives to conventional or common-sense views. Crucially, they may inhere in society, but are not necessarily reified. Social theory is principally a description of relations, and the relationships or structures identified have the same philosophical status as universals. At this point it is necessary to clarify the distinction between description and explanation.

The problem might be stated thus. Underneath the ostensible world of innocent meanings lies the more tangled 'social construction of reality' that may serve to transform the manner in which people 'know' their society. This latent content is exposed by structuralist authors such as Barthes (1973, 1979), Castells (1983) or Hall (1978), but the product is in Habermas' terms emancipatory in nature and serves well only in a descriptive mode. When looking for explanation the dangerous tendency is to locate a group of people best served by such transformations, to attribute to this group the propensity to effect these transformations, and then to dignify the group with the spurious unity of some omnibus, normally pejorative, title such as 'the bourgeoisie' or the 'power-bloc'.[24] Significantly, for Lévi-Strauss the business of locating such a force was considered irrelevant, just as if Darwin had tried to locate some teleological device behind evolution. Darwin's model works as a descriptive theory tied to a specific causal mechanism based on the situational logic of mutation and selection. It is not controlled by any entity in exactly the same sense as the mythologies explicated by Barthes are not, contrary to his own opinion, controlled by any one group; although, as he often demonstrates, the more perceptive and Machiavellian members of society may use the labyrinthine maze of synthetic meaning to their own advantage. This does not detract either from the realistic status of such structuring of the world or from the significance of power relations in social analysis.

This point is illustrated in Figure 12.5, which represents a composite model of some of the ideas of Castells (1983), Sivanandan (1982, 1983)

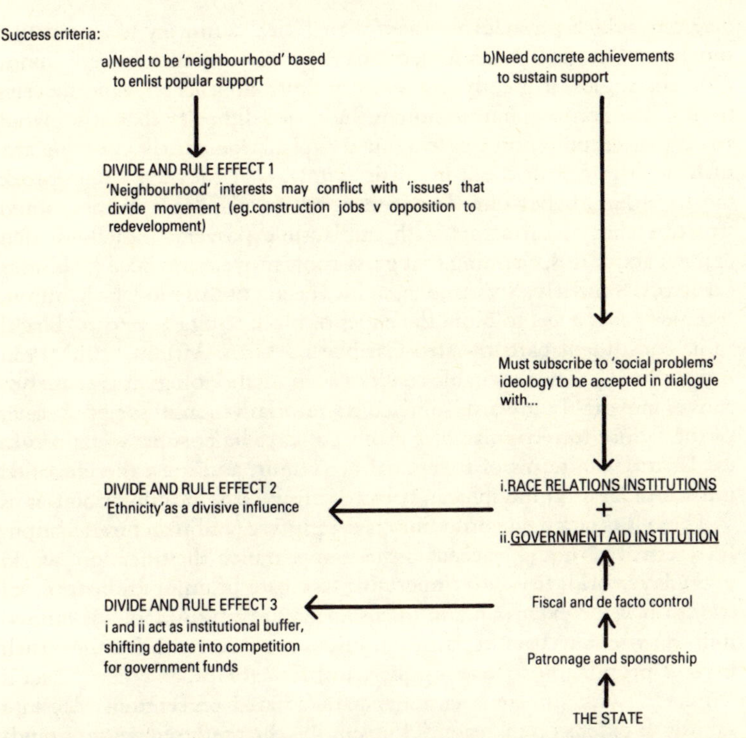

Figure 12.5 Systematic problems of 'popular movements' and 'ethnic' mobilization (adapted from Castells 1983, Sivanandan 1982, and Fisher 1985). The *description* in the flow diagram contains no *causal* mechanism. Not every effect has a premeditated cause – e.g. (i) and (ii) may act *in effect* as an institutional buffer, but are not necessarily designed to do so. Each element of the diagram cannot necessarily be treated as an 'historical actor' because of the relationship between intentionality and action.

and Fisher (1985). Because of its hybrid nature it is important to stress that this is only a personal view of their otherwise disparate work. The basic theme of the diagram is that in order to win widespread public support popular movements are often forced to subscribe to a set of conventional assumptions and social institutions that will, in the long run, hinder the very chances of success of these movements.[25]

There are two connected points that I want to make about such models. The first is that they are by their very nature synchronic, in that they describe a conjunction of structures at a particular point in time, described from one particular angle. It is a fundamentally descriptive diagram, which provides a powerful analytical summary of a particular situation, but just as an object may be photographed from many different angles, it is only one of an infinite number of valid descriptions.[26] The second point to note is the major difficulty that arises when analysts attempt to progress to a causal explanation, tacitly assuming that such descriptions define a monistic whole, that they are sufficient for understanding behaviour. The tendency is to transform the descriptive structure into social actors with implausible powers. Castells is quite explicit about this, claiming that grass-roots movements become historical actors. Similarly, Sivanandan, in his version of this model, claims that 'Ethnicity *was a tool* to blunt the edges of black struggle, return "black" to its constituent parts of Afro-Caribbean, Asian, African, Irish' (1983, my emphasis). The inevitable result of such methodology is that analysis moves inexorably towards some conspiratorial vision of society because of the failure to recognize that it must always be possible to rationalize explanations in terms of individual behaviour, and must therefore take into consideration the major problems of philosophical intentionality.

There is thus a dual constraint on descriptive (and structural) analysis. It is essential not to accept surface-appearance classification of the everyday world. It is also imperative to create meaningful behavioural sets, to balance extension and intension, in creating suitable descriptive units on which to base explanation. Just as extension and intension are in inverse proportion to one another, the probability of a shared 'social language' with similar intentional states (shared perceptions) decreases rapidly as extension increases.[27] Put simply, the preferred descriptions of actions and events are likely to be similar only when behavioural sets show maximum feasible intension and minimum feasible extension. This is not an unqualified call for intensive studies of the social world, but rather a suggestion that if explanation rests on descriptive sets of maximum extension then it will necessarily be precariously *a posteriori*, will not on its own be able to understand the *meanings* of events or actions, and must be rationalized in terms of unacknowledged conditions of action. All of these may be necessary and acceptable constrictions, although in the case of Castells in particular, they are often not considered.[28]

Rioting is a particularly germane example, where the basic analytical unit, 'rioter', can be defined by a series of averages calculated from those known to be 'on the streets', thus creating a set of extreme extension. Individuals in this set will interpret any one action in terms of widely differing intentional states. Poor explanations may select a couple of normatively derived salient characteristics from such meaningless averages and then 'subject' these 'straw men' to explanation. This is palpable nonsense, but it is important to recognize the constrictions that such considerations put on a hypothetically 'good' explanation. The problem of intentionality handicaps any simplistic exposition of the *feelings* of all the people involved in riots, while the relationship between theoretical and empirical domains prohibits the production of a single *definitive* account of events. It is possible to infer what some, or even many, of those involved considered they were doing, and it is possible constructively to argue a case for the most important features of the context of their behaviour, the material causes of action: but such a case can never be exhaustive. In such analysis the rôle of empirical evidence is not to falsify or verify hypotheses but to subvert ascendant theory or prevalent paradigms. This is not to accept some form of epistemological relativism, merely to acknowledge once again that 'objectivity is a direction not a terminus' (Harré 1979).

Cashmore & Troyna's study, with which this essay started, is a classic example of this problem. There is a tendency to identify a relevant concept, and then to use it like a favourite tool for every analytical job. The progression is from relevance to prominence, ascendancy and on to determination; the descriptive concept, be it 'class', 'culture', 'race', 'space' or any one of many neologisms, becomes increasingly powerful, is used to explain more and more, *extensio ad absurdum*. 'Black youth' is one valid descriptive unit but it is no *definitive* unit – and it certainly does not provide an explanatory model for all of the behaviour of the members of this group, precisely because the behaviour of a particular set of people is not wholly accounted for by their being young and black. The latter flaw is simply academic; the former much more insidious because if the category poses as reality rather than as one of many contingent descriptions of that reality, it implicitly recognizes, or even enhances, a form of racist stereotyping. It is this tendency to try to found explanation on descriptive models with implausible social actors that initiates the move towards an implausible structuralism, flawed in its conception because as Giddens puts it 'social systems have no purposes, reasons or needs whatsoever; only human individuals do so. Any explanation of social reproduction which imputes teleology to social systems must be declared invalid' (Giddens 1979).

It is this contingent value of social description, differing as it does from common-sense understandings of 'knowledge', that creates a major ethical contradiction in study of the social world. All descriptions are of limited use, yet as they accrue academic respectability, gaining status as part of received wisdom, the texts take on meanings of their own, set in the context of the history that follows their production, a normal characteristic of all literature. Indeed, for some the author must die that the text may survive,[29] an easy aesthetic ideal, free from responsibility. However, past pressures of scientific pretensions and present pressures of policy ramifications often result in academic studies of the social dismissing notions of explanation as puzzle-solving[30] and adopting positions of self-righteous certainty, selling 'truth' by the tome. But who is responsible for the path of the text through time? To acknowledge that the notion of value-freedom in the production of knowledge is spurious and fallacious is to lay the author open to personal as well as textual criticism, occasional polemical vilification, impugning individual motivation, alongside criticism of the literary product; moral judgements in hindsight rather than in context. Hence in one sense the text must die that the author may survive, bolstered by the kudos of the work, not harassed by its history. This is an ethical conflict of interest of major proportions.

Because of the mismatch between description and reality there exists a multiplicity of valid descriptions of the rioting in Britain in 1980 and 1981, yet this inevitable diversity has been used by analysts to justify *a priori* paradigms, circular proofs of validity based on manifestly biased significations of the phenomenon 'riot'. The riot is in this sense different from other social concepts only inasmuch as the scarcity of empirical evidence makes such distortion particularly easy. For only if theory is amenable to genuine modification via the empirical can worthwhile advances in understanding be made. Certainly a history of the riots is always a version, never a catalogue. This is the relevance of Wilde's suggestion that life imitates art, meaning that we compose the reality we perceive with mental structures that are cultural not natural in origin. Epistemological problems may well be profound, but in an era when naïve empiricism has been rightly discredited there is a danger that validation of theory will become incestuous rather than logical, based on the passwords of cliques and schools of thought, disparate vocabularies acting as the shibboleths of social science, the product of an academic Tower of Babel.

The half of the story that was rarely told in media coverage of the 1980 and 1981 riots was a tale of a conflict between police and black communities in Britain that was both deeply rooted historically and focused on powerfully evocative symbolic locations. These locations were to provide the dramatic stages for the transformation of common hostility into collective disorder. The violent and often horrific realization of conflict 'on the streets' is conditioned by both time and space.[31]

Notes

1 The Supplement to the *Oxford English Dictionary* (p. 998) defines these terms as follows: *Explicans* – The explanatory part of an explanation; in the analysis or explication of a concept or expression, the part that gives the

meaning. *Explicandum* – The fact, thing or expression to be explained or explicated.

2 'Social problem', 'black youth' and 'crisis' are in this context all value-loaded terms, not neutral descriptions.

3 For more detailed criticism of the 'cultural' model in general and the work of Cashmore & Troyna, as well as that of Lea & Young, in particular, see Gutzmore (1983). Lawrence (1982) also provides a more historical critique of black pathology and sociology.

4 The complex arguments and statistical difficulties in this field can in no way be summarized by a straightforward acceptance of 'black criminality'. Most of the empirical work that has been done in this field (e.g. Stevens & Willis 1979, Deutsch 1982) certainly does not accept such a notion.

5 See also Rex (1982), Cooper (1985), and several of the contributions to both Cowell, Jones & Young (1982) and Benyon (1984). Joshua & Wallace (1983) provide a more sophisticated explanation of the Bristol riot of 1980, but still stress the 'commonality of cause' behind collective disorder.

6 For a discussion of the difference between the collective as aggregate, the collective as structure and the collective as supra-individual, see Harré (1979, p. 140).

7 Rude (1967) has used similar methods to analyse *The crowd in history* and uses the phrase 'faces in the crowd'.

8 During the riots there were several cases reported of charges levelled against people being randomly chosen and arbitrarily changed midway through legal proceedings. Such cases clearly restrict the value of this particular sort of analysis. Nevertheless, if such cases are exceptional rather than normal, the link between charge and behaviour will remain, even if in individual cases this link is undoubtedly misleading.

9 For all analysis of the arrest data the police 'ethnic' coding is used for practical reasons. There is no normative comment implicit in this essay on the validity or ethics of such coding.

10 Descriptive terms such as 'rioters' and 'looters' are in this context loose analytical terms used to break down the arrest data. The precise relationship between such descriptive classes and reality is moot.

11 See, for instance, Howe (1981) and Gilroy (1982).

12 Again there is no claim that these data are an ideal source, only a suggestion that given the scarcity of alternatives, examination of such material is preferable to self-indulgent rationalism. For an examination of the flaws of the Home Office Statistical Unit data see LAG (1982).

13 Again it was necessary to rely on the police classification of 'ethnicity'.

14 The full data set consisted of 1050 arrests of which 503 (47.9 per cent) were 'white', 430 (40.95 per cent) were 'Afro-Caribbean', 99 'Asian' (9.43 per cent) and 18 of 'other' ethnic origins.

15 If the 27 incidents of public disorder in London in July 1981 (as defined by the Home Office) are ranked according to seriousness, as evaluated by either Wanderer (1974) or Spilerman (1971), there is a positive correlation of between 0.65 and 0.75 between riot seriousness and the proportion of those arrested that came from the 'Afro-Caribbean' group.

16 This tradition includes an illustrious assortment of pejorative descriptions of

the crowd that includes Taine's 'la canaille', Clarendon's 'dirty people without a name', Le Bon's 'hypnotized mob', Freud's 'psychologically disturbed, enlarged family' and Blumer's description of the 'crowd' as 'a herd'. For lucid histories of the crowd as myth see Rude (1967) and Berk (1974).

17 For fuller analysis of the mystification of the past by the transformation of history into nature, see Barthes (1973, 1979); and for what is in part one of the more successful attempts to use such ideas in British 'race relations sociology', see Hall *et al.* (1978).

18 For example, Hytner (1981), 33.6: 'the spark that led to the conflagration'; Scarman (1981), 8.9: 'the spark' that started the rioting.

19 See Austin (1962) for an accessible introduction to the philosophy of language, and Searle (1969, 1983, 1984) for discussion of 'speech acts' and the problems of intentionality.

20 See Barthes (1967) or Eco (1977) for two of the more lucid examinations of notions of 'langue' and 'parole'.

21 The problematic relationship between conscious intentional states and action is complicated yet further when preconscious and subconscious motivation is also considered. This complexity is rarely mirrored in geographical 'perception studies'. (See for example Downs & Stea (1977) for an introduction to a voluminous literature.)

22 See Sayer (1984) for a discussion of knowledge in context and the relationship between theory and knowledge.

23 'Usefulness' of an explanation can be assessed by satisfaction of conditions of practical adequacy. In Sayer's terms (1984, p. 66) 'To be practically adequate, knowledge must generate expectations about the world and about the results of our actions which are actually realised.'

24 The notion of 'power-bloc', derived from Laclau (1977), is often used by writers from the Centre for Contemporary Cultural Studies.

25 The notion of 'institutional buffering' is neither new nor restricted to the work of these authors – e.g. in British 'race relations sociology' see Ben-Tovim & Gabriel (1982).

26 Descriptions are theory-loaded but are not theory-determined. A description is useful if it is practically adequate for an explanation.

27 A human set of minimum extension would consist of 'the individual', whose characteristics are, *pace* subconscious influences, broadly unitary; hence the set is also representative of maximum intension. The number of characteristics that is shared by a group of humans must inevitably decrease as the size of the set increases and the parameters that define the set are relaxed. Hence intension decreases as extension increases. For fuller discussion of this see Harré (1979).

28 Castells (1983, p. 320): 'The movements become social actors by being engaged in a mobilisation towards an urban goal.'

29 See Eco (1985) on post-modernism and the novel.

30 Giddens (1979) advocates such an approach as preferable to naïve notions of explanation and 'the knowledge' thus produced.

31 The following references are not referred to in the text but are of major background significance to the chapter as a whole: Canetti (1973), Castells (1976), Giddens (1982) and Sivanandan (1973, 1976).

References

Ardrey, R. 1969. *The territorial imperative*. London: Fontana.

Austin, J. L. 1962. *How to do things with words*. Oxford: Clarendon.

Barthes, R. 1967. *Elements of semiology*. London: Cape.

Barthes, R. 1973. *Mythologies*. London: Paladin.

Barthes, R. 1979. *The Eiffel Tower*. New York: Hill & Wang.

Ben-Tovim, G. & J. Gabriel 1982. The politics of race in Britain 1962–1979. In *Race in Britain*, C. Husband (ed.) London: Hutchinson.

Benyon, J. (ed.) 1984. *Scarman and after*. Oxford: Pergamon.

Berk, R. A. 1974. *Collective behaviour*. Dubuque, Iowa: W. C. Brown.

Brixton Rastafarian Collective 1981. *Evidence to Lord Scarman's Enquiry*. Kew: Public Records Office.

Burgess, J. 1985. News from nowhere. In *Geography, the media and popular culture*, J. Burgess & J. R. Gold (eds). Beckenham, Kent: Croom Helm.

Canetti, E. 1973. *Crowds and power*. Harmondsworth: Penguin.

Cashmore, E. & B. Troyna (eds) 1982. *Black youth in crisis*. London: Allen & Unwin.

Castells, M. 1976. *The urban question*. London: Edward Arnold.

Castells, M. 1983. *The city and the grassroots*. London: Edward Arnold.

CCCS 1982. *The empire strikes back*. London: Hutchinson/Centre for Contemporary Cultural Studies.

Cooper, P. 1985. Competing explanations of the Merseyside riots of 1981. *British Journal of Criminology* **25**, 60–8.

Cowell, D., T. Jones & J. Young (eds) 1982. *Policing the riots*. London: Junction Books.

Deutsch, F. 1982. *Street crime in London 1981*. London: CRE.

Downs, R. M. & D. Stea 1977. *Maps in minds*. New York: Harper & Row.

Eco, U. 1977. *A theory of semiotics*. London: Macmillan.

Eco, U. 1985. *Reflections on 'The Name of the Rose'*. London: Secker & Warburg.

Fisher, G. 1985. *Ethnic participation, ideologies, strategies and realities*. Unpublished.

Giddens, A. 1979. *Central problems in social theory*. London: Macmillan.

Giddens, A. 1982. *Profiles and critiques in social theory*. London: Macmillan.

Gilroy, P. 1982. Police and thieves. In *The empire strikes back*. London: Hutchinson/Centre for Contemporary Cultural Studies.

Gutzmore, C. 1983. Capital, black youth and crime. *Race and Class* **25**(2), 13–31.

Habermas, J. 1971. *Theorie de Gesellschaft Oder Sozial Technologie*. Quoted in McCarthy, T. (1978) *The critical theory of Jurgen Habermas*. Oxford: Polity.

Hall, S., C. Critcher, T. Jefferson, J. Clarke & B. Roberts 1978. *Policing the crisis*. London: Macmillan.

Harré, R. 1979. *Social being*. Oxford: Blackwell.

Harré, R. & P. F. Secord 1972. *The explanation of social behaviour*. Oxford: Blackwell.

House of Commons 1972. *Report of the Select Committee on Police/Immigrant Relations*. London: HMSO.

Howe, D. 1982. Brixton before the uprising. *Race Today*, February/March, 62–9.

Husband, C. 1982. *'Race' in Britain*. London: Hutchinson.

Hytner, B. (chairman) 1981. *Report of the Moss Side Enquiry Panel to the Leader of the Greater Manchester Council*. Manchester: Council Publication.

James, C. L. R. 1981. Comments on the Scarman Report. *New Society*, 3 December.

Joshua, H. & T. Wallace 1983. *To ride the storm*. London: Heinemann.

Laclau, E. 1977. *Politics and ideology in Marxist theory*. London: New Left Books.

LAG 1982. Legal Action Group examination of problems of Home Office Statistical Unit Arrest Data. *LAG Bulletin*, November 1982.

Lawrence, E. 1982. In the abundance of water the fool is thirsty. In *The empire strikes back*. London: Hutchinson/Centre for Contemporary Cultural Studies.

Le Bon, G. 1896. *The crowd, a study of the popular mind* (reprint, 1960). London: Macmillan.

Lea, J. & J. Young 1982. The riots in Britain 1981. In *Policing the riots*, D. Cowell, T. Jones & J. Young (eds). London: Junction Books.

Murdock, G. 1984. Reporting the riots. In *Scarman and after*, J. Benyon (ed). Oxford: Pergamon.

Rex, J. 1982. The 1981 urban riots in Britain. *The International Journal of Urban and Regional Research* **6**(1), 88–113.

Rude, G. 1967. *The crowd in history*. New York: Wiley.

Runciman, W. G. 1983. *A treatise on social theory*. Cambridge: Cambridge University Press.

Russell, B. 1946. *History of Western philosophy*. London: Allen & Unwin.

Sayer, A. 1984. *Method in social science*. London: Hutchinson.

Scarman, Lord 1981. *The Brixton disorders 10–12 April 1981*. London: HMSO.

Searle, J. R. 1969. *Speech acts*. Cambridge: Cambridge University Press.

Searle, J. R. 1983. *Intentionality: an essay on the philosophy of the mind*. Cambridge: Cambridge University Press.

Searle, J. R. 1984. *Minds, brains and science*. London: BBC.

Sivanandan, A. 1973. Race, class and power: an outline for study. *Race* **14**, 383–9.

Sivanandan, A. 1976. Race, class and the state. *Race and Class* **17**(4), 347–68.

Sivanandan, A. 1982. *A different hunger*. London: Pluto Press.

Sivanandan, A. 1983. Challenging racism: strategies for the '80s. *Race and Class* **25**(2), 1–11.

Spilerman, S. 1971. The causes of racial disturbances: tests of an explanation. *American Sociological Review* **36**, 627–49.

Stevens, P. & P. Willis 1979. *Race, crime and arrests*. Home Office Research Unit Study number 58. London: HMSO.

Thackrah, J. R. 1985. Reactions to terrorism and riots. In *Contemporary policing*. Bramshill Police College.

Wanderer, J. J. 1974. An index of riot severity and some correlations. *American Journal of Sociology* **74**, 500–6.

13 *'A permanent possession'? US attitudes towards Puerto Rico*

Peter Jackson

This chapter is concerned with the way in which racist ideas and stereotypes are reproduced in public discourse by a variety of means. Racism, which can be personal or institutional, intended or unintended, concerns beliefs and practices that are based on the assumption of racial difference. Characteristically, racist beliefs reflect and actively reproduce structured inequalities between social groups in terms of their differential access to the material and symbolic rewards of status and power. Beliefs and practices are mutually reinforcing through the process of social reproduction. In dealing with racism as an ideology at the level of discourse, therefore, we are also implicitly addressing the question of racism as institutionalized in practice.[1]

Racist ideologies are expressed through a variety of media in both public and private domains. Political discourse represents a relatively public domain with privileged access to the mass media. Less public, but no less privileged in its own way, is the discourse of academic social science. Despite the appearance of 'political neutrality', 'academic detachment' and 'scientific objectivity', a closer reading of the evidence suggests that North American social science has consistently performed a highly ideological rôle in systematically perpetuating the structured inequalities of power that characterize the relationship between the United States and Puerto Rico.

In each domain, racism can be defined as an ideology in the sense of a set of beliefs and practices that conceal the interests of a super-ordinate group (Urry 1981). Like other ideologies, racism is revealed through its recourse to 'characteristic selectivities' or to unintentional biases (Williams 1981). Commonly, racist ideas take the form of 'common-sense' statements that are accepted and perpetuated without critical thought.[2] In other words, racist ideologies involve specific forms of 'unexamined discourse' (Gregory 1978). This chapter attempts to show how the discourse of social science has been no less ideological in its effects than the more public area of political discourse.

The chapter falls into three main sections. The first provides a brief discussion of the development of US interests in Puerto Rico; the second examines the public domain of political discourse about Puerto Rico; and the third discusses the ideological rôle of the social sciences in the context of US–Puerto Rican relations.

The United States and Puerto Rico

The United States invaded Puerto Rico at the end of the Spanish–American War in 1898 and took over the administration of the island from Spain after Puerto Rico had enjoyed a brief period of relative autonomy in 1897 (Lewis 1963, Maldonado-Denis 1972). While many Puerto Ricans welcomed their new governors, others began immediately to organize resistance. The Puerto Rican educator Eugenio María de Hostos gained an audience with President McKinley for this purpose but the United States' self-proclaimed Manifest Destiny inevitably prevailed. Indeed, the *New York Times* saw fit to advise the Puerto Ricans to welcome this opportunity to 'come at once under the beneficient sway of the United States' rather than engaging in 'doubtful experiments at self-government'.[3] This article also argued that 'the circumstances of the conflict for the enfranchisement of Cuba and Puerto Rico' fully entitled the United States to retain the island as a 'permanent possession'.

Direct military rule ended in 1900 and the Puerto Ricans became US citizens in 1917. It was not until 1947, however, that Luis Muñoz Marín became the first Puerto Rican Governor to be elected by popular vote. Shortly thereafter, Puerto Rico was given a new 'temporary' political status as a Commonwealth (*Estado Libre Asociado*), 'freely associated' with the United States. This status has persisted to the present day despite the efforts of those who advocate either statehood or independence. As a Commonwealth, Puerto Rico is officially a self-governing territory but depends on the United States for defence and other crucial areas of government including trade and immigration. The Puerto Ricans can migrate freely to the mainland as some two million have now done. While they remain on the island, however, they have no presidential vote and no seat in Congress.

Under Commonwealth government, the United States was able to transform the island from a poor but relatively self-sufficient peasant economy to an export-orientated cash-crop economy. The principal mechanism for this process was a programme of tax incentives, known familiarly as Operation Bootstrap, that encouraged mainland businesses to locate on the island. Despite the attainment of impressive rates of economic growth, the achievements of the programme have been greatly exaggerated. According to some observers, the 'stricken land' became a 'land of wonders' virtually overnight.[4] Other assessments reveal that the programme made little impact on the island's chronically high unemployment rate, which in 1982 still ran at 24 per cent of the labour force (Wagenheim 1983). Descriptions of the island's economic history as a 'showcase of democracy' therefore seem rather far-fetched.

Puerto Rico has been aptly described as an 'artificial economy' (García-Passalacqua 1984, p. 78). Until the Depression, Puerto Rico was practically a monocrop economy dominated by US sugar companies. Then, by means of Operation Bootstrap, it became equally dependent on US-based multinational investment. US companies have invested particularly heavily in petrochemicals and pharmaceuticals, in scientific and precision instruments, in banking and finance, all producing high rates of profit but making little impression on the island's chronic unemployment problem. In fact, with the closure of companies like the Commonwealth Oil Refining Company in Peñuelas in 1983, the problem is further exacerbated. Yet newspapers like the *International Herald-Tribune*, in a special supplement on Puerto Rico (July 1981), can still describe the island as 'the ideal second home for American business', with a highly skilled labour force, an attractive tax-incentive programme, duty-free entry into the US market, well-developed industry services, a prime location, and a sophisticated banking and financial sector with no exchange risk.

At present, the island is faced with an even more demeaning form of dependence on federal welfare payments. This has led to particularly severe consequences as a result of President Reagan's 'new federalism' and associated cutbacks in public spending. In July 1982, for example, Congress removed Puerto Rico from the federal Food Stamp programme at a time when some 50–60 per cent of the island's population was eligible for this particular form of assistance. In place of Food Stamps worth $875 million and Medicaid costs amounting to a further $45 million in fiscal 1981, Puerto Rico was awarded a block grant of $825 million in fiscal 1982 (*New York Times*, 3 January 1982). Other programmes, such as those administered under the Comprehensive Employment Training Act (CETA) have also been revoked in Puerto Rico.

Other recent federal initiatives have also had a deleterious effect on the island economy. Reductions in US corporate tax, for example, have diminished the comparative advantage of Puerto Rico for US investors. Uncertainties about the future of other fiscal instruments, such as

Figure 13.1 The Caribbean Basin Initiative, 1982. The stipple indicates beneficiaries of the CBI initiative.

Section 936 of the Internal Revenue Code, have also threatened existing investments, particularly in the pharmaceutical industry, besides their immeasurable effect in discouraging new investment. The Caribbean Basin Initiative (CBI), introduced to prevent the spread of Communism in the Caribbean, has reduced Puerto Rico's competitive edge even further by extending tax benefits to other areas in the Caribbean[5] (see Fig. 13.1).

The economic and political aspects of US imperialism in Puerto Rico are further exacerbated by a longstanding military presence concentrated in the offshore islands of Vieques and Culebra. Both islands have been in regular use by the US Navy since the 1940s for firing ranges although the military were forced to abandon Culebra in 1975 after fierce opposition from local people and *independentistas* (Maldonado-Denis 1972, pp. 262–73). In circumstances of such wide-ranging and deeply structured inequality as those that exist between the United States and Puerto Rico it is hardly surprising that many aspects of public discourse take on a distinctly ideological form.

Racism in political discourse

With this brief history of US–Puerto Rican relations in mind, I turn now to consider the way in which racist ideas are expressed in public discourse, especially in the political arena. I argue that such ideas serve to perpetuate the structured inequalities of power that have characterized the relationship between Puerto Rico and the United States. All political discourse about Puerto Rico must therefore be seen in the light of the evolving political and economic relationship between the United States and Puerto Rico and in the context of a growing Puerto Rican population in the United States (see Table 13.1).

Puerto Rico's Commonwealth status has been described as little more than 'a verbal disguise of colonial subjection' (Lewis 1974, p. 28). The status issue is still bitterly contested and continues to permeate every aspect of the island's relationship with the mainland. Political and economic imperialism inevitably spill over into the cultural domain. This is most clearly shown historically but remains no less true today. For example, a soldier who saw active service during the US campaign in Western Puerto Rico that followed on from the landing in Guánica Bay offered the following opinion concerning the most appropriate method of dealing with the typical Puerto Rican: 'A thick, stout cudgel or a bright, sharp axe will be more effective than honeyed words in helping him cheerfully to assimilate new ideas.' And sure enough, the American claim to come 'bearing the banner of freedom' was quickly

Table 13.1 Puerto Ricans in the continental United States, 1940–80.

	Born in Puerto Rico (first generation) (thousands)	Born in United States (second generation) (thousands)	Total (thousands)
1940	70.0	n.a.	70.0
1950	226.1	75.3	301.4
1960	617.0	275.5	892.5
1970	810.1	581.4	1391.5
1980	n.a.	n.a.	2013.9[a]

[a] Puerto Rican origin or descent.

Sources: A. J. Jaffe *et al.* (eds) *Spanish Americans in the United States: changing demographic characteristics*, New York: Research Institute for the Study of Man, September 1976; and New York Department of City Planning, *The Puerto Rican New Yorker*, December 1982.

belied by the imposition of a military government. The soldier's commentary is therefore most instructive in revealing the cast of mind that can rationalize such rampant imperialism. His views on race and class are also extremely apt:

> About one-sixth of the population in this island – the educated class, and chiefly of pure Spanish blood – can be set down as valuable acquisitions to our citizenship, and the peer, if not the superior, of most Americans in chivalry, domesticity, fidelity and culture. Of the rest, perhaps one half can be moulded by a firm hand into something approaching decency; but the remainder are going to give us a great deal of trouble. They are ignorant, filthy, untruthful, lazy, treacherous, murderous, brutal and black. (Hermann 1900; quoted in Gannon 1979, pp. 34–5.)

Such racist opinions were not confined to the military. The popular American magazine *Atlantic Monthly* produced an account in very much the same terms: 'The lower classes are destitute of moral perception, and disgusting in their habits of life . . . For arrant despicable cowardice, the world cannot produce their match' (Pettit 1899).

Even when a non-military government had been established, a favourite theme of American politicians was the Puerto Ricans' inability to govern themselves. William H. Hunt, the second Governor of Puerto Rico, was not alone in his opinion that 'The Puerto Ricans lack any and all capacity to govern their own destiny' (Andreu Iglesias 1984, p. 92). Press accounts were no more flattering, the *Morning-Sun* suggesting in 1902 that the Puerto Ricans were 'no more than savages who have replaced their bows and arrows with guns and knives' (ibid., p. 93). But the New York daily *Globe* kept up the usual theme, reporting in March 1904 that the 'natives of Puerto Rico' were not capable of governing themselves 'because they are a country that has not yet reached its maturity' (ibid., p. 95). As spokesmen for the Puerto Rican community in New York were quick to point out, however, these statements about the Puerto Ricans' inability to govern themselves were made at a time when they had not even been consulted about their national sovereignty. It might also be suggested that the 15 American Governors who administered the island over the 46-year period before a Puerto Rican was first appointed to the post did not prove themselves particularly adept at the job.

The Governorship has been described as 'a rich plum in the spoils system of American politics' (Lewis 1963, p. 119). The roster of American Governors includes one who sought to immortalize his name by signing a bill legalizing cockfighting with a tail feather and another who was rumoured to have decamped from La Fortaleza with a set of exquisite heirlooms on loan to an exhibition at the palace. Governor E. Montgomery Reilly was charged with a wide range of breaches of legal and political procedures. Blanton Winship was responsible for 20 deaths and scores of injuries during the Ponce Massacre in 1937 (ibid., pp. 119–21). Robert H. Gore's political career as Governor has been described by another historian as 'short but disastrous', fully justifying the soubriquet 'Gore's Hell' (Mathews 1960, pp. 58–116).

The cultural aspect of US imperialism is also clearly revealed in attitudes towards the Puerto Rican language both historically and today. From the earliest days of US occupation, a stereotype was created concerning the Puerto Ricans' debased Spanish. The first Commissioner of Education on the island asserted that the Puerto Ricans did not have 'the same devotion to their native tongue or to any national ideal that animates the Frenchman, for instance, in Canada or the Rhine Provinces' (Clark 1899, quoted in Wagenheim 1973, p. 141). 'A majority of the people', he continued, 'do not speak pure Spanish, their language is a patois almost unintelligible to the natives of Barcelona and Madrid. It possesses no literature and little value as an intellectual medium. There is a bare possibility that it will be nearly as easy to educate these people out of their patois into English as it will be to educate them into the elegant tongue of Castile.' Such an attitude towards the Puerto Ricans' use of Spanish was, of course, an ideal justification for the convenient American policy of introducing English as the language of instruction in Puerto Rican schools and the adoption of a system of education that had Americanization as one of its clear goals. This policy, which the historian Gordon Lewis (1963, p. 118) describes as 'egregious folly', continued until 1948.

Vestiges of such stark cultural imperialism persist in contemporary US attitudes towards the Puerto Ricans' 'language problem' on the mainland. Countless observers have described the Puerto Ricans as 'illiterate in two languages' or referred contemptuously to their blend of English and Spanish as 'Spanglish'.[6] Other neologisms, such as 'Neorican' or 'Nuyorican', have a less negative connotation and can be used as positive self-ascriptions (cf. Attinasi 1979). In some contexts, in fact, language has taken on a central significance in Puerto Rican resistance to American cultural hegemony. For example, in the political domain, bilingualism is now among the primary demands of most Puerto Rican organizations (e.g. National Puerto Rican Task Force 1977). Meanwhile, radical poets, like Pedro Pietri, have aptly summarizd the 'Nuyorican' experience in poems such as *The Broken English Dream* (1973):

To the united states we came
To learn how to mispell our name
To lose the definition of pride
To have misfortune on our side
To live where rats and roaches roam
in a house that is definitely not a home
To be trained to turn on television sets
To dream of jobs you will never get
To fill out welfare applications
To graduate from school without an education
To be drafted distorted and destroyed
To work full time and still be unemployed

The linguistic basis of Nuyorican poetry and its roots in the 'sobering reality of the New York ghetto' are even more clearly revealed in Tato Laviera's poem, *My graduation speech* (quoted in Flores *et al.* 1981, p. 214) which begins:

i think in spanish
i write in english

i want to go back to puerto rico,
but i wonder if my kink could live
in ponce, mayaguez and carolina
tengo las venas aculturadas
escribo en spanglish
abraham in espanol
abraham in english
tato in spanish
'taro' in english
tonto in both languages

Returning to the more explicitly political domain, it is clear that the current rhetoric about Puerto Rico that emanates from the White House is no less arrogant than presidential statements of the recent past. President Ford's suggestion in December 1976 that the moment had arrived for Puerto Rico to be fully assimilated as the nation's 51st state was interpreted at the time as no more than a characteristic gaffe (Rodríguez Beruf n.d.). President Carter's outspoken stance against US 'colonialism' in 1977 came to little more, despite his offer to let the Puerto Ricans decide their own future. As a presidential candidate, Ronald Reagan declared himself personally in favour of statehood while endorsing a referendum on the status question (*Wall Street Journal*, 11 February 1980). He later gave more unequivocal support to statehood in a statement that was widely interpreted as a reversal of his campaign pledge, if the Puerto Rican people chose it 'in a free and democratic election' (*New York Times*, 10 and 13 January 1982).

However, the Puerto Rican electorate has continued to be highly divided on the subject of statehood. The 1970s marked the 'end of consensus' in terms of the established electoral dominance of the Popular Democratic Party (PPD). The PPD registered its first defeat for 30 years in 1969 and its long-term leader, former-Governor Luis Muñoz Marín, died in 1980. The pro-statehood New Progressive Party (NPP) won the election in 1980 by a tiny margin (less than 1 per cent of the vote) but the PPD retained control of the Senate. The narrowness of the electoral margin led to a bitter stalemate that was only finally resolved by recourse to the federal courts, and the new governor, Romero Barceló, was forced to abandon his commitment to a plebiscite on statehood. The 'status soap opera' continues (Villamil 1984).

Before the 1984 election, the NPP again promised a referendum on statehood in 1985 if re-elected. The election was complicated by the advent of a new party, the Puerto Rican Renewal Party, led by the former mayor of San Juan, Hernán Padilla. The Renewal Party's platform was based on vociferous criticism of Reagan's economic policy and a moratorium on the status question until the island's economic problems were resolved. In the event, the PPD candidate, Rafael Hernández Colón, was elected governor, promising to improve the island's relationship with Washington, to restore tax exemptions and to abandon efforts to win greater autonomy for the Commonwealth (*New York Times*, 3 January 1985). Indeed, it is one of the ironies of Puerto Rico's relative powerlessness that successive Resident Commissioners in Washington are judged by the Puerto Rican electorate in terms of their ability to secure federal funds for the island (*El Nuevo Día*, 12 March, 12 June and 8 July 1981). Their success in this direction, of course, only exacerbates the problem of Puerto Rico's chronic dependence on the US.

Any decision the Puerto Ricans may themselves take about statehood or independence may not be acceptable to Washington. There are, in fact, a number of grounds on which the United States may feel reluctant to admit the island as the 51st state. First, there are the economic costs of administering an island that is several degrees poorer than Mississippi, presently the poorest state of the union. As a state, Puerto Rico would be eligible for all the benefits of the federal system. Although the Puerto Ricans would have to start paying federal income tax, they would also make disproportionately large demands on a welfare system that is currently being heavily curtailed. The political consequences of statehood are also far from negligible. As a state, Puerto Rico would command more votes in Congress than at least 30 of the existing states. Culturally, too, there are grounds for assuming that the United States would be less than eager to admit Puerto Rico to the union. Although Spanish is now widely spoken in New Mexico, Texas, California, Florida and New York, the perceived 'racial' difference of the Puerto Ricans may prove a less permeable barrier.

Political discourse in the United States includes occasional discussion of the nature of Puerto Rican migration to the mainland. Even in the early years of this century many observers felt that the migration could be attributed to Puerto Rico's 'overpopulation'. As Bernardo Vega recalls in his *Memorias*: 'Although the population [of Puerto Rico] had yet to reach one million, there were those who argued that "overpopulation" was the cause of all the misery' (Andreu Iglesias 1984, p. 88). Historians and social scientists later added their support to the view that, to quote one such instance, 'Puerto Rico's central problem since its annexation to the United States has been overpopulation' (Handlin 1959, p. 49).

The view that migration provided a 'safety valve' for the solution of Puerto Rico's population problems is still commonly expressed (e.g. Friedlander 1965). Some observers have even claimed that migration is beneficial to all concerned: migrant families themselves, the area to which they migrated, and the area from which they came (Senior & Watkins 1975). Only recently has the unfounded optimism of such views been critically exposed. Arguing from an explicitly Marxist perspective, the History Task Force of the Centro de Estudios Puertorriqueños at the City University of New York rejects the premise that population exerts an unbearable pressure on the productive system's resource base, necessitating large-scale out-migration. Instead, they argue that a system of production such as that employed in Puerto Rico exerts an intolerable pressure on the population (History Task Force 1979). Statistical analysis of migration and employment data clearly supports this more radical view, as the ebb and flow of migration and return is closely related to the fluctuating demands of US capital rather than simply reflecting the poverty of the island's resource base (Jackson 1984). Puerto Rico's resource base must in any case be seen in the context of specific processes of exploitation rather than as a purely natural endowment. From such a perspective, migration is better understood in terms of political economy than in simple demographic terms.

Racism in academic discourse

Far from combating the overt and implicit racism that pervades political discourse about Puerto Rico, the great bulk of social-science research emanating from the United States has actually tended to reinforce it. While the common-sense character of public discourse about 'race' is relatively understandable, however, the social-science profession has no such excuse for failing to adopt a more critical stance. This section attempts to demonstrate the ideological nature of much of the received wisdom of North American scholarship about Puerto Rico. It also uses predominantly Puerto Rican scholarship to indicate the possibility of a more critical rôle for social science in the context of the United States' continuing colonial relationship with Puerto Rico.

Even the most benign of US social scientists cannot escape the charge of unintentional racism in some of their writing about Puerto Rico. The Jesuit sociologist, Joseph P. Fitzpatrick, is a case in point. Fitzpatrick served for many years as the unofficial interpreter of Puerto Rican culture for mainland social-service agencies (e.g. Fitzpatrick 1971). On a practical level, his research and teaching helped prevent much misunderstanding and maltreatment of the Puerto Rican migrants (see, for example, Fitzpatrick 1960). At the same time, his writings remained wedded to an assimilationist perspective and served to perpetuate some of the most enduring myths about Puerto Rican culture.[7] For example, he argued that second-generation Puerto Ricans found themselves in a 'cultural no man's land', torn between two cultures, a condition which he proceeded to relate to various social problems such as mental illness and drug abuse (Fitzpatrick 1955). Such a simplistic account of the nature of cultural and social change tends to obscure the structural causes of such problems (cf. Jackson 1981a). Other examples show that similar forms of analysis have much less benign consequences.

The academic analysis of Puerto Rican mental illness provides an extreme case of cultural 'misunderstanding' that undoubtedly serves the interests of the US psychiatric profession better than those of its Puerto Rican clients.

During the late 1950s and early 1960s several mainland scientists became interested in the so-called *ataque* or 'Puerto Rican syndrome', a condition that was said to be rife among the early migrants in New York (Berle 1958). Rather than interpreting this condition as a culturally-specific response to situations of extreme stress, a variety of far-fetched theories were propounded, usually couched in extravagant medical jargon and with occasional reference to 'deficient child-rearing practices' and 'too frequent exposure to primal scenes' (Fernandez-Marina 1961; cf. Rothenberg 1964).

At the same time, several social scientists and medical practitioners were reporting on the alleged prevalence of schizophrenia among Puerto Ricans in New York (e.g. Malzberg 1956, Rogler & Hollingshead 1965, Rendon 1974). While some authors went so far as to posit an immediate connection between the experience of migration and the occurrence of mental illness, implying that the migrants were simply 'torn between two cultures', others were more cautious, suggesting that, in the absence of an adequate understanding of the social and cultural roots of schizophrenia, many of the migrants were being treated for mental illnesses they did not have (Fitzpatrick 1971, p. 162).

These are perhaps extreme cases.[8] But there are many other examples where North American social scientists have perpetuated unfounded myths about Puerto Rican culture and society with definite ideological effects. Two influential observers of the American scene, for example, describe Puerto Rican culture and family life as 'sadly defective', and take the prevalence of consensual unions among the Puerto Ricans in New York as evidence of 'an instability in the marriage form'. They declare that 'organizational life is not strong among the Puerto Ricans', that the migrants suffer from 'cultural schizophrenia', a condition that is presumably somewhat ameliorated by their alleged 'passion for music and dancing' (Glazer & Moynihan 1963). A team of sociologists from Columbia University led by C. Wright Mills reached similar conclusions, describing the Puerto Rican community in New York as 'a fragmented, disunited sphere for social living'. The Puerto Ricans were 'illiterate in two languages', had 'no effective leadership' and were generally without aspirations for the future (Mills *et al.* 1950). A later study from the same university concluded that it was not even possible to speak of a Puerto Rican culture in New York (Padilla 1958).

Thomas Cochrane's study of *The Puerto Rican businessman* (1959) provides a compendium of stereotypes about Puerto Rican 'culture'. Puerto Rican entrepreneurs tend to suffer in comparison with their US competitors, according to Cochrane, because of the 'highly individualistic' nature of the Puerto Rican business system, dominated by family proprietorship and control. Cochrane goes on to tell us that '. . . control based on seniority in the family tends to perpetuate the deterrants to entrepreneurial activity inherent in Spanish cultural traditions'. The value that the Puerto Rican businessman places upon 'secure and dignified living', together with their 'distrust of change', works against the expansion of enterprise. They are also, moreover, fearful of delegating authority, have little interest in technological innovation and prefer face-to-face contacts, preferably within their own families.

If Cochrane's own opinions were not enough, the sources that he quotes embody countless other myths about Puerto Rican 'culture'. A couple of examples will suffice. Cochrane quotes William L. Schurz on the Latin American's individualism: 'Though gregarious, he [sic] is averse to merging his personality in any group or to sacrificing its claims to any mass demands . . . he is not a good organization man. . . . He does not take kindly to the restraints of teamwork and resents the discipline of the group.' He quotes former-Governor Tugwell on the Puerto Ricans' pride, which is 'almost an obsession and which leads frequently to the substitution of fancy for fact'. Others write confidently about 'the Spanish race' and of *dignidad* and *personalismo* as innate characteristics of the Puerto Rican people.

Other characteristics glibly attributed to 'Puerto Rican culture' can be just as easily questioned. For example, there is a good deal of agreement among US social scientists about the Puerto Ricans' apparent apathy concerning elections on the mainland, compared with the extraordinarily high registration and turnout rates on the island (Estades 1978, Jennings & Rivera 1984). However, recent research on the early history of the Puerto Rican community in New York reveals a far more politicized and organized community than previously supposed, with more than 30 000 registered Puerto Rican voters during the 1940s, for example (Sánchez Korrol 1983, p. 184). The Puerto Ricans are also often described as an undisciplined and docile labour force, uncomplainingly resigned to constant exploitation. Again, evidence now exists with which to contradict this stereotype (e.g. Galvin 1979, Quintero Rivera 1976). Finally, there is a substantial literature that describes the nature of Puerto Rican communities in New York and other mainland cities as 'disorganized' or 'pathological' (e.g. Chenault 1938, Thomas 1967). This derogatory stereotype is readily countered and yet, as with each of the previous ideas, the mythical view consistently prevails.[9]

Given these examples it is no wonder that some Puerto Rican authors have despaired of the whole North American social-science profession. One Puerto Rican sociologist has commented, for example, that 'the vast social science research on Puerto Rico conducted by most North

American investigators has contributed very little to a real understanding of our society' (Nieves Falcón 1971, p. 9). Other Caribbeanists have expressed a similar fear that 'the structure of United States academic institutions and the career patterns of United States academies create the kind of scholar and type of research that has not been in the best interests of the area' (Lewis 1974, pp. 255–6). It is with particular interest, then, that one turns to the social-science literature produced by Puerto Ricans themselves.

To some extent, the literature produced by the Centro de Estudios Puertorriqueños at the City University of New York and by the Centro de Estudios de la Realidad Puertorriqueña at the University of Puerto Rico can be seen as an attempt to refute the stereotypes about Puerto Rican culture that characterize popular and academic accounts of the island and its migrant communities.

Insularity is one such feature, persistently attributed to Puerto Rican culture. It is a theme that Antonio S. Pedreira developed at length in his *Insularismo: ensayos de interpretación puertorriqueña* (1934). In that work, Pedreira ventured the opinion that the Puerto Ricans were a characteristically weak, complacent, ignorant and confused people, with a penchant for rhetorical excess, given to fits of lyrical melancholia, cowardly and passive in the face of adversity. It is no surprise, then, that Puerto Rican scholars are today seeking to liberate themselves from the consequences of this 'most burdensome typological cliché' (Flores 1980, p. 11). Flores himself provides a parody of Pedreira's negative evaluation of Puerto Rican culture: 'Paralyzed by an inclement climate, diminutive geography and a disjointed racial fusion, Puerto Ricans are condemned to isolation from the world around them, economically, politically, intellectually and culturally. On their haunches in the face of Destiny, and wills weakened by the tropical heat, they have recourse only to optimistic metaphors and overblown rhetoric with which to "sweeten the pill" of their historical misery' (ibid., p. 76).

Yet the insular stereotype persists, recurring as one of the central themes in the recent Twentieth Century Fund study of Puerto Rico (Carr 1984). In this study, the author describes the Puerto Ricans' cultural identity as 'confused, provincial, and ambiguous'. He writes of the Puerto Rican *jíbaro* as 'a brutalised and shifty peasant in a rural backwater devoid of intellectual distinction or social grace', and he dismisses everyday life in San Juan as 'vulgar and pretentious'. But this is as nothing compared to the comments he reserves for 'the pathology of island politics', such as 'the Puerto Rican disease' itself: 'their insistence on the unique importance of their problems perversely ignored by the world in general and by the United States in particular'. Other events in the island's recent history, such as the alleged political killings at Cerro

Maravilla, are dismissed as evidence of the Puerto Ricans' 'latent paranoia'. Their concerns over the issue of national sovereignty are rejected as 'obsessive and Byzantine discussions' that prove them to be a 'politically introverted small island community', caught up in 'the jungle of tribal politics'. Finally, the Puerto Ricans are said to indulge in 'tasteless verbal violence' and to be prone to 'periodic fits of self-doubt'. The metaphors have scarcely changed in the 50 years since Pedreira wrote his diatribe.

Likewise, the characteristic docility attributed to Puerto Rican culture in Rene Marqués' literary and psychological study *El puertorriqueño dócil* (1962) has remained a persistent theme in analyses of Puerto Rican culture, despite the efforts of a younger generation of Puerto Rican scholars to draw attention to the history of active resistance and organized struggle that has characterized Puerto Rican history (Silén 1971, Quintero Rivera 1976, Galvin 1979). Marqués argued that 'there is scarcely an area in Puerto Rican society where, scratching the surface a little, docility does not appear as a constant and determining trait' (Marqués 1976, p. 70). According to Marqués, the Puerto Ricans were submissive, weak and ignorant, the victims of a pathetic inferiority complex, and lacking confidence in their own intelligence, knowledge and strength.

By contrast, a more radical Puerto Rican analyst has acknowledged 'the silence, surface submission, and nonmilitancy that has marked much of our past', but asserts in its place the tenacity and capacity for survival that the Puerto Ricans have shown in their 'resolute avoidance of a capitulation on the cultural front' (Bonilla 1980, p. 370). Bonilla himself accepts the challenge of turning this symbolic resistance to surer political purpose ('beyond survival'). That purpose is enshrined in his belief that 'the option to struggle for full sovereignty in the island and power over our own mainland communities remains very much alive'. Whether that option will be exercised, however, remains to be seen.

In order to avoid giving the impression that Puerto Rican scholarship is a homogeneous body of work whose authors are in total agreement about the interpretation of recent Puerto Rican history and the need for change, a further comment may be necessary. The reality is, of course, more complicated. Even those who profess a Marxist approach are quite seriously divided on the question of Puerto Rico's hopes for liberation and its survival as a nation. James Blaut (1977, 1987), for example, has criticized the History Task Force of the Centro de Estudios Puertorriqueños for their 'badly flawed theory' of uneven development whose 'serious contradictions' lead to 'depressingly pessimistic, almost defeatist' conclusions about the island's future status. Far from being a unitary body of ideas, therefore, Puerto Rican scholarship is extremely diverse. The point that has been argued here, however, is that these ideas, together with their counterpart in North America, serve particular interests and cannot be divorced from the political and economic context in which they are articulated.

Conclusion

This chapter has attempted to interrogate a number of stereotypes about Puerto Rico and the Puerto Ricans. These stereotypes include the small size of the island and its relative lack of resources; the degree to which the island is assumed to be overpopulated; the alleged docility and insularity of the Puerto Rican people; and the supposed prevalence of a variety of social pathologies such as drug addiction and welfare-dependence together with a series of physical and mental diseases, such as tuberculosis and schizophrenia.

Each of these stereotypes, in one way or another, serves US interests in maintaining the entrenched inequalities of power that are enshrined in Puerto Rico's colonial relationship, to the continuing profit of American corporate capitalism. A small island with few resources can be said to be incapable of self-government on economic as well as on political grounds. 'Overpopulation' justifies the encouragement of mass migration at a time of labour shortage in the United States. A docile workforce can be readily exploited with few qualms about the periodic suppression of civil rights in cases of labour unrest and workers' struggles. And 'pathological' communities deserve to be treated like second-class citizens, at the mercy of the state's largesse and always vulnerable as the earliest victims of any cutbacks in federal, state and city services (Marcuse 1981).

Television and other mass media in the United States play a crucial rôle in perpetuating a racist stereotype of the Puerto Ricans as shiftless welfare-recipients whose teenage children are organized into gangs which roam the streets engaging in the most threatening types of criminal behaviour – rape, arson, drug-dealing and mugging. The news media report little about the Puerto Ricans that is not in some way connected with the violent activities of terrorist groups such as the FALN. But, as has been shown, their exaggerated images are fuelled by popular social-science accounts which concentrate on poverty, crime and sub-cultural 'deviance'. Oscar Lewis's studies of the 'culture of poverty' in San Juan and New York (Lewis 1965) are perhaps the best-known example of this tendency towards sensationalist analysis. He describes the Rios family in *La Vida*, for example, as

> closer to the expression of an unbridled id than any other people I have studied. They have an almost complete absence of internal conflict and of a sense of guilt . . . [They] show a great zest for life, especially for sex, and a need for excitement, new experiences and adventures. Theirs is an expressive style of life. They value acting out more than thinking out, self-expression more than self-constraint, pleasure more than productivity, spending more than saving, personal loyalty more than impersonal justice. They are fun-loving and enjoy parties, dancing and music. They cannot be alone; they have an almost insatiable need for sociability and interaction (Lewis 1965, p. xxvi).

The most infamous aspects of Lewis's 'culture of poverty' concept, of course, is the idea that it is perpetuated across the generations because of its effects on children. By the time they are six or seven, according to Lewis, they have absorbed the basic values and attitudes of the subculture and are 'not psychologically geared to take full advantage of changing conditions or increased opportunities which may occur in their lifetime' (ibid., p. xlv). Despite the extensive critique that has been made of this material (e.g. Valentine 1968, Rainwater & Yancey 1967) it remains a powerfully persistent theme in American thinking about Puerto Rico. For all that, 'its lurid and ugly story of a few families of prostitutes is no more typical of the island than *Tobacco Road* is of the mainland' (Steiner 1974. p. 508).

Only lately has the more direct experience of a growing Puerto Rican population in American cities been available as a possible antidote to these distortions. However, the extent to which the Puerto Ricans are segregated in the *barrios* of New York and other cities significantly reduces the probability for such an improvement in mutual understanding (see, for example, Jackson 1981b).

The prevalence of racist ideas and stereotypes in public political discourse about Puerto Rico is, then, closely mirrored in academic literature about the island and its migrant communities in the United States. This should not surprise us, given the extent to which academic social science is dominated by North American institutions and authors. While individual authors are still capable of adopting a critical perspective, the great bulk of the literature accurately reflects the interests and biases of its sponsors. The growing body of social-science research that is emerging from Puerto Rican authors, whether from the University of Puerto Rico or from the City University of New York, is therefore a highly significant feature in recent Puerto Rican intellectual history, providing a forceful critique of much of the received wisdom generated by mainland social science. Naturally, these sources have their own biases which reflect the context of Puerto Rico's abiding colonial status. I have argued, in fact, that every aspect of political and academic discourse about Puerto Rico bears the imprint of this fundamentally unequal relationship. It is my belief, however, that a critical analysis of Puerto Rican discourse offers at least one avenue for moving beyond the level of 'things said' to engage in the transformation of that relationship.

Notes

1 On the nature of discourse, see Foucault (1972).
2 The 'common-sense' nature of racist ideologies is discussed in Lawrence (1982).
3 The story is reprinted in Wagenheim (1973, pp. 106–10).
4 These phrases are taken from Tugwell (1947) and Hanson (1955), the second edition of which was entitled *Puerto Rico: land of wonders*. Puerto Rico's transformation 'from a scabrous slum ... to a shiny exhibit of democracy and free enterprise in action' is eulogized by Hancock (1960). Wells (1969) is only slightly more critical.
5 For further discussion of the CBI, see Carr (1984), Feinberg *et al.* (1983) and Vega (1985).
6 For a vigorous polemic that sets the issue of bilingualism in the context of 'Americanization', see J. Flores *et al.* (1981, pp. 193–217). The authors argue, in particular, that 'code-switching' is not a form of compensation for monolingual deficiency but a positive means of expanding communicative and expressive potential (p. 200).
7 Assimilationist ideologies are criticized in Blaut (1983).
8 For a general study of ethnic minorities and psychiatry, see Littlewood & Lipsedge (1982).
9 More positive interpretations of Puerto Rican community organization include Andreu Iglesias (1984), Sánchez Korrol (1983) and Colón (1982).

References

Andreu Iglesias, C. (ed.) 1984. *Memoirs of Bernardo Vega*. New York and London: Monthly Review Press.

Attinasi, J. 1979. Language attitudes in a New York Puerto Rican community. In *Bilingualism and public policy: Puerto Rican perspectives*, R. V. Padilla (ed.), 10–63. City University of New York: Centro de Estudios Puertorriqueños.

Berle, B. B. 1958. *Eighty Puerto Rican familes in New York*. New York: Columbia University Press.

Blaut, J. M. 1977. Are Puerto Ricans a national minority? *Monthly Review* **29**, 35–55.

Blaut, J. M. 1983. Assimilation versus ghettoization, *Antipode* **15**, 35–41.

Blaut, J. M. 1987. *The national question and colonialism*. London: Zed Press.

Bonilla, F. 1980. Beyond survival: por que seguiremos siendo puertorriqueños. In *The intellectual roots of independence*, I. M. Zavala and R. Rodríguez (eds), 357–71. New York and London: Monthly Review Press.

Carr, R. 1984. *Puerto Rico: a colonial experiment*. New York and London: New York University Press.

Chenault, L. R. 1938. *The Puerto Rican migrant in New York City*. New York: Columbia University Press.

Cochrane, T. C. 1959. *The Puerto Rican businessman: a study in cultural change*. Philadelphia: University of Pennsylvania Press.

Colón, J. 1982. *A Puerto Rican in New York and other sketches*. New York: International.

Estades, R. 1978. *Patterns of political participation of Puerto Ricans in New York City*. Universidad de Puerto Rico, Editorial Universitaria.

Feinberg, R. E., R. Newfarmer & B. Orr 1983. The battle over the CBI. *Caribbean Review* **12**, 15–48.

Fernandez-Marina, R. 1961. The Puerto Rican syndrome: its dynamics and cultural determinants. *Psychiatry* **24**, 79–80.

Fitzpatrick, J. P. 1955. The integration of Puerto Ricans. *Thought* **30**, 402–20.

Fitzpatrick, J. P. 1960. New York City and its Puerto Rican 'problem'. *Catholic Mind* **58**, 39–50.

Fitzpatrick, J. P. 1971. *Puerto Rican Americans: the meaning of migration to the mainland*. Englewood Cliffs, NJ: Prentice Hall.

Flores, J. 1980. *The insular vision: Pedreira's interpretation of Puerto Rican culture*. Centro de Estudios Puertorriqueños, CUNY, Working Papers No. 1 (revised edn).

Flores, J., J. Attinasi & P. Pedraza Jr 1981. *La Carreta* made a U-turn: Puerto Rican language and culture in the United States. *Daedalus*, 193–217.

Foucault, M. 1972. *The archaeology of knowledge*. London: Tavistock.

Friedlander, S. L. 1965. *Labor migration and economic growth: a case study of Puerto Rico*. Cambridge, Mass.: MIT Press.

Galvin, M. 1979. *The organized labor movement in Puerto Rico*. London: Associated University Presses.

Gannon, P. S. 1979. *The ideology of Americanization in Puerto Rico, 1898–1909*. PhD thesis, New York University.

García-Passalacqua, J. M. 1984. *Puerto Rico: freedom and equality at issue*. New York: Praeger.

Glazer, N. & D. P. Moynihan 1963. *Beyond the melting pot*. Cambridge, Mass.: MIT Press.

Gregory, D. 1978. *Ideology, science and human geography*. London: Hutchinson.

Hancock, R. 1960. *Puerto Rico: a success story*. Princeton, NJ: Van Nostrand.

Handlin, O. 1959. *The newcomers: Negroes and Puerto Ricans in a changing metropolis*. Cambridge, Mass.: Harvard University Press.

Hanson, E. P. 1955. *Transformation: the story of modern Puerto Rico*. New York: Simon & Schuster.

History Task Force, Centro de Estudios Puertorriqueños, City University of New York, 1979. *Labor migration under capitalism: the Puerto Rican experience*. New York: Monthly Review Press.

Jackson, P. 1981a. A transactional approach to Puerto Rican culture. *Revista/Review Interamericana* **11**, 53–68.

Jackson, P. 1981b. Paradoxes of Puerto Rican segregation in New York. In *Ethnic segregation in cities*, C. Peach, V. Robinson & S. Smith (eds), 109–26. London: Croom Helm.

Jackson, P. 1984. Migration and social change in Puerto Rico. In *Geography and ethnic pluralism*, C. G. Clarke, D. Ley & C. Peach (eds), 195–213. London: Allen & Unwin.

Jennings, J. & M. Rivera (eds) 1984. *Puerto Rican politics in urban America*. Westport, Conn.: Greenwood Press.

Lawrence, E. 1982. Just plain common sense: the 'roots' of racism. In *The empire strikes back: race and racism in 70s Britain*, Centre for Contemporary Cultural Studies (eds), 47–94. London: Hutchinson.

Lewis, G. K. 1963. *Puerto Rico: freedom and power in the Caribbean*. New York: Monthly Review Press.

Lewis, G. K. 1974. *Notes on the Puerto Rican revolution*. New York: Monthly Review Press.

Lewis, O. 1965. *La vida: a Puerto Rican family in the culture of poverty – San Juan and New York*. New York: Random House.

Littlewood, R. & M. Lipsedge 1982. *Aliens and alienists*. Harmondsworth: Penguin.

Maldonado-Denis, M. 1972. *Puerto Rico: a socio-historical interpretation*. New York: Vintage Books.

Malzberg, B. 1956. Mental disease among Puerto Ricans in New York City, 1949–51, *Journal of Nervous and Mental Disease* **123**, 262–9.

Marcuse, P. 1981. The targeted crisis: on the ideology of the urban fiscal crisis and its uses. *International Journal of Urban and Regional Research* **5**, 330–55.

Marqués, R. 1962. *El puertorriqueño dócil*. Translated as: *The docile Puerto Rican*, 1976. Philadelphia: Temple University Press.

Mathews, T. 1960. *Puerto Rican politics and the New Deal*. Gainesville, Florida: University of Florida Press.

Mills, C. W., C. Senior & R. K. Goldsen 1950. *The Puerto Rican journey*. New York: Harper.

National Puerto Rican Task Force on Educational Policy 1977. *Toward a language policy for Puerto Ricans in the United States: an agenda for a community in movement*. New York: National Puerto Rican Task Force on Educational Policy.

Nieves Falcón, L. 1971. Puerto Rico: a case study of transcultural application of behavioural science. *Caribbean Studies* **10**, 5–17.

Padilla, E. 1958. *Up from Puerto Rico*. New York: Columbia University Press.

Pettit, W. V. 1899. Porto Rico. *Atlantic Monthly* **83**, 635–7.

Pietri, P. 1973. *Puerto Rican obituary*. New York and London: Monthly Review Press.

Quintero Rivera, A. 1976. *Workers' struggle in Puerto Rico*. New York and London: Monthly Review Press.

Rainwater, L. & W. L. Yancey 1967. *The Moynihan report and the politics of controversy*. Cambridge, Mass.: MIT Press.

Rendon, M. 1974. Transcultural aspects of Puerto Rican mental illness in New York. *International Journal of Social Psychiatry* **20**, 18–24.

Rodríguez Beruf, J. (n.d.). *Puerto Rico: 51st state?* London: Committee for Puerto Rican Independence.

Rogler, L. H. & A. B. Hollingshead 1965. *Trapped: families and schizophrenia*. New York: Wiley.

Rothenberg, A. 1964. Puerto Rico and aggression. *American Journal of Psychiatry* **120**, 962–70.

Sánchez Korrol, V. E. 1983. *From colonia to community: the history of Puerto Ricans in New York City, 1917–1948*. Westport, Conn.: Greenwood Press.

Senior, C. & D. O. Watkins 1975. Toward a balance sheet of Puerto Rican migration. In *Status of Puerto Rico: selected background studies prepared for the United States–Puerto Rico Commission on the Status of Puerto Rico*, 689–795. New York: Arno Press.

Silén, J. A. 1971. *We, the Puerto Rican people: a story of oppression and resistance*. New York and London: Monthly Review Press.

Steiner, S. 1974. *The islands: the worlds of the Puerto Ricans*. New York: Harper & Row.

Thomas, P. 1967. *Down these mean streets*. New York: Knopf.

Tugwell, R. G. 1947. *The stricken land: the story of Puerto Rico*. New York: Doubleday.

Urry, J. 1981. *The anatomy of capitalist societies*. London: Macmillan.

Valentine, C. A. 1968. *Culture and poverty: critique and counter-proposals*. Chicago and London: University of Chicago Press.

Vega, B. 1985. The CBI faces adversity. *Caribbean Review* **14**, 18–43.

Villamil, J. J. 1984. The status soap opera. *Caribbean Review* **13**, 3–44.

Wagenheim, K. 1973. *The Puerto Ricans: a documentary history*. New York: Praeger.

Wagenheim, K. 1983. *Puerto Ricans in the US*. New York and London: Minority Rights Group Report No. 58.

Wells, H. 1969. *The modernisation of Puerto Rico*. Cambridge, Mass.: Harvard University Press.

Williams, R. 1981. *Culture*. London: Fontana.

14 *Racist and anti-racist ideology in films of the American South*

JOHN SILK

In this chapter the treatment of 'race' in the cinema, restricted here to those films set in the American South, is considered in the light of a number of factors: (i) the continually changing context of relations between blacks and whites, particularly the level and nature of struggle against oppression mounted by the former; (ii) the way in which these struggles related to the economic organization of the motion-picture industry, and to censorship; and (iii) the impact of economic and political relations between the South and other sections or regions of the United States, and the overall 'laws of motion' of capitalist competition and accumulation within the United States and in the capitalist world as a whole.

I am concerned with the rôles that films play in the relationships between the political and cultural, and the economic and cultural, spheres. These rôles are considered primarily in terms of dominant ideology, which is constituted by ideas, themes, stereotypes and images which tend, whether by conscious intention or not, to justify and reinforce the interests of one class at the expense of another (or others). This may be done by misrepresenting or obscuring the class issues and conflicts involved. In capitalist societies, this means the justification of the exploitation and oppression of the proletariat by the bourgeoisie. For example, individualism tends to promote feelings of self-congratulation for success and self-blame for failure, rather than an examination of systematic or structural features of society. Nationalism, on the other hand, appeals to and fosters loyalties which are apparently above conflicts between exploiter and exploited, oppressor and oppressed. Both ideological factors are evident in the films to be considered, but are subordinate to the shifting portrayal of blacks and of black–white relations in films of the American South. Throughout, I argue that changes in portrayals of the South in general, and of black–white relations in particular, can be related to class interests. As well as examining the way in which the films reinforce dominant ideology, based on class, and related ideologies which support racial oppression, I shall be looking for ways in which films have been used by oppressed and exploited groups to counter such ideologies and I shall estimate the extent of their success.

Economic and political context – the 19th-century legacy

The crucial issue over which the American Civil War had been fought was the future of a mode of production based on slavery in the American South, and the political power that this gave to the slave-owning class both in the Southern states and in Washington. As the 19th century progressed, slavery became a fetter to continuing capitalist development in the United States. The problem was to be solved by the Civil War, which ended with a military victory for Northern interests in 1865. To secure political power in the South after the war, an alliance of black freedmen, Southern poor white farmers (scalawags) and Northern businessmen (carpetbaggers) was formed by the Radical wing of the Republican Party. A small army of occupation was established, and Constitutional Conventions were set up in which blacks and some poor whites actively participated. Some Southern war leaders were disqualified from voting. Progressive measures in education and social welfare were enacted by many Conventions and there was also strong pressure for land reform. This period of Radical Republican dominance in Southern legislatures was known as (Radical) Reconstruction. Opposition to Reconstruction came from boycotts by Democratic party politicians and many white voters, and also from guerilla warfare carried out by the Ku Klux Klan and other white secret societies against blacks and sympathetic whites. Once the South was 'safe for capital', in both territorial and political terms, the coalition of 1866 could finally be disbanded by 1877 – in particular, blacks had served their purpose.

The counter-revolutionary forces which were to 'redeem' the South from black and carpetbag rule mounted a campaign to promote a 'New South' based on massive investment in industry by Northern industrial and banking interests (Gaston 1970; Woodward 1971, Ch. 6). It was argued that such investment would guarantee ultimate prosperity for the South based on its vast natural resources and reserves of cheap labour.

The New South remained little more than a dream – despite rapid expansion of the railway network after the Civil War, and the movement of industry to the region, the former was entirely controlled by Northern bankers by the 1890s, and the latter typified by low wages. The region showed all the signs of a colonial economy (Woodward 1971, Ch. 11). The greater part of the raw or crudely processed materials of the region left to be fabricated in the North or abroad, and the economy was dominated by absentee owners, the net result being a region of low wages and poverty. Any threats to the North–South business alignment were severely resisted, as in the case of the Populist movement of the 1890s. Squabbles over the black vote and, more fundamentally, the pressure to counter any working-class resistance and to keep costs low in an 'internal colony' contributed to federal government and supreme court actions which encouraged enactment of a whole gamut of segregation or 'Jim Crow' laws by the turn of the century, depriving blacks of civil rights, and allowing lynch mobs to operate unmolested (Woodward 1974).

Birth of a Nation and the end of an era

When it first appeared D. W. Griffith's *Birth of a Nation* (1915) incorporated many of the devices employed in popular fiction of the 1880s and 1890s that dealt with Southern themes (Buck 1937, McKeathen 1980), as well as being more directly based on two 'race-baiting' novels by Dixon (1903, 1905). During Reconstruction, the industrial capitalists formed the ruling class in the North and in the United States as a whole, but were 'dominant' rather than 'leading' or 'hegemonic', particularly in the South, ruling by force rather than consent (Silk & Silk 1985). They had also to win the ideological battle to achieve a greater degree of social stability, and to justify their activities on three major issues, the first being reconciliation with Southern ruling groups, the second being the disfranchisement and subjugation of blacks in the South. For both tasks, the prime 'audience' was the mass of the population in the North and West. The third issue involved persuading the Southern ruling class, and Southern whites in general, that the period beginning with Redemption represented one in which all whites, North and South, had the same material interests, and in which white Southerners, because of their knowledge of how to deal with blacks, whether under slavery or Jim Crow, had snatched a moral victory from the jaws of military defeat.

The policies of the Redeemers were justified in terms of a number of inter-related themes (Buck 1937, Silk & Silk 1985). An over-arching notion was that of national reconciliation or national unity. The most important single theme was the racist portrayal of blacks, which was counterposed to the 'brotherhood' of all whites based on white supremacy or Aryan superiority. Blacks were portrayed either as active or passive, the former being realized in characters who were either beasts or militants, the latter in terms of child-like creatures or menials. As incorporated in films, Bogle (1973, Ch. 1) identifies five denigratory black stereotypes. Of the passive characters, the 'tom' and the 'coon' were male, the former being a socially acceptable Good Negro character, 'who never turn(s) against white massas' (Bogle 1973, pp. 4–6). 'Coons' are shown as amusement objects and black buffoons, important variants being the 'pickaninny' and the 'uncle remus'. However, the 'pure coon' is the most degrading of the passive black stereotypes, being a lazy, subhuman, chicken-stealing creature who butchers the English language (Bogle 1973, p. 8). Such stereotypes are also asexual beings. Black female stereotypes include the 'tragic mulatto', and the 'mammy', the latter being shown as fiercely independent (within rather strict limits) and usually as big, fat and cantankerous. The 'aunt jemima' or 'handkerchief head' represents a calmer version. The active stereotype is that of the beast or 'brutal black buck', being a male who was either aggressive and violent, or more sinisterly, oversexed and savagely lusting after white flesh (Bogle 1973, p. 13). These stereotypes were used to show that blacks are harmless, and happy, only when kept in their place.

The justification for keeping blacks in their place was the rôle of 'Southern white womanhood' in perpetuating white supremacy. White women were placed on a pedestal, symbolizing the purity of the Anglo-Saxon race. Any threat of violation was dealt with in terms of 'Southern chivalry', and it was implied that lynch mobs and the Ku Klux Klan were the modern equivalents of the presumed knightly ancestors of Southern slave-owning aristocrats and engaged in a crusade to protect Southern women.

In *Birth of a Nation*, Griffith wished to portray accurately the Civil War and surrounding events from the Southern point of view (Armour 1981, p. 14). The film shows the Old South as a happy place based on an agrarian lifestyle and a social structure characterized both by tradition and by well-to-do well-mannered whites. The Southern family, the Camerons, live in a modest house with classical white columns. 'Toms' and 'mammies' staff the hall and kitchen or work contentedly in the cotton fields. 'Coons' amuse themselves and white onlookers during the extravagant two-hour lunch break normally granted to slaves – a marked contrast with the experience of most Northern industrial workers. The arrival from the North of the visiting Stonemans sets the scene for shy flirtation or good-natured horseplay between the offspring from both families. Griffith cleverly uses normal family behaviour, together with humour, to enlist the audience's sympathy for the whites. Such humour is used in an altogether different manner when, after the Civil War, the Radical Republicans are shown to take advantage of Lincoln's assassination to punish the South. At the Constitutional Convention in South Carolina, black delegates are shown swilling alcohol and sitting with bare feet up on desks. 'Aunt jemimas', like one of the loyal maidservants, remark that 'Dem free-niggers f'um de N'of am sho' crazy', while Lydia, a mulatto, hates whites and refuses to be treated as an inferior. It is typical of popular fiction and films that deal with Reconstruction from this viewpoint that the contrast between active and passive blacks should be displayed concretely, either within the same scene, or by cutting from one scene to another.

Southern white womanhood, in the form of Lucy Cameron, literally leaps to her death from a pedestal, in this case a cliff top, rather than submit to a black 'buck', the freedman Gus. The mulatto Silas Lynch tries to force Elsie Stoneman into marriage. One of the Cameron brothers leads a group of upright Southern men in a posse of the Ku Klux Klan to rescue Elsie, break the power of blacks, carpetbaggers and scalawags, and restore white supremacy and all else the South had lost. The film closes with split-screen shots showing two honeymoon couples – one being a Cameron (South), the other a Stoneman (North). In each case, there are glimpses of happy crowds wearing classical Greek costume, and a monstrous allegorical figure of Mars who dissolves to 'The Prince of Peace' (Cripps 1977, p. 51) – such is the meaning of Southern Redemption.

The film had an enormous impact. It undoubtedly played a rôle in fostering the growth of the Ku Klux Klan which had been revived as a national, not just a Southern, organization in 1915. Protests came from blacks, particularly from the National Association for the Advancement of Colored People (NAACP), and from white liberals. Picketing, 'race riots' and mob action accompanied its presentation in many cities. It was banned for some time in New York, and refused a licence in at least eight states.

The controversy generated by the film was one of the reasons why the active black, whether as beast or militant, almost totally vanished from the cinema screen until the late 1960s. Black characters, except for films with predominantly black or all black casts, became asexual. The typical black character in 1920s Hollywood was a 'jester' (Bogle 1973, p. 19). However, the portrayal of blacks as passive, stupid and non-threatening simply meant removing the most controversial aspects of a racist image of blacks from the screen. The general practice of showing whites as the most important and valued people, whether in films set in the South or elsewhere, and of excluding blacks altogether from most films, reinforced the negative self-image of a group that continued to suffer appalling *de jure* oppression in the South and *de facto* oppression elsewhere. The message for whites was that blacks were unimportant and not worthy of respect and, by implication, that they belonged at the bottom of the socio-economic ladder. Treatment of the South, both in the cinema and the literary field, flagged from 1916 to 1928. Apart from Buster Keaton's *The General* (1926)[1] and yet another (sixth!) version of *Uncle Tom's Cabin* (1927), no well-known film appeared (Kirby 1978, p. 44).

There were other reasons, both within and outside Hollywood, for avoiding controversy centred on issues of 'race' or class (Sklar 1975, pp. 82–5). Since 1907 middle-class reformers and moralists had seen the cinema as likely to promote immoral behaviour among the masses unless appropriate supervision could be exercised by their social superiors. Scandal in Hollywood in 1921–2 and a fall in cinema attendance were thought to be causally related, although the latter was due to competition from commercial radio and the new leisure activity of motoring. The alarmed producers successfully countered with mild self-censorship, imposed by the Hays Office,[2] prohibiting ethnic name-calling and miscegenation ('inter-racial' sexual relations) with an eye to the moralists and a gesture towards liberalism. The studio heads could also point to the disastrous effect on takings in Southern (white) box offices if any more-favourable images of blacks were portrayed, and to outright censorship of any treatment of integration, desegregation or 'passing' (attempts by light-skinned blacks to pass as whites). Reorganization of the film industry along oligopolistic lines during World War I also reinforced conservative tendencies, as did the increasing reliance on outside capital from the banks and, later, from the communications industry to meet the enormous cost of conversion to sound.

Blacks returning from war service in Europe expected definite moves towards first-class citizenship, whereas many whites were equally determined to prevent it (Franklin 1974, Ch. 19). In the 'Red Summer' of 1919 blacks fought back when attacked by whites, the pattern being repeated during the next few years. Moderate reform organizations like the NAACP failed to secure mass black support chiefly because they represented the interests of upper-class blacks dependent upon the goodwill of white liberals and terrified of black direct action or violence under any circumstances. The enormous popular following that Garvey enjoyed with his 'back-to-Africa' movement, in which he also exalted everything black and declared that blacks had a noble past in Africa, showed how important it was to improve the self-image of blacks. An interest in current social and economic problems, including those of 'race', among many white writers was paralleled by a movement amongst blacks who produced a crusading, protest literature as part of the Harlem Renaissance (Franklin 1974, Ch. 20). There was also considerable labour unrest after the War, and a 'red scare' during which many communists and socialists were rounded up, and imprisoned or deported. Under these conditions, it was too much to expect other than caution on such issues by film makers who were 'deeply committed to capitalist values, attitudes and ambitions. Any new options they offered would clearly avoid breaking away from the fundamental economic and social mould' (Sklar 1975, Ch. 2).

Partly as a direct response to *Birth of a Nation* some Afro-Americans turned to the cinema as a means of expressing the 'black point of view'. These efforts depended upon a growing black ghetto audience in the cities, particularly in the North and West, as labour moved off the land into services and manufacturing. One of the traditional responses of the black petite bourgeoisie and bourgeoisie to racism, establishing a parallel black economy to serve black needs, was therefore adopted in the production of 'race movies' (Cripps 1977, p. 71). Lack of capital, hectic competition, insufficient black spending power and the postwar depression killed off many companies, but over 100 appeared during the 1920s to serve 700 ghetto cinemas. The enormous costs of conversion to sound, and the Depression, liquidated most black independents in the 1930s, although Oscar Micheaux produced films until 1948. From the early 1930s, however, most 'race movies' were made almost entirely by whites (Bogle 1973, pp. 107–8).

Whoever made the films, their plots did little to bolster a strong black self-image. Few, if any, dealt with the specific problems of blacks in the ghetto or in the South. There was a strong middle-class preoccupation with 'passing', and the general emphasis on 'the lighter the better' encouraged an interest in hair straightening and skin lightening liable to promote a negative self-image verging on self-hate amongst blacks. In this sense white oppression was unwittingly reinforced by the black oppressed. Another important feature of many films was the boosting of capitalist values through the portrayal of 'Black Horatio Alger' characters. Such themes propagated the view that all would be well if only more blacks were allowed to join the middle and upper echelons of white capitalist society, taking on their cultural and economic values. Many 'race movies' were also poorly produced and difficult to follow, and Cripps (1977, p. 326) argues that most blacks after 1932 watched films produced by Hollywood.

Hollywood's Golden Age and *Gone with the Wind*

Mythical views of the South had been increasingly questioned during the 1920s, and the entire social mores of the region ridiculed by commentators like H. L. Mencken and by reaction to the 'Monkey Trial'[3] in Tennessee in 1925 (Tindall 1980, Ch. 3). More attention was being paid to the poverty and racism of the 'embarrassing New South' (Kirby 1978, Ch. 3) in both serious and popular literature. 'Sharecropper realism' and the 'Southern Gothic School', the latter including works by Faulkner and Caldwell, exploited these themes. Southern scholars like Odum (1936) and Vance (1932, Ch. 17) supported the contention that the region's colonial status persisted and that the Southern states propped up the American league table on virtually all measures of economic health and social welfare.

Such views had some impact on Hollywood as part of the sudden turn to, and interest in, social realism in 1930 which took the form of films emphasizing gangster violence, sex and political melodrama. *Cabin in the Cotton* (1932) deals with the class struggle between sharecroppers and the merchants to whom their crops are in lien, but the characters are all white, even though most sharecroppers were black. However, treatment of class militancy was daring for Hollywood, let alone struggle based on 'race' and class (Kirby 1978, pp. 46–7). *I am a Fugitive from a Chain Gang* (1932) concentrated on the convict lease system and *They Won't Forget* (1938) on racism and lynching, but the major protagonists are white with few, relatively minor, parts for blacks. Such critical social comment was welcome in a medium which mostly showed the South, Old or New, as an escapist alternative during the Depression, but the indirect treatment of 'race' shows that expediency had its ideological limits when this involved challenging practices of what Marable (1983, p. 10) has termed a racist/capitalist state.

As in the 1920s, the middle classes and the 'authorities' were far less concerned about the impact of the written word than that of the visual moving image. By 1934, the studios were forced to give in to massive pressure exerted by the Catholic Church through the Legion of Decency and to tighten censorship. The Production Code Administration prohibited a variety of acts and verbiage on the screen, mostly related to sex, and many of the most important contemporary moral, social and political issues vanished from the screen (Sklar 1975, pp. 173–4). The New Deal Administration of 1932 wanted to foster patriotism and commitment to national values. Loss of confidence in 'the American Way', brought about by the experience or observation of ruin for those who had done all the right things, deferring gratification and working long and hard, had to be restored. Film moguls found opportunities for proft-making by supporting traditional American culture and values (Sklar 1975, p. 175). Stressing national unity and reconciliation was a well-tried way of seeking to counter a classic crisis of capitalism and the heightened struggles between classes, 'races' and sections that it was liable to trigger off.

Where blacks featured at all prominently, as in a number of films with Southern settings in the 1930s, they did so as servants or menials (Bogle 1973, Ch. 3). In escapist films which temporarily insulated audiences from Depression conditions, a pre-industrial South was portrayed in which relatively simple and unproblematic ways of life and social hierarchies existed. They continued to validate the post-Reconstruction settlement, obscuring external control over the Southern economy and confirming domination of black by white. Films like *Jezebel* (1938) and *Mississippi* (1935) continued the myth of the Old South. Bill 'Bojangles' Robinson appeared as an urbane 'tom' figure with Shirley Temple in two films with Southern settings, *The Littlest Rebel* (1935) and *The Little Colonel* (1935). In the latter, he acts as her 'Uncle Billy', exuding contentment as the film wends through a thin plot which includes a North–South marriage and the foiling of two swindlers from 'out West'. The device of using a child, not yet fully socialized, and a socially marginal black, in order to show affectionate and asexual black–white relationships, is highly effective. Two other black film stars, Stepin Fetchit and Hattie McDaniel, play respectively an archetypal 'coon' and 'mammy' in *Judge Priest* (1934).

Gone with the Wind (1939) raises the black servant tradition to its peak (Bogle 1973, p. 92). The film repeats the sequence of Old South, Civil War and Reconstruction shown in *Birth of a Nation*, but with significant variations. The most provocative forms of Griffith's racism are toned down, as is the vicious racism in Mitchell's (1936) novel, partly because Selznick, the producer, wanted to avoid controversy but also because the NAACP and other groups fought to get the most offensive scenes removed or softened (Reddick 1975, p. 15). For example, the clash between beast and child, militant and menial, is concretized in the scene where Sam, the loyal ex-slave, rescues Scarlett from attack by a black, but only after he has first saved her from assault by a poor white. Ashley Wilkes is the Southern gentlemen who says at one point that he 'would have freed all slaves anyway', war or no war. There is no explicit reference to the Ku Klux Klan, although its 'political meetings' are whitewashed by the attendance of so exemplary a character as Wilkes. However, there are also suggestions of the embarrassing New South. After the defeat, Scarlett joins the scalawags and runs her husband's lumber mill using convict labour, despite the protestation of other

Southerners who had fallen on hard times but who 'kept their honour and kindness'.

Nevertheless, the overall survival of the mythical South symbolized by Scarlett's eventual return to the plantation at Tara provided an example of American determination – '. . . survival and recovery were not only possible but probable, given the regenerative strength of American native character. The picture, therefore, was a national epic of contemporary meaning' (Campbell 1981a, p. 119) as the Depression persisted and the run-up to World War II continued.

However, the meaning to many blacks and left-wing radicals was only too clear. Black leaders condemned .the film, and in *The New Masses* an article criticized *Gone with the Wind* as 'vicious', 'reactionary', 'inciting to race hatred', 'slander of the Negro people' and 'justifying the Ku Klux Klan' (Reddick 1975, p. 16).

The black response to Hollywood stereotyping of characters and issues was conflict-ridden. It was recognized that black film stars were highly-skilled performers, but blacks were unhappy with the rôles that had to be played to achieve success. In Hollywood itself, criticism was muted because of the fear that even minor gains would be lost. Such ambivalence is very evident in Bogle (1973) and well documented by Cripps (1977).

The number of blacks who looked for more radical solutions, whether inside or outside the film industry, was very small. Of those who turned to the left, Richard Wright and Paul Robeson are among the best known. Robeson placed much faith in opportunities in Europe after the eclipse of 'race movies' in America (Cripps 1977, Ch. 12). Unfortunately, he was unable to exercise much control over the form of the final product and in films like *Sanders of the River* (1936) found himself unwittingly justifying British imperialism, much to his own outrage and that of the black press in America. This factor, together with remoteness from the American black experience and the hostility of American exhibitors to any foreign film, accounted for his failure.

From World War II to the late 1960s

World War II marked a turning point in black–white relations in the United States, and the beginning of the end of the Jim Crow system in the South, even though such changes took more than 20 years to complete. As the United States entered the war, the obvious contradiction between racism in the Southern states and elsewhere in the US and the fight against fascism abroad embarrassed the government. Pressure for change came from black militants such as those led by A. Philip Randolph, who threatened a march on Washington in 1941 if the government did not promise a fair deal on employment for blacks in war industries, and from the Federal government itself. The Office of War Information (OWI) became the arbiter and censor of racial themes, and exerted pressure to increase opportunities for blacks in the film industry. NAACP leaders and Hollywood studio heads met and the latter promised to liberalize their depiction of blacks. This produced little result at first. The profitable Old South themes, 'darkened and sweetened' (French 1981, p. 248) were retained in *Dixie* (1943), in which smiling, singing darkies and black minstrels feature, but without any reference to slavery or black subservience. Disney's *Song of the South* (1946) was enthusiastically received, attracting only some protest and adverse critical comment (Campbell 1981a, pp. 153–4).

Nevertheless, the hegemonic rôle assumed by the United States in the capitalist world after 1945 continued to conflict with the oppression of blacks within the country and especially with the Jim Crow system. As leader of the free world against Communism, the US government wanted as many allies as possible in unaligned and Third World countries. Increasingly, the leaders of such nations were to be black nationalists who owed their position to anti-colonial movements. Blacks in the South contrasted their own position with that of black Africans, the example of Ghana in 1957 being particularly influential, and a younger generation also compared their own social and economic position under the Jim Crow system with an ideology that promised freedom, democracy and opportunity for all. Further massive migration from the South to Northern and Western cities after 1945 also increased the political influence of blacks – for example, their votes were critical, and they were assiduously courted by Kennedy in the presidential election of 1960 which he barely won. Federal government and Supreme Court actions and decisions, which had bolstered racism in the interests of national reconciliation and capitalist expansion in the late 19th century, now favoured an ideological climate in which anti-racism, or at least a sensitivity to the feelings of blacks, came to predominate. Once the anti-communist witch-hunt of the McCarthy era had ended, and the historic Supreme Court decision decreeing an end to public-school segregation on racial lines had been reached in 1954, conditions were more favourable than they had been since Reconstruction for a successful black struggle for freedom and equality.

American films reached the peak of their popular appeal in 1946, but audience research showed that competition from radio and newspapers was important. Audiences fell in 1947 and the decline accelerated with the impact of television from 1949 to 1950. Harassment by the House

Un-American Activities Committee, a major anti-trust suit and economic retaliation against the industry by foreign governments (especially that of Britain), added to their troubles. Survey research had also showed that increases in film-going between 1935 and 1945 were attributable largely to middle- and higher-income groups. The initial reaction of studio heads was to go for 'good pictures' which moved sharply away from the escapist themes of the 1930s and presented controversy in realistic terms. Films on suitable Southern themes were often drawn from the work of 'quality' writers like Faulkner or Tennessee Williams, contributing to a 'Southern Gothic School' in the cinema which lagged 15–20 years behind its literary counterpart. *Intruder in the Dust* and *Pinky*, the former based on a story by Faulkner, were among five that appeared in 1949. These 'problem movies' (Cripps 1975) died out by 1954.

Other films were adapted from the work of 'serious' authors, like Tennessee Williams' *A Streetcar Named Desire* (1951) and *Suddenly, Last Summer* (1959), and Caldwell's *God's Little Acre* (1958). As the battle with television intensified, distributors and exhibitors increasingly took on the censors, realizing that shock and titillation based on liberal doses of sex and violence were likely to attract audiences. Plenty of cheap, clean material was available on television which, as family entertainment, now attracted far more attention from censors. Hollywood's period of self-censorship gradually faded after 1954, aided by a series of Supreme Court decisions in obscenity cases and liberalization of views on such issues within the Catholic church (Sklar 1975, pp. 295–6). Advertising for the new Southern Gothic films emphasized corruption and decadence, presenting them as 'simmering stories of life in the Deep South, steamy with sex and laced with violence and bawdy humour' (Campbell 1981a, p. 159). Such films concentrated almost entirely on whites. *Band of Angels* (1957), set in the slavery era, was no portrayal of the Old South. The cruelty and lust of a slave-owner, and tensions between him, the overseer and slave, are highlighted.

As the Civil Rights struggle reached its height in the early 1960s, integration was still taboo on film and Hollywood mostly avoided dealing openly with such controversy. Johnson (1975, pp. 167–8) says the industry 'would hesitate to release a fiction film based upon the true-life horrors experienced by white and Negro civil rights workers in the backward counties of Mississippi, Georgia and Alabama. To make such films today would be inflammatory and raise cries of anarchy'. However, *The Intruder* (1961), filmed in a Southern town, includes a sequence in which black high-school students are led by a courageous white editor as they approach an all-white school for their first day's attendance (the film had still not been generally exhibited in the US by 1964). *Gone are the Days* (1963) attempts to satirize the stereotyped view of the black held by white Southerners, but given events in the South during that year such humour tended to fall flat. Finally, *Nothing but a Man* (1964) shows day-to-day life for two young blacks who marry and struggle to live peacefully in small-town Alabama in the 1960s. The humanity and nobility of their bearing in the face of discrimination, as shown in the film, is a frequently used device when dealing with the oppressed and exploited. Such people bear up stoically, their infinite reasonableness and patience standing for and justifying the Gandhi-style tactics and demands of black leaders like Martin Luther King.

To Kill a Mockingbird (1963) and *In the Heat of the Night* (1967) are both anti-racist liberal fantasies set in small Southern towns. *Mockingbird*, set in Alabama in 1935, skilfully uses the devices of seeing events through the eyes of children. In a dramatically superb, but totally unrealistic, scene they shame a lynch mob into disbanding, and (for a while) save Tom Robinson, a black who is unjustly accused by a poor white teenage girl, Mayella Ewell, of attempted rape. In the film, no reason is given for the apparently unjust and irrational behaviour of most whites, including the segregated courthouse and farcical trial. The childrens' ability to differentiate clearly between right and wrong is reinforced by Boo Radley, a simpleton who kills Mayella's father because the latter attacks them. Ewell hates the children's father because the latter defended Tom Robinson in court. *In the Heat of the Night* provides a strong plot in which a black homicide detective, Tibbs, from Philadelphia is unwittingly drawn into a murder investigation in Mississippi. The murdered man, Colbert, is a northern industrialist who is to build a factory in the town. It is only because Tibbs solves the murder case that Colbert's wife agrees to go ahead. This portrayal of the North as rational and forward-looking, and as an example to and potential saviour of an irrational, seedy and backward-looking South, is hypocritical but not unusual and is symbolized here by the rôles played by Tibbs and the Colberts. In this film and *Mockingbird* many of the pre-1945 images of the South, whether Old or New, are inverted. This applies also to the rôle of white women, as in each film the 'gutter' stereotype triggers off many of the problems to be resolved.

From the late 1960s to the early 1980s

After victory in the battle for Civil Rights legislation in 1965, the urban riots which occurred in many American cities in the latter half of the 1960s, and the rise of the Black Power movement, blacks turned more and more to the promotion of a positive self-image in which, apart from largely rejecting traditional racial stereotypes, they also moved towards a rejection of white culture and mores. 'Black is beautiful' and Afro hairstyles became popular as hair straighteners and skin lighteners were forgotten. There was no longer any attempt to discourage an emphasis on black values and attitudes for fear that it might appear to support Jim Crow and white supremacy.

The recession that Hollywood suffered in the mid-1960s provided an opportunity for many black film producers, performers and technicians (Murray 1973, p. 118). A number of independent companies sprang up to fill the gap as the major studios drastically cut back production, among them being those run by black film-makers like Van Peebles, who made *Sweet Sweetback's Baadasssss Song* (1971), and Parks, who made *Shaft* (1971). By Hollywood standards, both films were made on extremely low budgets but grossed $11 million and $18 million respectively in box-office receipts. A number of other 'black blockbusters' followed (Michener 1975). They appealed strongly to black audiences, starring black heroes and heroines, being set in ghetto areas like Harlem and Watts and portraying characters like the black stud in *Sweetback* who successfully rebelled against white society. They also included plenty of sex and violence.

Monaco (1984, p. 187) points out that the 'Hollywood Renaissance' of 1968–70 was based to a considerable extent on Black films – written, directed and acted by blacks, and sometimes even financed and produced by them as well. This success had also been made possible by continued black migration to Northern and Western cities, and the white flight to the suburbs, so that large downtown cinemas were increasingly patronized by blacks – a weekend visit to the cinema was a popular alternative for people whose regular entertainment was radio in cramped tenements and who could not afford a television set (Murray 1973, p. 248).

Once Hollywood realized the economic potential of black audiences, many talented blacks suddenly found that they were being offered rôles in films. 'Blaxploitation', in which white action genres – urban private eye, cop, drug and caper movies – were reworked to include blacks, often reversed racial stereotypes and at times even showed some sensibility to blacks (Monaco 1984, p. 191). It soon came to be realized that black audiences also watched many non-black films, and that whites might be attracted to films starring blacks in black stories, but directed and written by whites. Apart from a few independent productions, the 'crossover' film replaced the black film by the mid-1970s. As Monaco (1984, pp. 187, 188) comments, 'the virtual disappearance of Black film in the mid-seventies has been the greatest feature of the American film business in recent years . . . Blacks in film, as elsewhere, have in a way been co-opted, as Black aspirations have been trimmed, modified and channelled by the industry to serve its own ends'.

The most striking example in a Southern setting of a blaxploitation crossover film is *Mandingo* (1975), together with its sequel *Drum* (1976), the former ranking as the 18th most profitable film in 1975. *Mandingo* results from a strange mixture of liberal attempts to correct the stereotypes of the Old South, gestures to greater black self-awareness and pride, and the need to make profits. Slave-breeding practices are highlighted, as are the sexual appetites of plantation owners and overseers for young black females. Floggings are common, and slave resistance – whether in the form of illicitly learning to read, running away, or occasional open defiance – is portrayed. The physical trappings of the Falconhurst plantation, and the social atmosphere, are seedy and depressing – no great civilization of ladies and gentlemen here! The marriage of the slave-owner's son Hammond, to Blanche, is a transaction that mirrors the buying and selling of slaves. Blanche does not initially appear on a pedestal, and falls decisively further once she orders Mede, the 'fighting black buck', to sleep with her. Such breaking of taboos leads to disaster when Hammond finds out – 'his' son is born with a black skin – and the film ends on a note of black defiance. As Campbell (1981b, p. 113) points out, publicity for *Mandingo*, *Drum* and other blaxploitation films emphasized 'rape, mayhem, murder and assorted other ingredients necessary to new films about the slave South', but they nevertheless contained much new information (for films) and gave the lie to the formerly predominant portrayals of such a system as simple and benign.

Apart from recognizing the blaxploitation aspects of such films, some critics feared that scenes of black rebellion might be seen to condone or encourage renewed and widespread urban rioting by blacks and other oppressed groups. While this could be the case (although at the time of writing it has not yet occurred), I think it more likely that the films, in a distorted way, reflect the black struggles and rebelliousness of the 1950s and 1960s, and the tendency of blacks (and some whites) to re-examine black history, seeing it not as an unending period of passivity by an oppressed group that only struggled in the period after 1945, but rather as an unending struggle which bore little fruit except when other circumstances were favourable. This is part of a more general process by which exploited and oppressed groups – whether they be ethnic groups, women or the working class – come to realize that, in order to suit the interest of exploiters and oppressors, they have been 'hidden from history', and so attempt to provide an alternative historical account which does their own cause justice.

The setting of particularly violent confrontations between black and white well in the past is also a distancing device commonly used when someone wishes to comment on a contemporary issue which is too contentious to be dealt with directly. A film that lies totally outside the blaxploitation tradition is *Freedom Road* (1979), based on a novel written in 1944 by a (white) member of the American Communist Party, Howard Fast. In a film which would have been regarded as totally unacceptable until the late 1960s, the relationships between 'race' and class in the Reconstruction era are examined. Not only is Reconstruction shown in a positive light, but so also is the possibility of an alliance between black freedmen and poor white farmers. Both in its structure and in the artistic devices it employs, the film is the antithesis of the kind of account given in *Birth of a Nation* and *Gone with the Wind* and in bourgeois cultural forms generally. The ideal-typical individuals with whom we identify (for black audiences there is an added draw of Muhammed Ali playing the major part) are all killed, but the collectivity survives, as do the ideals for which they fought, because someone passes on an account of their actions in an oral, literary or other tradition of communication (Silk & Silk 1986). We finally note here that the chief visual medium responsible for presenting black history through popular culture has been television, notably through *Roots* (1977), based on Haley's novel of the same name, and *King* (1981). *Roots* traces Afro-American history from its origins in West Africa, through slavery, the Civil War and up to the end of World War I. *King* traces the progress of the Civil Rights struggles of the 1950s and 1960s in relation to the leadership of Martin Luther King.

Apart from films dealing explicitly with black–white relations or black history, a noticeable sub-genre since the late 1970s deals chiefly with the experiences of working-class Southern whites both in the factory and on the farm. In *Norma Rae* (1979), set in 1978, black–white relations are shown as good but a scuffle breaks out when management try to incite racial antagonisms as Norma battles to set up a labour union in a textile mill (see also Adams 1981). *The Coal Miner's Daughter* (1980) is a rags-to-riches story, based on the life of Loretta Lynn and set initially in a Kentucky mining town. Battles against the elements and against exploiters figure prominently in these films, and also in *The Doll Maker* (1980). Parallels with the 'sharecropper realism' films of the early 1930s suggests that recession in the US economy in the late 1970s triggered renewed interest by some directors and producers in the plight of the poor and unorganized. Rôles for strong female characters in all the above-mentioned films, and a close examination of what it means to be poor, without a man but with children, in *Raggedy Man* (1981), also suggest a feminist influence.

A notable absence in these films, and all the others we have discussed, is the strong black woman. There has been a renaissance in literature by black women, many of them Southerners, during the past 15 years (Evans 1984) and the recent release of the film version of Alice Walker's (1983) *The Color Purple* (1985) starts to redress the balance.

Conclusion

The economic super-exploitation and colonial position of the South from the late 19th century, and the oppression of blacks under Jim Crow, were justified, confirmed and obscured by the ideological themes and devices which proliferated in popular and serious literature about the region from the 1880s. Racist ideology formed the most important single cluster of themes, and also helped justify American imperialism.

Birth of a Nation was the first film to make use of all these themes in dealing with the major periods of Southern history. It was the last to deal so crudely and contentiously with them because of the opposition it provoked, the ultra-conservatism of Hollywood producers, and the censorship pressure exerted by middle-class groups. Southern blacks as beasts or militants disappeared from the screen. Instead, loyal and contented menials in a refined and civilized 'Old South' of moonlight and magnolias or in an agreeable post-1877 setting of rural stability came to predominate, despite a small number of films on the 'embarrassing New South' that appeared in the 1930s. Political protest over the specific position of blacks and the more general plight of the poor during the Depression which generated movements in other media failed to register in 'race movies' which instead showed a strong sense of black inferiority and a desire to attain superior class positions in the existing capitalist socio-economic structure. Robeson's attempts at radical protest in the medium failed.

Major changes in the world position of the United States government and capital which had occurred by the end of World War II, together with internal developments, provided the most favourable conditions for black struggles in the South since Reconstruction. These factors combined to produce an ideological climate in which anti-racism became predominant, although direct treatment of such conflicts on film was tentative. Black urban rebellion and the rise of Black Power in the 1960s were accompanied by a new pride in black history and culture, and black separatist sentiment made a significant appearance for the first time since the 1920s. Black resistance and brutal black 'bucks' appeared in films in a form that audiences were expected to applaud; and formerly idyllic images of the South, whether Old or New, were rudely inverted to portray a backward, uncultured and irrational society, populated by seedy planters, vicious overseers and white men who raped black women. Competition with television, falling audiences and relaxation of censorship ensured that the treatment and advertising of such films was sensationalist and titillating, relying heavily on sex and violence.

Black consciousness and unity have not weathered well the recession of the late 1970s and the rightward swing since Reagan's 1980 election, with increased assimilation into mainstream society (for some) and the persistence of institutional racism (Jewell 1985). Parent (1985) also comments that blacks accept the 'American idea' and above all want to be allowed to participate in the system, accepting an ideology of individualism which, if they fail, leads to self-blame. There is resistance to practices associated with structural features of society which foster exploitation and oppression (Marable 1985), but films sympathetic to such practices in a Southern setting deal almost entirely with rural poor whites.

Notes

1 Where known, the date of first release of a film is given in brackets on its first appearance in the text.
2 The Hays Office was the name by which the Motion Picture Producers and Distributors Association, formed in 1922, came to be known, after its president Will H. Hays. As a former Postmaster General in Harding's Administration he was well equipped to act as public relations man and lobbyist on behalf of the industry.
3 At the 'Monkey Trial' or 'Scopes Trial', held in Dayton, Tennessee, a schoolteacher, John Scopes, was tried for having taught the Darwinian theory of evolution in Tennessee public schools. The trial attracted nationwide publicity and, from nearly all quarters, ridicule.

Films mentioned in the text

Band of Angels (1957)
Birth of a Nation (1915)
Cabin in the Cotton (1932)
The Coal Miner's Daughter (1980)
The Color Purple (1985)
Dixie (1943)
The Doll Maker (1980)
Drum (1976)
Freedom Road (1976)
The General (1926)
God's Little Acre (1958)
Gone are the Days (1963)
Gone with the Wind (1939)
I am a Fugitive from a Chain Gang (1932)
The Intruder (1961)
Intruder in the Dust (1949)
In the Heat of the Night (1967)
Jezebel (1938)
Judge Priest (1934)
King (1981) (television)
The Little Colonel (1935)

The Littlest Rebel (1935)
Mandingo (1975)
Norma Rae (1979)
Nothing but a Man (1964)
Pinky (1949)
Raggedy Man (1981)
Roots (1977) (television)
Sanders of the River (1936)
Shaft (1971)
Song of the South (1946)
A Streetcar named Desire (1951)
Suddenly, Last Summer (1959)
Sweet Sweetback's Baadasssss Song (1971)
They Won't Forget (1938)
To Kill a Mockingbird (1963)
Uncle Tom's Cabin (1927)

References

Adams, M. 1981. 'How come everybody down here has three names?' Martin Ritt's Southern films. In *The South and film*, W. French (ed.), 143–55. Jackson, Miss.: University Press of Mississippi.

Armour, R. A. 1981. History written in jagged lightning: realistic South vs. romantic South in *Birth of a Nation*. In *The South and film*, W. French (ed.), 14–22. Jackson, Miss.: University Press of Mississippi.

Bogle, D. 1973. *Toms, coons, mulattoes, mammies and bucks*. New York: Viking Press.

Buck, P. 1937. *The road to reunion*. Boston: Little & Brown.

Campbell, E., Jr 1981a. *The celluloid South*. Knoxville: University of Tennessee Press.

Campbell, E., Jr 1981b. 'Burn, Mandingo, Burn': the plantation South in film, 1958–1978. In *The South and film*, W. French (ed.), 107–16. Jackson, Miss.: University Press of Mississippi.

Cripps, T. R. 1975. The death of Rastus: negroes in American films since 1945. In *Black films and filmmakers*, L. Patterson (ed.), 53–64. New York: Dodd & Mead.

Cripps, T. R. 1977. *Slow fade to black*. New York: Oxford University Press.

Dixon, T., Jr 1903. *The leopard's spots*. New York: Doubleday & Page.

Dixon, T., Jr 1907. *The clansman: an historical romance of the Ku Klux Klan*. New York: Doubleday & Page.

Evans, M. (ed.) 1984. *Black women writers (1950–1980): a critical appraisal*. New York: Anchor.

Fast, H. 1979. *Freedom road*. London: Futura.

Franklin, J. H. 1974. *From slavery to freedom* (4th edn). New York: Knopf.

French, W. (ed.) 1981. *The South and film*, Jackson, Miss.: University of Mississippi Press.

Gaston, P. 1970. *The New South creed*. Baton Rouge: Louisiana State University Press.

Haley, A. 1976. *Roots*. New York: Doubleday.

Jewell, K. S. 1985. 'Will the real black, Afro-American mixed, coloured negro please stand up'? Impact of the Black Social Movement, twenty years later. *Journal of Black Studies* **16**(1), 57-75.

Johnson, A. 1973. The negro in American films: some recent works. In *Black films and filmmakers*, L. Patterson (ed.) 153–81. New York: Dodd & Mead.

Kirby, J. T. 1978. *Media-made dixie*. Baton Rouge & London: Louisiana State University Press.

McKeathen, L. H. 1980. *The dream of Arcady: place and time in Southern literature*. Baton Rouge & London: Louisiana State University Press.

Marable, M. 1983. *How capitalism underdeveloped black America*. London: Pluto.

Marable, M. 1985. *Black American politics*. London: Verso.

Michener, C. 1975. Black movies. In *Black films and filmmakers*, L. Patterson (ed.), 235–46. New York: Dodd & Mead.

Mitchell, M. 1936. *Gone with the wind*. London: Macmillan.

Monaco, J. 1984. *American film now*. New York: Plume.

Murray, J. P. 1973. *To find an image*. Indianapolis and New York: Bobbs-Merrill.

Odum, H. 1936. *Southern Regions of the United States*. Chapel Hill: University of North Carolina Press.

Parent, W. 1985. A liberal legacy: blacks blaming themselves for economic failures. *Journal of Black Studies* **16**(1), 3–20.

Reddick, L. 1975. Of motion pictures. In *Black films and filmmakers*, L. Patterson (ed.), 3–44. New York: Dodd & Mead.

Silk, C. P. & J. A. Silk 1985. Racism, nationalism and the creation of a regional myth: the Southern states after the American Civil War. In *Geography, the media and popular culture*, J. Burgess & J. Gold (eds),165–91. London: Croom Helm.

Silk, C. P. & J. A. Silk 1986. *Against the grain: oppositional ideology in Howard Fast's* Freedom Road. Unpublished manuscript, Department of Geography, University of Reading.

Sklar, R. 1975. *Movie-made America*. London: Chappell.

Tindall, G. B. 1980. *The ethnic Southerners*. Baton Rouge: Louisiana State University Press.

Vance, R. B. 1932. *Human geography of the South*. Chapel Hill: University of North Carolina Press.

Walker, A. 1983. *The color purple*. London: The Women's Press.
Woodward, C. Vann 1971. *Origins of the New South*, 2nd edn. Baton Rouge: Louisiana State University Press.
Woodward, C. Vann 1974. *The strange career of Jim Crow*, 2nd edn. New York: Oxford University Press.

Index